Society of Economic Geologists Annual Conference (SEG 2024)

AA005390

Sustainable Mineral Exploration and Development

Windhoek, Namibia
27-30 September 2024

ISBN: 979-8-3313-0729-5

Printed from e-media with permission by:

Curran Associates, Inc.
57 Morehouse Lane
Red Hook, NY 12571

Some format issues inherent in the e-media version may also appear in this print version.

Copyright© (2024) by Society of Economic Geologists (SEG)
All rights reserved.

Printed with permission by Curran Associates, Inc. (2025)

For permission requests, please contact Society of Economic Geologists (SEG) at the address below.

Society of Economic Geologists (SEG)
7811 Shaffer Parkway
Littleton, CO 80127-3732
USA

Phone: +1(720) 981-7882
Fax: +1(720) 981-7874

www.segweb.org

Additional copies of this publication are available from:

Curran Associates, Inc.
57 Morehouse Lane
Red Hook, NY 12571 USA
Phone: 845-758-0400
Fax: 845-758-2633
Email: curran@proceedings.com
Web: www.proceedings.com

TABLE OF CONTENTS

ORAL SESSION 02

ORAL SESSION 03

ORAL SESSION 04

TRIFECTA 2

ORAL SESSION 05

ORAL SESSION 06

ORAL SESSION 07

ORAL SESSION 08

ORAL SESSION 09

ORAL SESSION 10

TRIFECTA 3

ORAL SESSION 11

ORAL SESSION 12

ORAL SESSION 13

ORAL SESSION 14

POSTERS 000-025

POSTERS 026-050

POSTERS 051-075

POSTERS 076-100

POSTERS 101-125

POSTERS 126-150

POSTERS 151-175

POSTERS 176-200

POSTERS 201-225

POSTERS 226-250

POSTERS 251-276

SPEED TALKS 1

SPEED TALKS 2

SPEED TALKS 3

SEG 2024

Sustainable Mineral Exploration and Development

Windhoek, Namibia | September 27–30, 2024

CONFERENCE PROGRAM

seg2024.org

B2GOLD
NAMIBIA

Progressive rehabilitation of the Otjikoto waste rock dump

Planning with THE END IN MIND

PLANNING FORMS A CRUCIAL PART OF ANY PROJECT, NO MATTER WHAT SIZE. PLANNING OF REHABILITATION RISKS AND CRITERIA IN THE EARLY PHASES OF THE LIFE OF MINE CAN IMPROVE DECISION MAKING TO REDUCE OPERATION COSTS AND MAXIMIZE SUSTAINABLE BENEFITS.

At Otjikoto Mine, rehabilitation planning has been part of our responsible mine planning strategy long before operations commenced. Long-term plans have been developed for the type of landscape we will leave behind at the end of life of mine. A comprehensive progressive rehabilitation plan has been developed for Otjikoto Mine and took into account various considerations. The plan addresses issues that include: hydrology, geotechnical geochemical, landform design, rehabilitation approaches, flora, revegetation techniques, engagements with communities, and regulatory compliance.

The plan includes resource identification and allocation to ensure that sufficient resources are available for the duration of the life of mine. From the start of operations, it was planned to harvest and stockpile topsoil from disturbed areas for future rehabilitation activities. Some of our oldest topsoil stockpiles date back to 2013 and are being used in the rehabilitation of previously disturbed areas. As part of continuous planning, the stockpiles are regularly assessed to determine the bio-physical and chemical properties of the soils. This is to ensure their full functionality in the rehabilitation process.

Planning has allowed Otjikoto Mine to implement a good progressive rehabilitation program where the operation has been able to;

- Develop an environmental monitoring program that keeps track of any changes in environmental aspects.
- Continuously improve on rehabilitation approaches and develop success completion criteria.
- Make informed decisions on financial provision for rehabilitation.
- Detect rehabilitation risks and develop corrective actions early.

Planning has further provided Otjikoto Mine the opportunity to take a leading stance in mine rehabilitation approaches by developing an integrated and progressive rehabilitation program that ensures that areas are rehabilitated while resources are still available. Additionally, the operation has adopted leading international practices where there is opportunities for long-term sustainable benefits for B2Gold and Namibia.

Topsoil collection and stockpiling for rehabilitation

Conference Hotel
Safari Conference Centre at the Mercure / Mövenpick Hotels
Corner of Auas and Aviaton Road | Windhoek, 9000 | Namibia
Phone: +264 61 296 8000

Wi-Fi details at registration

The information in this program is current through **September 1, 2024**. Please refer to the online platform for updated information:

seg2024.org/virtual

Speaker Ready Room (Gamsberg)

Thursday, September 26: 3:00 pm – 6:00 pm
Friday, September 27: 7:00 am – 5:00 pm
Saturday, September 28: 7:00 am – 5:00 pm
Sunday, September 29: 7:00 am – 5:00 pm
Monday, September 30: 7:00 am – 5:00 pm

Parent's room: Huab

Registration

Friday, September 27: 6:30 am – 6:00 pm
Saturday, September 28: 6:30 am – 6:00 pm
Sunday, September 29: 6:30 am – 6:00 pm
Monday, September 30: 6:30 am – 12:00 pm

Exhibits and Posters (Exhibit Hall, Poster and Speed Talk Tent)

Date	Time	Event
Thursday, September 26	3:00 pm - 6:00 pm	Exhibit Move In (Poster and Exhibitor Only Access)
Friday, September 27	8:00 am - 3:00 pm	Exhibit Move In (Poster and Exhibitor Only Access)
	6:00 pm - 6:30 pm	Exhibitor Only Access
	7:00 pm - 9:00 pm	Posters and Exhibits Open - Welcome Reception
Saturday, September 28	9:00 am - 10:00 am	Exhibitor Only Access
	10:00 am - 7:30 pm	Exhibits Open
	5:30 pm - 7:30 pm	Exhibit and Poster Reception
Sunday, September 29	9:00 am - 10:00 am	Exhibitor Only Access
	10:00 am - 7:30 pm	Poster and Exhibits Open
	5:30 pm - 7:30 pm	Exhibit and Poster Reception
Monday, September 30	9:00 am - 10:00 am	Exhibitor Only Access
	10:00 am - 2:00 pm	Posters and Exhibits Open
	2:00 pm - 4:00 pm	Exhibits Dismantle (Exhibitor Only Access)

All sessions and events listed are at the conference hotel.
www.seg2024.org/venue

FRIDAY

9:30 am-4:00 pm
Student and Early Career Professionals Events

4:30 pm-5:00 pm
Opening Performance – African Cultural Diversity Dance

5:00 pm-5:15 pm
Welcoming Remarks
Inge Zaamwani-Kamwi

5:15 pm-6:30 pm
Panel Discussion
Geoscience Capacity Building for the Future - the African Perspective

6:30 pm-7:00 pm
Presidential Address
Steve Piercey, SEG President 2024

7:00 pm-9:00 pm
Welcome Reception

SATURDAY

8:00 am-8:15 am
Opening Comments
Anna Nguno and Laurence Robb, Conference Chairs
Jennifer Craig, SEG Executive Director

8:15 am-10:00 am
Plenary Sessions:
- Fostering Sustainable Development – Opportunities for the Extractive Industry
- Enhancing Perceptions and Building Trust in the Mining Industry
- The Impact of the Pan-African Reporting Code on the Future of Mining in Africa

10:00 am-11:00 am
Posters and Speed Talks

11:00 am-12:30 pm
Trifecta 1: Gold in Africa

12:30 pm-2:00 pm - Lunch

2:00 pm-3:30 pm
Oral Session 01
Gold: Enhanced Discovery and Development I

Oral Session 02
Innovative Technology Developments in Mineral Deposit Science I

3:30 pm-4:00 pm - Break

4:00 pm-5:30 pm
Oral Session 03
Gold: Enhanced Discovery and Development II

Oral Session 04
Innovative Technology Developments in Mineral Deposit Science II

5:30 pm-7:30 pm
Exhibit and Poster Reception, Speed Talks

7:30 pm-10:00 pm
SEG Awards Ceremony (Ticket Required)

SUNDAY

8:00 am-10:00 am
Trifecta 2: Central African Copperbelt

10:00 am-11:00 am
Posters and Speed Talks

11:00 am-12:30 pm
Oral Session 05
Resource Development: ESG from Exploration to Remediation
Oral Session 06
Vital High-Volume Base Metals I

12:30 pm-2:00 pm - Lunch

2:00 pm-3:30 pm
Oral Session 07
Africa's Iconic Ore Deposits

Oral Session 08
Vital High-Volume Base Metals II

3:30 pm-4:00 pm - Break

4:00 pm-5:30 pm
Oral Session 09
New Discoveries and Developments I
Oral Session 10
Vital High-Volume Base Metals III

5:30 pm-7:30 pm
Exhibit and Poster Reception, Speed Talks

7:30 pm-10:00 pm
Industry Outlook Dinner (Ticket Required)
Mfikeyi Makayi, CEO KoBold Metals Zambia

MONDAY

8:00 am-10:00 am
Trifecta 3: The Bushveld Complex

10:00 am-11:00 am
Posters and Speed Talks

11:00 am-12:30 pm
Oral Session 11
The Energy Transition: Metals of the Future I

Oral Session 12
New Discoveries and Developments II

12:30 pm-2:00 pm - Lunch

2:00 pm-3:45 pm
Oral Session 13
The Energy Transition: Metals of the Future II
Oral Session 14
Specialty Metals and Materials I

3:45 pm-4:15 pm - Break

4:15 pm-4:45 pm
Motivating Future Geoscientists: NGEA Survey Insights

4:45 pm-5:15 pm
Conference Reflections and Future Perspectives (Panel Discussion)

5:15 pm-6:00 pm
Closing Ceremony and Concluding Remarks
Anna Nguno and Laurence Robb, Conference Chairs
Anne Thompson, 2025 SEG President
Rick Valenta and Mark Noppé, SEG 2025 Conference Chairs

6:00 pm-8:00 pm
Closing Social (Ticket Required)

Welcome to the SEG 2024 Conference!

Thank you for joining us as we explore this year's theme, Sustainable Mineral Exploration and Development. The SEG 2024 Conference will feature an engaging in-person technical program highlighting the need for metals of the future, the logistical challenges of exploration and development, innovative developments and exploration discoveries, and the critical significance of world-class ore districts.

The technical program promises to showcase groundbreaking research and facilitate vibrant discussion. Each day will begin with a trifecta session that focuses on a key topic: Gold in Africa, the Central African Copperbelt, and the Bushveld Complex. Technical sessions will then delve into eight key themes with oral contributions from academia and industry speakers representing a wide diversity of background and geography. Conference attendees will also have the opportunity to explore a record number of poster presentations.

In addition to this exceptional technical program, pre- and post-conference field trips will explore key areas in Namibia, Zimbabwe, and Zambia and highlight important resources such as diamonds, rare earth elements, base metals, gold and iron oxide copper-gold, copper, lithium, uranium, pegmatites, and chromite and platinum group metals. Workshops, a robust social program, and networking opportunities specifically geared toward students and early career professionals round out the offerings.

The venue is the Mercure (Safari) Hotel Conference Centre in Windhoek, Namibia—a country renowned for spectacular geology, unique ore deposits, and leadership in mineral resource sustainability. SEG 2024 is organized in collaboration with the Society of Economic Geologists, the Geoscience Council of Namibia, and the Geological Society of Namibia, and we are grateful for their partnership.

We look forward to this chance to network with friends and colleagues old and new!

Anna Nguno and Laurence Robb
Co-Chairs, SEG 2024 Organizing Committee

Jennifer D. Craig, *SEG Executive Director*

Anna Nguno

Laurence Robb

Jennifer D. Craig

Conference Information

Conference Organizing Committee

Conference Chairs:	Anna Nguno, Laurence Robb
SEG Executive Director:	Jennifer D. Craig
Advisory, Logistics:	Gabi Schneider
Technical Program:	Judith Kinnaird, Paul Nex, Ester Shalimba
Poster Program:	Richard Manyanga, Claudia Mutongolume
Early Career Program:	Halleluya Ekandjo, Absai Vatuva
Workshops:	Mary Barton, Samuel Nengola
Field Trips:	Roger Swart, Robert Kaemba, Patrick Redmond
Social Program:	Kaarina Ndalulilwa
Sponsorships and Exhibitors:	Stephen Frindt, Vilho Shafombabi
Marketing and Media:	Inge Mbidi, Josephine Uushona

SEG Contact Information
7811 Shaffer Parkway
Littleton, CO 80127-3732
Tel.1.720.981.7882
E-mail: seg@segweb.org

The information in this program is current as of September 1, 2024, and is subject to change before conference events begin. Please check the SEG 2024 app for updated information.

seg2024.org/virtual

Table of Contents

Organizational Partners

SEG 2024 is organized in partnership with the Society of Economic Geologists, the Geoscience Council of Namibia, and the Geological Society of Namibia.

The Society of Economic Geologists (SEG) is an international scientific association devoted to advancing the science, discovery, and responsible development of mineral resources. Our members come from industry, academia, and government institutions, creating a unique, diverse network of professionals.

segweb.org

The Geoscience Council of Namibia (GSCNam) was established to serve as a regulatory juristic body for geoscientists in Namibia. GSCNam works to promote the geoscience profession in Namibia and to ensure that geoscientists are fully engaged in the transformation of their profession; a profession that is increasingly relied upon by the public to provide expert opinion and service, and to safeguard the public interest.

geocouncil.org.na

The Geological Society of Namibia aims to promote the study of geosciences in Namibia and do all things conducive to the advancement of geological science. The Society hosts regular talks on geological topics from across the world and organizes field excursions to mines and geologically interesting sites across Namibia.

geolsocnamibia.org

SEG Awardees for 2024

The SEG Awards ceremony will be Saturday, September 28, 2024, at 7:30 pm. The event will be held at the Safari Conference Centre.

R.A.F. PENROSE GOLD MEDAL

Holly J. Stein
AIRIE Program

SEG SILVER MEDAL

Massimo Chiaradia
University of Geneva

SEG WALDEMAR LINDGREN AWARD

Bertrand Rottier
Université Laval

RALPH W. MARSDEN AWARD

Douglas J. Kirwin
Independent Consultant

BRIAN J. SKINNER AWARD

Federico Cernuschi
Eclectic Rock

HONORARY FELLOW

Reid Keays
Monash University

SEG DISTINGUISHED LECTURER

Adam C. Simon
University of Michigan

INTERNATIONAL EXCHANGE LECTURER

Moira T. Smith
Liberty Gold

SEG THAYER LINDSLEY LECTURER

Sarah A.S. Dare
Université du Québec à Chicoutimi

REGIONAL VICE PRESIDENT LECTURER

Reimar Seltmann
Natural History Museum

Student and Early Career Program

Thursday, September 26

8:00 am-2:00 pm Matchless Mines Field Trip

6:00 pm Joe's Beerhouse Social Event—Sponsored by iCRAG

Friday, September 27

9:30 am-4:00 pm Student and Early Career Professionals Events

Pre-Conference Field Trips

September 19–25

Diamond Excursion – A Visit to the Namibian Diamond Mega-Placer

Leaders: Gottfried Grobbelaar, John Ward, Jana Jacob, and Roger Swart

September 21–26

REE Deposits of Northern Namibia

Leaders: Rainer Ellmies and Pete Siegfried

September 21–26

Base Metal Deposits of Southern Namibia

Leaders: Patrick Redmond and Eckhart Freyer

September 21–26

A Transect Through Gold Deposits and Prospects of the Damara Belt, Namibia

Leaders: Emmanuel Shilongo, Josia "PJ" Shilunga, Jeremie Lehmann, and Alexander Kisters

September 21–26

Zimbabwe - Lithium

Leaders: Adam Moodley and Gayle Hanssen

September 22–26

Zambian Copperbelt - Classic Copperbelt

Leaders: Murray Hitzman, Elijah Mwelwa, and Robert Kaemba, and Timothy MacIntyre

Pre-Conference Workshops

September 26

Technical Writing for Emerging Geoscience Professionals

Presenter: Jacob Hanley

September 27

Structural Geology and Exploration Insights Applied to Orogenic Gold Systems

Presenter: Miguel Nassif

Post-Conference Field Trips

October 1

The Iron Oxide Copper-Gold (IOCG) Vein System of the Onganja Mine, Windhoek District

Leader: Robert Carr

October 1–7

Uranium, Pegmatites, and Critical Raw Materials in Namibia

Leaders: Gabi Schneider, Judith Kinnaird, and Paul Nex

October 2–5

Zimbabwe - Great Dyke

Leaders: Caston Musa and Andrew Du Toit

October 2–6

Zambian Copperbelt – Domes Region

Leaders: Murray Hitzman, Elijah Mwelwa, Tim MacIntyre, and Robert Kaemba

Post-Conference Workshops

October 1

DroneSOM - Mineral Exploration Mapping Tools Using Machine Learning and Drone Geophysics

Presenters: Ina Storch, Andreas Knobloch, Johanna Pesonen, Longying Xiao, Ari Saartenoja, and René Forsberg

October 1

Critical Minerals: Geological Perspectives for Discovery and Recovery

Presenters: Richard Herrington, Maria Boni, Reimar Seltmann, Frances Wall, and Francesco Putzolu

October 2

Fluid and Melt Inclusions in Ore Systems: Fundamentals and Practical Applications

Presenters: Jacob Hanley and Tobias Fusswinkel

IVANHOE MINES

Mining with a greater purpose.

Technical Program

DAY 1 – FRIDAY, SEPTEMBER 27, 2024

Opening Event

4:30 pm - 5:00 pm	Opening Performance – African Cultural Diversity Dance
5:00 pm - 5:15 pm	Welcoming Remarks *Inge Kameeja Zaamwani (Invited Speaker)*
5:15 pm - 6:30 pm	Geoscience Capacity Building for the Future - the African Perspective (Panel Discussion) *Anne Thompson (Chair), Kurt House, Benjamin Mapani, Kombada Mhopjeni, Franck Twite*
6:30 pm - 7:00 pm	Presidential Address *Stephen Piercey (2024 SEG President)*
7:00 pm - 9:00 pm	Welcome Reception

DAY 2 – SATURDAY, SEPTEMBER 28, 2024

Welcome and Plenary Sessions

8:00 am - 8:15 am	Opening Comments *Jennifer Craig, Anna Nguno, Laurence Robb*

8:15 am - 8:45 am	Fostering Sustainable Development – Opportunities for the Extractive Industry *Thabisile Phumo (Invited Speaker)*
8:45 am - 9:15 am	Enhancing Perceptions and Building Trust in the Mining Industry *John Roos (Invited Speaker)*
9:15 am - 9:45 am	The Impact of the Pan-African Reporting Code on the Future of Mining in Africa *Deng Ngang Deng (Invited Speaker)*
9:45 am - 10:00 am	Q&A
10:00 am - 11:00 am	Poster Viewing and Speed Talk Session

TRIFECTA 1: GOLD IN AFRICA *Lynnette Greyling, Chair*

11:00 am - 11:30 am	First Major Ore-Grade Gold Concentration on Earth's Surface: Why, When, and Where? *Hartwig Frimmel (Invited Speaker)*
11:30 am - 12:00 pm	Gold in Africa – History, Endowment, and Discovery *Mike Robertson (Invited Speaker)*
12:00 pm - 12:30 pm	Beyond the State-of-the-Art Geophysical Surveys for Mineral Exploration: Where Do We Go Next? *Musa Manzi (Invited Speaker)*
12:30 pm - 2:00 pm	Lunch Break

DAY 2 – SATURDAY, SEPTEMBER 28, 2024

	Oral Session 01 **Gold: Enhanced Discovery and Development I**	**Oral Session 02** **Innovative Technology Developments in Mineral Deposit Science I**
2:00 pm - 2:15 pm	Large Gold Systems of Central Asia: Geodynamics and Metallogeny Revisited *Reimar Seltmann*	The Transformative Impact of Enhanced Geoscience Data Collection and AI-Powered Modeling on the Mining and Exploration Industry *Patrick Redmond*
2:15 pm - 2:30 pm	Permeability Evolution Within an Orogenic Disseminated Gold Deposit: Evidence from Liba Gold Deposit, Central China *Pengcong Zhang (Student)*	Synergies of Drone-Borne and Satellite Data for Noninvasive Mineral Exploration in Namibia *René Booysen*
2:30 pm - 2:45 pm	New Genetic Type of Gold Deposit for Timok, Serbia: the High-Grade Gold Skarn Mineralization at Coka Rakita Project *Mladen Zdravkovic*	Geochemical Signatures of Magmatic Hydrothermal Alteration in Carbonate Rocks Around the Basin Porphyry in the Drum Mountains of West-Central Utah *Michael Kirschbaum (Student)*
2:45 pm - 3:00 pm	Insights into the Metallogenic Evolution of the Timmins Gold Camp Using Quartz-Hosted Melt Inclusions from the Porcupine Intrusive Suite *Sarah Speight*	Development of New Exploration Vectors by Quantitative Mineralogical Analysis at the Spremberg-Graustein-Schleife Cu-Ag Kupferschiefer Deposit, Lusatia, Germany *Yonghwi Kim*

3:00 pm - 3:15 pm	Source of Fluids in Orogenic and Intrusion-Related Gold Deposits: Insights from Trace Elements in Pyrite *Dominique Genna*	Do Mineral Systems Have Fingerprints? (Yes, and They Can Be Used to Find World-Class Orebodies!) *Paul Hodkiewicz*
3:15 pm - 3:30 pm	Orogenic Gold Grain Coarsening from Colloidal Precursor by Ostwald Ripening *Bjorn von der Heyden*	Influence of Mineral Microinclusions on Vectoring and Fertility Assessment Tools in the Green Rock Environment: A Case Study from Batu Hijau Porphyry Cu-Au Deposit, Indonesia *Axel Cima (Student)*
3:30 pm - 4:00 pm	Break	
	Oral Session 03 **Gold: Enhanced Discovery and Development II**	**Oral Session 04** **Innovative Technology Developments in Mineral Deposit Science II**
4:00 pm - 4:15 pm	Petrography, Ore Mineral Paragenesis, and Gold Deportment of the Twin Hills Gold Deposit, Central Namibia *Johannes Sales*	Mineral Systems: A Discussion on the Continuing Challenges and Emerging Opportunities *Joel Hrominchuk*

DAY 2 – SATURDAY, SEPTEMBER 28, 2024

4:15 pm - 4:30 pm	U-Pb Geochronology of the Archaen Tati Greenstone Belt Gold deposits, NE Botswana: Evidence of Two Post-Archean and Post-Metamorphic Mineralization Events. *Charles Geraud Kono (Student)*	Hydrothermal Alteration and Volcanic Stratigraphy at the Paleoproterozoic Kay Mine Volcanogenic Massive Sulfide Deposit, Black Canyon City, Arizona USA *Aaron Adsit (Student)*
4:30 pm - 4:45 pm	Intrusion-Associated Au Mineralization in the Archean Mooshla Intrusive Complex, Doyon-Bousquet-LaRonde Mining Camp, Abitibi Greenstone Belt, Québec, Canada *Kevin Neyedley (Student)*	DrillDown: An Open-Source Python Package for "Drilling Down" Into Ore Deposits Datasets *Robert Collar (Student)*
4:45 pm - 5:00 pm	Valley – A High-Grade Reduced Intrusion Related Gold System in Yukon, Canada *Thomas Branson*	Perkins Discontinuities: Structurally Controlled Grade Patterns Diagnostic of Epigenetic Mineralization at the Deposit-Scale *Jun Cowan*
5:00 pm - 5:15 pm	Distribution and Partitioning of Trace Elements During Hydrothermal Sulfidation of Magnetite by Pyrite, KCD Gold Deposit, DRC *Yann Waku Mpaka (Student)*	Mineral Prospectivity Mapping on Point Data using Graph Deep Learning Regression: Case Study of Copper Occurrence in Northern Ireland *Felix Sihombing (Student)*

5:15 pm - 5:30 pm	Bi-Pb Melts as Highly Efficient Agents for Au-Ag Enrichment in Epithermal Ore Systems *Néstor Cano (Student)*	One Person's Trash is Another's Treasure; Extracting Additional Information on Fluid Rock Interaction from Disturbed Isotopic Systems *Lucy Mathieson (Student)*
5:30 pm - 7:30 pm	Exhibit and Poster Reception, Speed Talks	
7:30 pm - 10:00 pm	SEG Awards Ceremony (Ticket Required)	

DAY 3 – SUNDAY, SEPTEMBER 29, 2024

TRIFECTA 2: CENTRAL AFRICA COPPERBELT *Murray Hitzman (Session Chair)*

8:00 am - 8:15 am	Morning Comments
8:15 am - 8:45 am	Mitigating the Streetlight Effect: Stepping Into the Darkness is an Important Step in Understanding the Copperbelt *David Selley (Invited Speaker)*
8:45 am - 9:15 am	Using Mg-Carbonates, Kyanite, and Immobile Elements as Pathfinders for Technology Metals in the Central African Copperbelt *Anna Bidgood (Invited Speaker)*

DAY 3 – SUNDAY, SEPTEMBER 29, 2024

9:15 am - 9:45 am	Provenance of the Katangan Basin in Zambia and the Implications for Basin Evolution and Exploration Search Space *Munshya Zimba (Invited Speaker)*	
9:45 am - 10:00 am	Q&A	
10:00 am - 11:00 am	Poster Viewing and Speed Talk Session	
	Oral Session 05 **Resource Development: ESG from Exploration to Remediation**	**Oral Session 06** **Vital High-Volume Base Metals I**
11:00 am - 11:15 am	Redefining Viability: Integrating Social and Environmental Dimensions into Mineral Exploration Success *Matthew Grimshaw*	Framboids, Fluids, and Field Mapping: Understanding Sedimentary-Rock-Hosted Metal Prospectivity Across 11 Orders of Magnitude *Elizabeth Turner (Invited Speaker)*
11:15 am - 11:30 am	Charting Sustainable Resource Development: ESG Perspectives in the DRC Mining Sector *Innocent Mufungizi (Student)*	Evolution of the Neoproterozoic Katangan Basin in the Kipushi Mine Area, Southern DRC and Northern Zambia: The Importance of Halokinesis, Global Glacial Events, and Tectonics *Helen Twigg (Student)*

11:30 am - 11:45 am	An Effective Formalization of Artisanal Mining Will Improve the Public Acceptability of the Mineral Extraction Industry in Mozambique *Manuel Nopeia*	Exploring Geological Complexities of the Kakula Cu Deposit, Democratic Republic of Congo *Franck Twite*
11:45 am - 12:00 pm	Exploration, Mining and Biodiversity in Angola: A New Business Model *Luis Chambel*	Beyond Kamoa and Kakula: Exploration in the Western Foreland *Yves Ilayao*
12:00 pm - 12:15 pm	Effective Bottom-Up Social and Environmental Strategies for Exploration Teams: Lessons from an Orphaned Mine *Robin Armstrong*	Geological Characterisation, Stratigraphic Correlation and 3D Modelling of the Musinga Copper Exploration Play, Northeast Kabompo Dome, Democratic Republic of Congo *George Aldiss*
12:15 pm - 12:30 pm	The Benign Nature of Tailings from the Tumas Uranium Project, Namibia: Aligning Environmental Safeguards with Economic Outcomes *Alex Otto*	Camp Creek Porphyry Alteration Mineralogy and Geochemistry: Exploring for Blind Copper Mineralization in Northern British Columbia, Canada *Maxwell Porter (Student)*
12:30 pm - 2:00 pm	Lunch Break	

DAY 3 – SUNDAY, SEPTEMBER 29, 2024

	Oral Session 07 Africa's Iconic Ore Deposits	Oral Session 08 Vital High-Volume Base Metals II
2:00 pm - 2:15 pm	Bringing 20 Years of Research and 500 Reef Intersections Together on Ivanhoe's Tier 1 Platreef Deposit: A Synthesis from the Northern Limb of the Bushveld Igneous Complex *Tim Dunnett*	Discovery of the Oak Dam West IOCG Deposit, South Australia *Kathy Ehrig*
2:15 pm - 2:30 pm	The Formation of High-Grade Iron Formation-Hosted Iron Ores of the Northern Cape Province of South Africa: Insights from Hematite Ages *Albertus Smith*	Triple-Halogen (Cl-Br-I) Signatures of Salt Melt Inclusions in Kiruna-Type Magnetite-Apatite Ore Systems – Proof of Formation Via Evaporite Anatexis *Tobias Fusswinkel*
2:30 pm - 2:45 pm	Geological and Metallogenic Characterisation of Manganocopper Mineralisation in the Tizi N'Isdid Deposits (Ounein Massif, High Atlas, Morocco) *Imad Aflla (Student)*	Using Geochemical and Petrophysical Properties to Target Sediment-Hosted Copper Deposits *J Porter*

2:45 pm - 3:00 pm	Primary-Hosted Uranium Resources, Geology, and Mineralisation in the Damara Orogen, Namibia *Guy Freemantle*	Linking the Nature and Timing of Diagenesis of Source Rocks with Hydrocarbon Migration and Copper Mineralisation Events in the Neuquén Super Basin, Argentina *Richard Herrington*
3:00 pm - 3:15 pm	Cu and Fe Isotope Contrasts Between the Paleoproterozoic Haib Porphyry Copper Deposit and Representative Phanerozoic Porphyry Systems *Bibi Ayesha Jogee*	Geometallurgical Characterization of Potential By-Products in the Spremberg-Graustein-Schleife Kupferschiefer Deposit *Zahra Nourizenouz (Student)*
3:15 pm - 3:30 pm	Petrogenesis and Structure of the UIS Pegmatites, Namibia: A New Source of Battery Metals in Africa *Laurence Robb*	Ore-Forming Processes at the Rosh Pinah Zn-Pb-Ag Deposit, Southern Namibia: Evidence from Mineralogy, Geochemistry, and Metal Distribution *Halleluya Ekandjo (Student)*
3:30 pm - 4:00 pm	Break	
	Oral Session 09 New Discoveries and Developments I	**Oral Session 10 Vital High-Volume Base Metals III**
4:00 pm - 4:15 pm	Epithermal Gold Discoveries in the New Khundii Metallogenic Province, Southwest Mongolia *Jeffrey Hedenquist*	Breccias in Magmatic-Hydrothermal Environments *David Cooke*

DAY 3 – SUNDAY, SEPTEMBER 29, 2024

4:15 pm - 4:30 pm	Structural Controls on Williams Brook Gold Mineralization, New Brunswick, Canada *Stefan Kruse*	Insights into Enhanced Brownfield Exploration Techniques at Morenci for Improved Ore Characterization and Mineral Reserve and Resource Addition *Paul Albers*
4:30 pm - 4:45 pm	Constraining Mineralization and Alteration Through New Geologic Mapping at Spring Peak Low-Sulfidation Epithermal System, Mineral County, NV *Marcus Angus (Student)*	Geological and Geochemical Evolution of the Supergiant High-Grade Resolution Cu-Mo Deposit, Arizona *Takeshy Coaquira (Student)*
4:45 pm - 5:00 pm	B2Gold's Antelope Discovery – Fresh Insights Lead to Exploration Success at Otjikoto *James Bursey*	Anatomy of a Moderated Telescoped Porphyry–High-Sulfidation System: Insights of a 2-Km Vertical Profile of Alteration at Valeriano Cu-Au Deposit, Chile *Jaime Osorio*
5:00 pm - 5:15 pm	The Highbury Lithium Deposit, KwaZulu - Natal, South Africa *Eugene Van den Heever*	Prospectivity Modeling and Metallogeny of Pre-Orogenic and Orogenic Base and Precious Metals of the South Nabitah Belts, Arabian Shield *Arnaud Fontaine*
5:15 pm - 5:30 pm	Predicting Giants: The Role of Geoscientists and Mineral Systems in our World of Emerging AI *Nicole Januszczak*	

5:30 pm - 7:30 pm	Exhibit and Poster Reception, Speed Talks
7:30 pm - 10:00 pm	Industry Outlook Dinner (Ticket Required) Human & Social Capital Development of the World's Future Mines *Mfikeyi Makayi (Invited Speaker)*

DAY 4 – MONDAY, SEPTEMBER 30, 2024

TRIFECTA 3: BUSHVELD *Marina Yudovskaya (Session Chair)*

8:00 am - 8:15 am	Morning Comments
8:15 am - 8:45 am	The Bushveld Complex and Its Mineralization, South Africa *Judith Kinnaird (Invited Speaker)*
8:45 am - 9:15 am	Processing of the Bushveld Ores *Kirsten Youlton (Invited Speaker)*
9:15 am - 9:45 am	Platinum Group Metals: It Begins and Ends in the Market *Richard Stewart (Invited Speaker)*
9:45 am - 10:00 am	Q&A
10:00 am - 11:00 am	Poster Viewing and Next Generation Explorers Award (NGEA) Speed Talk Session

DAY 4 – MONDAY, SEPTEMBER 30, 2024

	Oral Session 11 **The Energy Transition: Metals of the Future I**	**Oral Session 12** **New Discoveries and Developments II**
11:00 am - 11:15 am	National-Scale Critical Mineral Assessments to Support Accelerating Discovery in Australia *Arianne Ford*	The Nosib Polymetallic Mineral Deposit in the Otavi Mountain Land of Namibia – New Deposit Type or Variation on a Theme? *Jonathon Dugdale*
11:15 am - 11:30 am	Transtensional Tectonics Control Emplacement of Lithium-Tin Granites in the Eastern Erzgebirge (Germany/Czech Republic) *Jan Cerny*	The Eureka Gold Deposit: Latest Gold Discovery in an Emerging Gold District in the Northern Zone, Damara Belt, Namibia: Alteration, Mineralization and Structural Controls *Wynand Slabbert*
11:30 am - 11:45 am	Structural Controls on the Emplacement and Mineralization of Lithium-Bearing Pegmatites *Lot Koopmans (Student)*	A Potential Mafic-Ultramafic Magmatic Phase of the Kunene Complex in Angola and Namibia and Its Ni-Cu-PGE Fertility *Ben Hayes*
11:45 am - 12:00 pm	End-Orogenic, Granite-Hosted Uranium Mineralisation: A Solution from Namibia *Paul Nex*	The Geology of the Mingomba Deposit in the Zambian Copperbelt *Sitwala Mulonda*

12:00 pm - 12:15 pm	Evolution of the Magmatic System Associated with a Supergiant Porphyry Cu Deposit from Trace Element Profiles in Zircons *Massimo Chiaradia*	The Discovery and Geology of the Boundary Zone Zn-Pb-Ag Deposit, Yukon, Canada: New Perspectives on Mineralizing Processes in Sediment-Hosted Base Metals Deposits *Jack Milton*
12:15 pm - 12:30 pm	Mineralogy and Petrology of the Petalite-Subtype Prof Pegmatite, Revelstoke, British Columbia, Canada *Catriona Breasley (Student)*	Diamonds from a New Witwatersrand-Like Gold Deposit Show Evidence of Archaean Subduction in the Slave Craton, Canada *Rory Changleng (Student)*
12:30 pm - 2:00 pm	Lunch Break	
	Oral Session 13 **The Energy Transition: Metals of the Future II**	**Oral Session 14** **Specialty Metals and Materials**
2:00 pm - 2:15 pm	Iron Oxide-Copper-Gold Deposits: A One-Size-Fits All Genetic Model Does Not Work *Adam C. Simon (SEG Distinguished Lecturer)*	Primary Rare Earth Elements Mineralization of the World-Class Mt Weld Carbonatite Deposit, Western Australia *Arthur de Oliveira (Student)*

DAY 4 – MONDAY, SEPTEMBER 30, 2024

2:15 pm - 2:30 pm	Iron Oxide-Copper-Gold Deposits: A One-Size-Fits All Genetic Model Does Not Work, cont'd *Adam C. Simon (SEG Distinguished Lecturer)*	A Revised Model for High Grade Ni-Co-As-Ag-U-Sb-Bi Deposits: The Product of Heating and Oxidation of Marine Evaporite Brines and Hydrocarbons from Intracratonic Basins *Jacob Hanley*
2:30 pm - 2:45 pm	Is There a Link Between Orogenic Collapse and Lithium-Bearing Pegmatites? *Tarryn Cawood*	Paragenetic Evolution of Sn-dominant Lode Systems in the Pool Mining District, Cornwall *Kiara Brooksby (Student)*
2:45 pm - 3:00 pm	Petrography and Geochemistry of the Okohongo Copper-Silver Deposit, Kaoko Belt, Kunene Region, Namibia *Abner Nghoongoloka (Student)*	Micas as Tracers of Tungsten Mineralization Case Study: Santa Comba and Fontao Deposits NW Spain *Candela Pita Díaz (Student)*
3:00 pm - 3:15 pm	A Recipe for Copper-Rich Fluids: The Importance of Sulfide Resorption and Water-Rich Magmas *Olivia Hogg (Student)*	Controls on Germanium Enrichment in Zn-Pb Sediment-Hosted Mineral Deposits *Juan Bello Rodriguez (Student)*

3:15 pm - 3:30 pm	Temporal Evolution of Miocene-Pliocene Magmatic Suites Fertile for Porphyry Cu Deposits: Preliminary Results from the Tres Cerrillos Porphyry Cu Prospect, Ecuador *Hugo Carrasco (Student)*	Indium in Mine Wastes: Geochemistry, Mineralogy, and Microbiology *Olivia Mejías (Student)*
3:30 pm - 3:45 pm	The Birth of a World-Class Komatiite-Hosted Ni District: 35 Ma of Crustal Fertilisation, Gestation, and a Violent Short-Lived Labour *Anne Virnes (Student)*	Critical Metals in Base Metal Mineralization of the Otavi Mountain Land and the Kunene Region, Namibia *Ester Shalimba*
3:45 pm - 4:15 pm	Break	

CLOSING EVENTS

4:15 pm - 4:45 pm	Motivating Future Geoscientists: NGEA Survey Insights *Tim Dobush*
4:45 pm - 5:15 pm	Conference Reflections and Future Perspectives in Memory of Brian G. Hoal (Panel Discussion) *Richard H. Sillitoe (Chair)*
5:15 pm - 6:00 pm	Closing Ceremony and Concluding Remarks *Anna Nguno, Laurence Robb, Anne Thompson, Richard Valenta, Mark Noppé*
6:00 pm - 8:00 pm	Closing Social (Ticket Required)

Core scanning to mine planning

GeologicAI

Resource Modeling Solutions / A GeologicAI Division

Enhance Resource Characterization

Quantify and Manage Uncertainty

Improve Data Consistency

Optimize Efficiencies

geologicai.com

resourcemodelingsolutions.com

Invited Speakers

Anna Bidgood is an exploration geologist with KoBold Metals working on their copper discovery team. Anna was previously a Fulbright Scholar at the Carnegie Institution for Science and a postdoctoral researcher at the Ireland Research Centre in Applied Geosciences (iCRAG). Her research explores metamorphism, mineral growth, and mineralization in the Central African Copperbelt, a large sedimentary basin that contains economic concentrations of critical elements required in the transition from hydrocarbon- to renewable energy-based technology. Anna finished her Ph.D. in metamorphic geology and Himalayan tectonics at the University of Oxford in 2020.

Deng Ngang Deng is a South Sudanese-Canadian geologist based in Australia and the founder of Target Mining Resources, a mining services and training provider. He specializes in the mining and exploration of gold and base metals, with expertise in geochemical mapping. Deng is a member of the Australian Institute of Geoscientists and a coauthor of the Pan-African Resource Reporting Code (PARC), contributing to the development of resource reporting standards across Africa. His academic background includes a diploma of project management from Western Sydney Institute, an M.S. in geology from Laurentian University, and an Honours B.S. geology specialist from the University of Toronto.

Hartwig Frimmel, who obtained his Ph.D. in geology at the University of Vienna, is a professor and Chair of Geodynamics and Geomaterials Research at the University of Würzburg, Germany. He is also associated with the University of Cape Town, where he previously climbed the academic ladder from lecturer to associate professor. He was leader of the Earth Science subprogram within the South African National Antarctic Program, president of the Society for Geology Applied to Mineral Deposits (SGA), and director of Lithoscope consultancy. He has served on several editorial boards (including *Mineralium Deposita* for the past 25 years) and as a consultant to numerous national research funding agencies, mining/exploration companies, and government bodies.

Lynnette Greyling has an M.Sc. for a study on MVT-type Pb-Zn deposits in South Africa and a Ph.D. from the University of the Witwatersrand characterizing the hydrothermal fluids of the Zambian Copperbelt. As an exploration geologist at Randgold Resources, she worked on Archaean and Paleoproterozoic orogenic gold deposits in Tanzania and West Africa. She is interested in how structures create pathways and traps for hydrothermal fluids, leading to precious and base metal mineralization. This was the focus of her postdoctoral research on Birimian-hosted orogenic gold deposits in northern Cote d'Ivoire at the University of Cape Town, where she became a lecturer in economic geology. She has been a vice president of exploration for Goldrange Resources, with projects in Cote d'Ivoire and Tanzania. She is currently based in the United Kingdom and works as an independent consultant.

Murray Hitzman is a professor in the School of Earth Sciences, University College Dublin, and director of the Science Foundation Ireland Research Centre in Applied Geosciences (iCRAG). He previously served as associate director for energy and minerals at the U.S. Geological Survey, Charles Fogarty Professor of Economic Geology at Colorado School of Mines, a policy analyst in the White House Office of Science and Technology Policy and the U.S. Senate, and an exploration geologist conducting mineral exploration worldwide. Hitzman has B.A. degrees in anthropology and geology from Dartmouth College, an M.S. in geology from the University of Washington, and a Ph.D. in geology from Stanford University. He currently serves as a technical advisor for KoBold Metals.

INVITED SPEAKERS

Kurt House is the cofounder and CEO of KoBold Metals, an AI-driven mineral exploration company that integrates the expertise of top-tier mineral explorers with a world-class team of data scientists and software engineers from Silicon Valley giants such as Apple and Google. KoBold's mission is to accelerate the discovery of essential materials for the energy transition, and over the last six years, KoBold has established a diverse portfolio of over 60 projects across four continents, some of which are in partnership with industry leaders like BHP and Rio Tinto.

Judith Kinnaird, a former president of SEG, is internationally known for her work on ore deposits in Africa. She and a team of researchers have worked on copper mineralization in the Central African Copperbelt of Zambia and the DRC; pegmatite-hosted tin, lithium, and uranium deposits in the Damara belt of Namibia; tin, niobium, and tantalum in the Younger granites of Nigeria; lithium and gem-bearing pegmatites in Zimbabwe and Somaliland; chromitite and platinum group metals in the Bushveld Complex of South Africa and Malopo Farms Complex in Botswana; lead-zinc deposits in Ireland; and Archaean gold mineralization in South Africa and Mozambique. Currently, she is focusing on vanadium, platinum group elements, uranium, lithium, and rare earth elements needed for the green economy. Judith is codirector of a National Centre of Excellence (CIMERA) for Minerals and Energy Research Analysis, cohosted at the University of the Witwatersrand, Johannesburg, and funded by the NRF/DST of the South African government.

Mfikeyi Makayi is an engineer and seasoned mining professional with over 15 years' experience in industry. She worked on one of Zambia and Africa's largest copper mines, First Quantum's Kansanshi copper-gold mine, in mining operations and in the global internal audit division covering Australia, Finland, Mauritania, and Panama's operations. She was also the country manager and director of one of Zambia's largest mining and construction equipment supply and services companies, Barloworld Equipment, supplying Caterpillar products to various industries. Her portfolio covered First Quantum's Kansanshi and Sentinel mines, Barrick's Lumwana mine, Mopani copper mines, and Gemfield's Kagem mine, the world's largest emerald mine.

Musa Manzi completed his Ph.D. degree in geophysics in 2013 at the University of the Witwatersrand. He is an associate professor, founding director of the Wits Seismic Research Centre, and a scientific coordinator of the Eramin FuTuRe Project, an international consortium project funded through a pan-European network of research funding organizations. Manzi's innovative and outstanding contributions to geosciences are concerned with the development of novel seismic exploration methods to gain a deeper understanding of Earth processes, particularly in exploring for mineral, oil, and gas reservoirs in African basins and beyond.

Benjamin Mapani has made significant contributions to academia, including helping to build an international master's program in exploration geology at the University of Zimbabwe and building a geology program at the University of Namibia. In 2020, he was offered a professorial position in mining and sustainable development at the Namibia University of Science and Technology. Benjamin's research has focused on geodynamics, economic geology, groundwater, climate change, and environmental management of producing and abandoned mines. He has a combined experience of 35 years in geology—from mapping to structural geology, geochemistry, and applied geostatistics—and believes field work is the anchor for improving emerging AI technologies. He has consulted for a number of mines and organizations and has published extensively in the field.

Kombada Mhopjeni is the national program officer for natural sciences at the UNESCO Windhoek office. She leads the delivery of UNESCO's science programs in Namibia and coordination of the geoscience program in Southern Africa, drawing on over 18 years of scientific experience in the government, including various roles in geosciences and management. She served as the national representative for the International Union of Geological Sciences and the cosecretary general of the IUGS Commission for the Management and Application of Geoscience Information. Kombada is an executive member of the Geological Society of Namibia and founding national representative of the Young Earth Scientists Network Namibia chapter. She holds an M.Sc. in geology from the University of Western Australia.

Thabisile Phumo is an accredited public relations and communication specialist and Public Relations Institute of Southern Africa (PRISA) fellow with 28 years' experience in the public and corporate sector. She is a past president of PRISA and executive committee member of the World Communication Forum Association. She is currently the executive vice president and head of stakeholder relations at Sibanye-Stillwater and has worked in various capacities in the corporate affairs environment at the University of South Africa, the United Nations High Commissioner for Refugees, the Commission for Gender Equality, and Anglo American Platinum Ltd. She is co-editor of Strategic Communication: South African Perspectives, a member of the Academic Advisory Board for Strategic Communication at the University of Johannesburg, and chair of Group Technical Security Management (Pty) Ltd. Thabisile was recognized in Mining Elites in Africa 2022 and received the Alumni Dignitas Award 2023 from the University of Johannesburg.

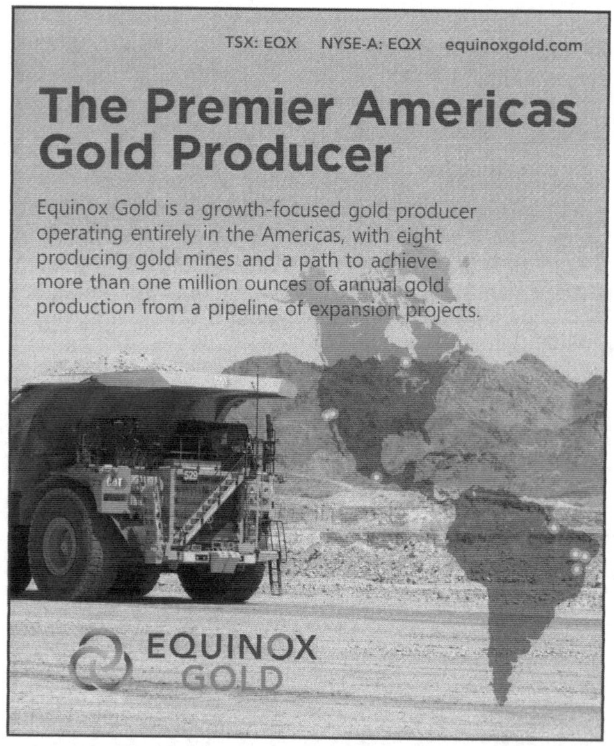

KoBold Metals

Finding the materials of the future with AI and HI

koboldmetals.com

Mike Robertson is a principal consultant with The MSA Group based in Johannesburg, South Africa. With an M.Sc. in the structural controls of gold mineralisation at the Sheba mine, Barberton greenstone belt, he has a particular interest in the application of applied structural geology to understanding controls on ore genesis and exploration targeting. This has led to work on orogenic gold deposits through southern, east, west, and northeast Africa, Saudi Arabia, Yemen, Western Australia, Russia, and Kazakhstan. He has been involved in mineral exploration, both in industry and in a consulting role, for 34 years.

John Roos is the country manager and director of B2Gold Namibia, the nation's largest gold mining company, which is 90% owned by B2Gold Corporation and 10% by EVI Mining Namibia. In his role, John oversees the company's long-term strategy and corporate development. Before his appointment as country manager in November 2022, John led the corporate finance function and managed various green energy projects within the company. John is also actively involved in the company's corporate social investment (CSI) initiatives, spearheading various projects. He is a registered chartered accountant (SA) and a member of the executive committee of the Chamber of Mines of Namibia. Additionally, John serves on the boards of the Save The Rhino Trust and Nakayale Private Academy.

David Selley is an explorer and researcher with expertise in structural geology and basin analysis. His first visit to Africa was in 1999—a regional targeting of the Birimian in Côte d'Ivoire and Mali as a very green exploration geologist for North Ltd. The continent is now his second home, having worked throughout most of Africa's countries and geologic provinces. The highlight has been a near 25-year involvement with the Central African Copperbelt, where he has had the privilege to study and explore for the most wonderful ores, all the time working with and being mentored by stellar research, exploration, and production geologists. The province boasts superb data sets, generated over 100 years of systematic exploration and mining—these underpin the concepts that will be presented. The scale and complexity of the mineralizing system means that each new discovery generates new challenges and a need to continually develop and refine our deposit models.

Adam C. Simon is the Arthur F. Thurnau Professor of Earth and Environmental Sciences at the University of Michigan, USA. He earned degrees in geology and geochemistry from the University of Maryland and Stony Brook University and was a postdoctoral fellow at Johns Hopkins University. His scholarly work focuses on the global flow of energy and mineral resources with an emphasis on the geologic availability of energy critical metals. Adam has coauthored two textbooks and published 100 papers in the field of energy and mineral resources. Adam has done research and given invited lectures on all seven continents, including a TEDX talk in 2022, and was awarded the University of Michigan Provost's Teaching Innovation Prize for his transdisciplinary pedagogy.

Richard Stewart has 25 years' experience in South Africa's geologic and mining industries, is a fellow of the Geological Society of South Africa and a Registered Natural Scientist. Richard joined the Sibanye Group in 2014 and has held the position of chief regional officer: southern Africa since 2022, subsequent to being the group chief operating officer (COO) from 2020. Prior to being COO, Richard was the executive vice president: business development at Sibanye-Stillwater, where he contributed significantly to the group's growth and acquisition strategy.

dPA 6658

Sustainability success starts with exploration

BARRICK

www.barrick.com | NYSE : GOLD · TSX : ABX

INVITED SPEAKERS

Elizabeth Turner has published extensively on sedimentary rock-hosted ore systems in northern Canada and the Central African Copperbelt. Her expertise in ore-hosting sedimentary systems, ore-forming fluids and events, carbonate and siliciclastic lithofacies and diagenesis, stratigraphy, basin analysis, paleobiology, early Earth-surface environments, fossil reefs, early animal and eukaryote evolution, and hydrothermal systems is founded in >30 years of remote field research in Proterozoic and Paleozoic sedimentary basins in northern Canada and central Africa, together with extensive microanalytical studies on the nature and evolution of ore-forming fluids and ore-hosting rocks.

Franck Twite was born in Lubumbashi, DRC Copperbelt, where he developed an interest in geology and mining. He pursued a Bachelor of Science in geology at the University of Lubumbashi, DRC, and a master's degree at the University of the Witwatersrand, South Africa. Franck has been a key member of the Kamoa copper project geology team since 2011. The work of this geology team led, among other things, to the discovery of the Kakula deposit (625 Mt at 2.73% Cu). Franck is currently accountable for the on-site geology functions including exploration, resource definition, and grade control. Franck was the leading author in two papers published in the *Journal of African Earth Sciences* and *Ore Geology Reviews*. In pursuit of further economic geology education, Franck recently enrolled as a Ph.D. student at the KU Leuven University in Belgium.

Kirsten Youlton is currently a senior mineralogist at SGS South Africa, where she specializes in automated scanning electron microscopy (autoSEM) of a range of commodities across various stages in the resource pipeline from exploration to optimization and even remediation. She routinely conducts mineralogical investigations into Bushveld ores in collaboration with the metallurgical team in order to understand, optimize, and improve processes. Prior to this, she was a senior scientist in the mineralogy division at Mintek.

Marina A. Yudovskaya is a graduate of the Lomonosov Moscow State University, where she completed her Ph.D. research on volcanogenic massive sulfide deposits of her native Rudny Altai. Upon completion of the Ph.D. degree, she went to work at IGEM in Moscow, with research focused on orogenic gold deposits in black shales, metal transport in fluids of active volcanoes, and unconventional platinum group element (PGE) mineralization. Since 2007, Marina has been involved in various projects on the Bushveld Layered Complex with affiliation at the University of the Witwatersrand (Johannesburg, South Africa). Her current projects also include studies on the intrusions and associated Cu-Ni-PGE deposits of the Norilsk region in the Russian Arctic.

Ingenesia "Inge" Zaamwani is an accomplished legal expert and executive with an extensive professional history in the mining and energy industries. She has held several government advisory positions, and she was CEO of NamDeb Diamond Corporation from 1999 to 2015, during which time she helped elevate the company to national significance and played an instrumental role in navigating the global financial crisis of 2008. Currently, Inge holds the position of senior presidential advisor, constitutional affairs and private sector interface, as well as five directorships and an honorary membership on the board of directors for the Namibia Chamber of Commerce and Industry.

Munshya Zimba is a principal geologist at First Quantum Minerals (FQM) in Zambia. His role focuses on regional generative exploration and support to the global exploration group. He has been working for over 16 years in the mining and mineral exploration industry with a focus on the Zambian Copperbelt. During this period, he has been involved in various phases of mining and exploration from generative to resource definition to production. He holds an M.S. in geochemistry from the University of Leeds, United Kingdom, and a B.S. in geology from the University of Zambia. He has a passion for the acquisition, integration, and interpretation of relevant regional geochemical, geophysical, and geologic data sets to improve understanding of the Katanga basin and subsequent mineral exploration targeting.

Posters

Posters will be on display during regular conference hours in the Poster and Speed Talk Tent. Titles and authors of posters are listed below; the presenter's name is underlined. For schedule updates, please go to www.seg2024.org.

Africa's Iconic Ore Deposits

P1.001 Investigating the Role of Country-Rock Assimilation on the Mineralisation of the Platreef, Turfspruit Farm, Northern Bushveld Complex, South Africa
Alessia Barbieri (Student), Katie McFall, Richard Herrington, Iain McDonald, Bothepha Phetlhu, Tim Dunnett

P1.002 Geology, Geochemistry and Zonation of the Tsumeb Polymetallic Deposit
Rob J. Bowell, Ester Shalimba, Abner Nghoongoloka, Matthew Leybourne, Daniel Layton-Matthews, Stephanie G. Lohmeier, Hartwig E. Frimmel, Helke Mocke

P1.003 Application of Conventional and Unconventional Exploration Methods for Sapphire Discovery in Southeastern Madagascar: From Regional to Project Scale
Evelyn Caiza (Student), Dibya J. Baral

P1.004 The Mineralogy of PGE in the Deep Critical Zone at Sandsloot, Northern Bushveld Complex, South Africa: PPGE Reef Underlain by an IPGE + Rh-Rich Base Metal Zone
Kate Canham (Student), David Holwell, Iain McDonald, Hannah Hughes, Katie McFall, Andy Lloyd, Erin Thompson, Kofi Acheampong, Lara Du Preez, Matthew Loader

P1.005 Lithostratigraphic Distribution and Geochemistry of Carrollite in the World-Class Tenke-Fungurume Cu-Co Mining District, Democratic Republic of the Congo
Malena Cazorla Martínez, Lingli Zhou, Murray Hitzman

P1.006 Mapping Under Cover: The Kalahari Copperbelt in Eastern Namibia
Branko Corner, P.R. Hildebrand, M. Kumar

P1.007 Geochemical and REE Data for Sheeted Leucogranites at the Primary Uranium Deposits of the Damara Orogen, Namibia
Guy G. Freemantle

P1.008 The Antelope Deposit, Otjikoto, Namibia: An Integrated Geochemical and Mineralogical Study
Charlie B. Hamblet (Student)

P1.009 Nickel Potential of the Historical Okiep Copper District: An Overlooked Target in a Convergent Margin Setting
John Paul Hunt, Louw van Schalkwyk, Errol Smart, Clive Benhura

P1.010 The Role of Exploration on an Active Mine: Case of Kamoa Copper Mine
Richard Ilunga, Franck Twite

P1.011 Geochronology and Assessment of the Age(s) of Khoemacau Cu-Ag Mineralization Event(s) in the Ghanzi-Chobe Belt Portion of the Kalahari Copper Belt
Tebogo Kelepile (Student), Thierry Bineli Betsi, Elisha Shemang

P1.012 The Growth and Expansion of Kansanshi, an Iconic Ore Deposit in the New Copper-Belt, NW Zambia
Nommia Kolala (Student)

P1.013 Petrography and Depositional Features of the Lower Argillaceous Series; Dunstan Farm, Chimanimani: Possible Charleswood Diamondiferous Grits Extension
Gerald S. Kupeta

P1.014 Geology of the Southwestern Part of Kakula: Structural Complexities and Evidence of Tectonic Inversion
Serge M. Makenda, Franck Twite

P1.015 Fingerprinting Mineralization in the Sediment-Hosted Cu-Ag System: Constraints from the Khoemacau Zone 5 Cu-Ag Deposit in NE Botswana's Portion of the Kalahari Copper Belt
Tefo Masena (Student), Thierry Bineli Betsi, Tebogo Kelepile

P1.016 Reassessing the Bulk Geochemistry and Origin of Hematitic Iron Ores from the Northern Cape Province of South Africa
Neo G. Moabi (Student), Bjorn von der Heyden, Janne Blichert-Toft, Fanny Thibon, Harilaos Tsikos

P1.017 Stratigraphic Position Change of the Mineralization at the Kamoa Deposit
Innocent Mushobekwa, Franck Twite

P1.018 The Effect of the Paarl Fault on the Main Sulphide Zone, Selukwe Subchamber, Great Dyke
Takudzwa Mwagura (Student)

P1.019 A Supergene-Hydrothermal Origin of the BIF-Hosted High-Grade Iron Ores in the Mbarga Prospect, Mbalam Iron Ore District, Southern Cameroon, Congo Craton
George L. Ngiamte (Student), Eleanor Green, Olugbenga A. Okunlola, Roland Maas, Alan Greig, Cheo E. Suh

P1.020 From Deposition to Diagenesis and Beyond – Tracing the Origins of Manganese-Lutites of the Kalahari Manganese Field
Valentin Oge (Student), Louis L. Coetzee, Bradley M. Guy, Albertus J. Smith, Jens Gutzmer

P1.021 Sequence Stratigraphic Correlation and Depositional Environment Mapping in the Katangan Basin of Central Africa: Implications for Mineral Exploration
Martin Purkiss, Michael C. Daly

P1.022 The Origin of Transgressive, Magnetite-Rich Gabbro at the Kameelhoek Chromite Mine on the Western Bushveld Complex
David E. Russo (Student), Rais Latypov, Grant M. Bybee, Sofia Chistyakova

P1.023 The World-Class Shamva Gold Deposit, NE Zimbabwe
Ahmet K. Sener

P1.024 The Implication of Metamorphisim and Deformation of the Vioolsdrif Domain on the Haib Porphyry Copper-Molybdenum Deposit, Namibia
Shafiishuna Silvanus (Student)

P1.025 Magmatic Stratigraphy of the Deep Platreef at Sandsloot, Northern Bushveld Complex: Carbonate Contamination and Controls on Ni-Cu-PGE Mineralisation
Erin S. Thompson (Student), David A. Holwell, Marc Reichow, Iain McDonald, Thomas Blenkinsop, Hannah Hughes, Katie McFall, Kate R. Canham, Matthew A. Loader, Lara Du Preez, Kofi Acheamponga, Andy Lloyd

P1.026 Geometallurgical Evaluation of the Lowermost, Low-Grade Manganese Bed in the Kalahari Manganese Field, South Africa
Louis van Eeden (Student)

P1.027 Timing Diagenetic Processes in the Zambian Copperbelt: Insights from U-Pb Dating of Detrital Ilmenite and Authigenic Hematite
Victor I. Vincent, Koen Torremans, Darwinaji Subarkah, Sarah Gilbert, Juraj Farkas, Aileen L. Doran, Alan S. Collins, Simon Jones, Jon Stacey, Murray Hitzman

P1.028 EPMA, Raman Spectroscopy and EBSD Analysis Track Physiochemical Changes During Hypogene to Supergene Mineralization in the Lubambe-Mingomba Cu-Co deposit
Victor I. Vincent, Koen Torremans, Richard Unitt, Christian Burlet, Aileen L. Doran, Simon Jones, Jon Stacey, Murray Hitzman

P1.029 The Hidden Talents of Quartz and Mica: Indicating Metallogenic Potential in A-Type Granitic Systems
Leo Vonopartis, Paul Nex, Judith Kinnaird, Laurence Robb

P1.030 Hypabyssal and Volcanic Rocks of the Western Foreland Shelf, Democratic Republic of Congo
James Yates, Amanda M. Stoltze

P1.031 The Significance of In Situ Sr-Nd in Apatite for Constraining Magma Sources and the Origin of Platinum Mineralisation in the Bushveld, South Africa
Peace Zowa, Ben Hayes, Grant M. Bybee

Gold: Enhanced Discovery and Development

P2.032 Gold Remobilization in Gossans of the Usakos Dome in the Southern Central Zone (sCZ) of the Damara Orogen, Namibia
Abia Amukongo (Student)

P2.033 Invisible Gold Mineralization within Korivipalle Dolostone and Dhone Dolerites: Implications from EPMA and LA-ICP-MS Analysis of Pyrite
Ramesh C. Behera (Student), Sahendra Singh, K S V. Kumar, Vinod K. Tirkey, Ramakrishna Reddy, S Shalivahan

P2.034 Liquid Bismuth Collector as a Gold Concentration Mechanism in a Late-Variscan Gneiss Dome: Martinamor Case of Study
Daniel Bermejo-López (Student), Lorena Ortega, Santos Barrios, Pedro Castiñeiras, Elena Crespo, Juan Gómez-Barreiro

P2.035 Structural Evolution of the Magino Gold Mine, Wawa Subprovince, Ontario: An Overprinted Archean Intrusion-Related Deposit
Ian Campos (Student), Bruno Lafrance, Ross Sherlock, Stefan Kruse

P2.036 Geochemistry, Petrography, and Mineralogy of Intrusive Metamorphic Dykes and Their Relation with Gold Occurances in the South of Minas Gerais, Brazil
Gabriela O. Carvalho (Student), Gustavo L. Pires, Silvia R. Medeiros

P2.037 The Buritica Gold Deposit, Middle Cauca Metallogenic Belt, Colombia
Daniel A. Castano Madrigal (Student)

P2.038 Geneses of Gold Mineralization in the Mugomo and Chifumbazi Prospects of the Muládzi Gold Field, Tete Province, Northern Mozambique
Euclesia P. Cossa (Student), Andrea Agangi, Manuel Nopeia, Vicente Manjate

P2.039 Chlorite Composition as a Tracer for Mineral Exploration of the Disseminated Gold Deposit
Tao Cui (Student), Kunfeng Qiu, Lejun Zhang

P2.040 Morphochemical and Isotopic Analysis (δ65Cu) of Gold from the Serra Dourada and Bananal Placer Deposits in the Aguapeí Belt, Southwest of the Amazonian Craton, Brazil
Kelvin dos Santos Alves (Student), Juan G. Barreiro, Santos B. Sánchez, Ryan Mathur

P2.041 Geochemical Analysis of Native Gold from the Roman Mining Area of Pino del Oro (Zamora, Spain)
Vanessa L. Escobar Duche (Student), Santos Barrios Sánchez, Kelvin dos Santos Alves, Juan Gómez Barreiro

P2.042 Geological Mapping and Reinterpretation of the Gold Prospect "Descoberto" Located Near the World-Class Cuiabá Orogenic Gold Deposit, Minas Gerais, Brazil
Beatriz A. F. Andrade (Student), Carolina P. N. Moreto, Ricardo O. A. Mabub

P2.043 Tellurium and Selenium in and Epithermal Gold Deposit in the Chatkal-Kurama Region (Uzbekistan)
Danis I. Filimon (Student), Rustam A. Khalmatov, Anna Gioncada

P2.044 Characterization of the Gold Hill Low Sulfidation Epithermal Au-Ag Vein Deposit, Northern Nye County, Nevada
Silas L. Goetz (Student), Simon Jowitt

P2.045 Understanding Microstructure of Gold Particles in Placers: A New Vector for Transport Intensity?
Juan Gomez-Barreiro, Santos Barrios Sánchez, Inéz Puente-Orench, Juan Morales Sánchez Migallón

P2.046 Chemical Composition of Garnets and the Relation to Mineralizing Fluids in the Tucano Gold Deposit, Amapá State, Brazil
Deniro F. Gonçalves Costa (Student), Gabriel A. Rodrigues Soares, Rosaline C. Figueiredo e Silva, Mahyra F. Tedeschi, Steffen G. Hagemann

P2.047 Genesis and Mineralogy of the Golddigger Property in British Columbia, Canada
Randall Karcher (Student), Simon Kocher, Holly Stein, Roger Rosmus, Simone Pujatti, Manuele Lazzarotto, Thomas Monecke

P2.048 An Assessment of the Grade Control Spacing at QKR Mine and the Effect on Au Grade Estimation
Ndinelao N. Kashidinge

P2.049 Studying the Internal Structure of Native Gold Using the EBSD Method
Aleksandr Kozin

P2.050 Mg Isotopes of Carbonate from the Shuiyindong Carlin-Type Gold Deposits of the Youjiang Basin (Southwest China) Provide Insights into Gold Genesis
Jialong Liu (Student), Xiaoye Jin, Xiaodong Deng, Jianwei Li

P2.051 Relationships Between the Tectonic Evolution and the Formation of Polymetallic Vein Deposits Along the Caledonian-Appalachian Belt
Taija Torvela, Rob Chapman, James Shaw, Georgian Manuc, Shane Webb, Lucia Savastano

P2.052 The Structural Setting of Navachab Gold Mine and Satellite Orebodies: Insights for Exploration
Richard R. Manyanga

P2.053 Characteristics of the Ore-Forming Fluids of the Giant Karagba-Chauffeur-Durba (KCD) Gold Deposit, Northeastern Democratic Republic of the Congo (DRC)
Rowin P. Marais (Student), Bjorn P. von der Heyden

P2.054 Mapping Cryptic Telescoping Events in Porphyry-Epithermal Systems Through Integrated Micro-textural-Chemical Pyrite Analysis at the Bonanza-Type Brucejack Au-Ag Deposit
Duncan F. McLeish (Student), Anthony E. Williams-Jones, James R. Clark

P2.055 Minor and Trace Elements in Native Gold: Controls and Metallogenic Implications
Julian D. Melo Gómez (Student), Evan Hastie, Harold Gibson

P2.056 Textural and Mineralogical Characterization of Gold Deposits of the Vumba Greenstone Belt of Northeastern Botswana
Lebogang Mokane, Andrea Agangi, Ryohei Takahashi, Pearlyn Manalo, Kelebogile Phili, Manuel Nopeia, Thierry Bineli Betsi, Tebogo Kelepile, Zibisani Bagai

P2.057 Links Between Mafic Volcanism and Porphyry Gold Mineralization: Insights from the Maricunga Belt, Chile
Allyson N. Murray (Student), Hannah C. Baker, J. Tomás Ovalle, Adam C. Simon

P2.058 Epithermal Au-Ag Mineralization at Galim-Legalgorou, Cameroon Volcanic Line (CVL): Constraints from Ore Mineralogy, Mineral Chemistry of Electrum and Fluid Inclusion Data
Terence Cho Ngang (Student), Cheo Emmanuel Suh, Thomas Wagner, Tasin Godlove Bafon, Tobias Fusswinkel, Akumbom Vishiti, Eric-Claude Essono Owona

P2.059 Provenance of the Placer Gold-Bearing Banket Conglomerate, Iduapriem, Ghana: Constraints from Morphology and Oxygen Isotope Characteristics of Quartz Pebbles
Edwin A. Obeng (Student), Kofi A. Ansah

CRITICAL PRODUCTIONS

P2.060 A Reappraisal of the Structures, Alteration and Tectonic Setting of the Rambi Gold Target, Western Kenya
Maryce N. Obindah (Student)

P2.061 Assessing Structural Impact on Gold Mineralization Using Aeromagnetic, Radiometric Data Sets and Field Observations: AngloGold Ashanti Iduapriem Mine, Ghana
Newman T. Odonkor (Student), Prince O. Amponsah, Samuel Nunoo, Marian S. Sapah

P2.062 Geochemistry, Sulfur Isotopes, and Fluid Inclusion Characteristics of the Neoarchean Gold Deposits of the Tati and Vumba Greenstone Belts of NE Botswana
Kelebogile Phili, Ryohei Takahashi, Andrea Agangi, Pearlyn Manalo, Lebogang Mokane, Manuel Nopeia, Hinako Sato, Thierry Bineli-Betsi, Tebogo Kelepile, Nigel Blamey, Tavis Enno, Zibisani Bagai, Kemmonye Baliki

P2.063 Precious and Base Metal Occurrence in Amphibolites within the Buenos Aires Complex, Tandilia System, Argentina
Belén Marone, Federico J. Saponara, María F. Lajoinie, Mabel E. Lanfranchini

P2.064 Evidence of Sulfide Melting in Pegmatite Veins Containing Precious and Base Metals in the Paleoproterozoic Basement of the Tandilia System, Buenos Aires, Argentina
Federico J. Saponara (Student), Belén Marone, María F. Lajoinie, Mabel E. Lanfranchini

P2.065 Chlorite Thermometry and Its Implication to the Genesis of Gold Mineralization at Paramanahalli, Dharwar Craton, India
Manju Sati (Student), Rajagopal Krishnamurthi, Sakthi S. Chinnasamy, Ajit K. Sahoo

P2.066 Vein Textures at the Moss Low-Sulfidation Epithermal Gold Deposit, Arizona: Constraints on the Processes of Mineral Deposition
Greg Seitter (Student), Thomas Monecke, Patrick Wood

P2.067 The Discovery Under Cover of the 1.3-Moz Dokwe Gold Project, SW Zimbabwe
Ahmet K. Sener

P2.068 The Karibib Region as an Emerging Gold District in the Damara Belt, Namibia: Linking Local Fluid Sinks to Regional Fluid Flow
Josia Shilunga (Student), Alexander Kisters

P2.069 Geology and Gold Mineralization at the Leeville Underground Complex, Carlin, Nevada
Koray Tasbicen (Student), Elizabeth A. Holley, Kendle Fraley, Elizabeth Stock, Carl Pollard, Joseph Becker

P2.070 Kinetics of Silica Deposition as a Control on Silica-to-Ore Ratio in Mineralized Epithermal Veins
Lauren R. Terry (Student), Thomas Monecke, Cliff Buck

P2.071 What Drove the Leaching of Gold from the Mesoarchen Land Surface: A Global Climate Modelling Approach
Lisa Wasitschek, Hartwig E. Frimmel

P2.072 Direct Dating of the Mineralization by the Single Grain (U,Th)-He Pyrite Geochronology
Olga Yakubovich

P2.073 Hydrothermal Alteration and Mineralization Footprint of the Čoka Rakita High-Grade Gold Skarn Deposit, Timok, Serbia
Jelena Zivanovic, Dragana Davidovic, Mladen Zdravkovic, Nutu Groza, Ivan Strmbanovic, Dragan Milosevic, Snezana Janjic, Olivera Dukic-Mijucic, Aleksandar Sestak, Vuk Milic, Ivana Suzic, Vaso Kitanovic, Dragana Drobnjak, Bojana Kotlajic, Tijana Mitic, Janja Mihajlovic, Nikola Novicic, Stefan Metodiev, Kuncho Kunchev, Kalin Kouzmanov, Aleksandar Pacevski, Paul Ivascanu, Istvan Marton

Innovative Technology Developments in Mineral Deposit Science

P3.074 Toward Automatic Airborne Magnetic Mapping – VAMPair Colibri
Laurent Ameglio, Francis Galluser

P3.075 Using InSAR to Detect Soil Expansion: A New Way to Explore for Kimberlites
Kamogelo Baloyi (Student), Stephanie Enslin-Scheiber

P3.076 MOSMIN: Multiscale Observation Services for Mining Related Deposits
René Booysen, Sandra Lorenz, Moritz Kirsch, Richard Gloaguen

P3.077 Hydrothermal Alteration Zones Estimated by ASTER Satellite Image and ASD TerraSpec on the Eastern Zone of Cerro de Pasco, Perú
Evelyn Caiza (Student)

P3.078 Deep Learning-Based Predictive Modeling of Mineral Prospectivity with a Class-Balanced Focal Loss Function for Tackling Training Data Inequality
Zhiqiang Zhang, Gongwen Wang, Emmanuel John Carranza

P3.079 Application of BEMD in Extracting Gravity Anomaly Components Showing Deep Ore-Forming Dynamic background of the Giant Jiaodong Gold Cluster Region
Yongqing Chen, Xu Zhu, Chen Chen

P3.080 How do We Determine and Use Geochemical Background in Big-Data-Based Exploration?
Mason L. Dykstra, Thomas-Louis de Lophem

P3.081 Enhancing Platinum Group Mineral Recovery from Flotation Tailings: A Process Mineralogy Approach for Alternative Pretreatment
Borbor Gibson (Student), Nonkuselo Madlakana, Glen Nwaila, Yousef Ghorbani, Eligiusz Gugala, Sehliselo Ndlovu, Jochen Petersen

P3.082 Idaho's Proterozoic-Hosted Copper-Cobalt Belt: Characteristics, Challenges and New Geologic Insights
Virginia S. Gillerman, Reed S. Lewis, Darin Schwartz, Katharina Pfaff, Daniel Schmidt, George King, William Stone

P3.083 Reuse of Granite Quarry Waste for REEs Recovery Along the Industrial Production Chain: Results from Preconcentration Tests
Silvia Gioiello (Student), Licia Santoro, Alberto Cazzaniga

P3.084 Geometallurgical Evaluation of a Tailing Storage Facility to Address Environmental Concerns and Maximize the Extraction of Contained Resources
Petya Gutzmer, Kai Bachmann, Raimon Tolosana-Delgado, Laura Tusa, Cecilia Contreras, Philipp Buettner, Jens Gutzmer

P3.085 Muon Attenuation Radiography and Tomography in the Polymetallic St. Christoph Skarn Deposit, Ertzgebirge, SE Germany
Marko Holma, Pasi Kuusiniemi, Gergely Surányi, Gábor Nyitrai, László Balázs, Lars Starke, Eric Hohlfeld, Jens Metschurat

P3.086 A GIS and Remote Sensing Based Approach for the Classification of Mine Tailings in South Africa
Gaone Joseph (Student)

P3.087 Critical Mineral Exploration with Compiled Data: Joint Imputation and Clustering for Improved Analytics
Rebecca M. Montsion, Gerhard Visser, Carsten Laukamp

P3.088 Application of Machine Learning Algorithms to Predict Rock Types Using Geochemical Data: A Case Study from the Obuasi Gold District, Ghana
Abdullah B. Muhammed (Student), Emmanuel D. Sunkari

P3.089 A Process Mineralogy Approach for Direct Leaching of Precious Metals from the UG2 Ore in the Bushveld Igneous Complex
Nonhlanhla Ndlovu (Student)

P3.090 Satellite-Based Mapping for Lithium Pegmatite Mineral Exploration in Namibia: A Criteria-Based Approach
Olivia Rhind, Christopher Lambert, Alfie Baines, Mark Broadley

P3.091 Drone-Based Electromagnetic Survey System for Mineral Exploration
Ari J. Saartenoja, Markku Pirttijärvi, Pekka Korkeakangas

P3.092 Tin and Zinc Stable Isotope Variations in Cassiterite and Sphalerite from the Bolivian Tin Belt
Evan C. Hirsh, Ryan Mathur, Adam C. Simon, Wayne G. Powell

P3.093 Hyperspectral Outcrop Surveys for 3-D Geological Mapping
Laura Tusa, Samuel Thiele, Erik Herrmann, Cecilia Contreras

P3.094 Precompetitive Support of Exploration and Development in the Critical Minerals Era
Richard K. Valenta

P3.095 Innovative Narrow Reef Mining (NRE) Technology for Ultrathin Subhorizontal Ore Deposits
Sibila Borojevic Sostaric, Ema Vokic (Student), Vjecislav Bohanek, Luka Petro, Paulo Plese

P3.096 Modeling Field Strength and RQD to Interpret Potentially Poor Ground Conditions in Southwest Wau Graben, Papua New Guinea
Farzi Yusufali, Daniel Mogilny, Yush Rajcoomar

P3.097 Application of Anomaly Components of Multi-Element Combination Extracted from Stream Sediments Data by BEMD in Mapping Ore Prospectivity Within the JGCR, Eastern China
Aoyue Zheng

P3.098 Evaluating the Importance of Ore-Related Features and Favorability of Ore-Forming Spaces by Machine Learning Prediction Combined with Dynamics Simulation
Liangming Liu, Feifu Zhou

P3.099 Using Machine Learning to Integrate 3D Geological Modeling with Geodynamics Simulation to Predict Potential Ores: An Example from Dongguashan Orefield, Eastern China
Feihu Zhou (Student), Liangming Liu

P3.100 The Prediction of Uranium Occurrence Using GIS Based Geostatistical Algorithms in Erongo Region
Silas T. David

P3.101 EIS: Exploration Information System – Linking of Mineral Systems and Mineral Prospectivity Mapping
Andreas Knobloch, Vesa Nykänen, Hafsa A. Munia, Tobias E. Bauer, Guillaume Betrand, Juha Kaija, Joy Cremesty, John M. Carranza

P3.102 Mineral Exploration Mapping Tools Using Machine Learning and Drone Geophysics
Andreas Knobloch, Sami Niemi, Jochen Kamm, Rene Forsberg, Ari Saartenoja

New Discoveries and Developments

P4.103 Mapping Nigeria's Mineral Frontier: AI-Driven Prospecting for Investment Opportunities
Luis Pizano, Abba Usman, Abdulrazaq Garba, Pablo Martin-Cocher, Andreas Barth, Andreas Brosig, Andreas Knobloch

P4.104 Exploring for Cu-Au Alkalic Porphyry Deposits from the Fringes
Fabrizzio Abello (Student), Farhad Bouzari, Shaun Barker, Devin Wade

P4.105 Unlocking Gold Potential: Remote Sensing Insights and Applied Analysis of the Qena-Safaga Area in Egypt's Eastern Desert
Mohamed Badawi (Student)

P4.106 Trace Element Systematics of Auriferous Sulfides and Its Implications on the Gold Mineralization Within the Vempalle Dolostone of the Cuddapah Basin, India
Sahendra Singh, Vijay K. K S, Ramesh C. Behera, Vinod K. Tirkey, Shalivahan S

P4.107 Rare Earths and Other Minor Elements in Hungarian Coal Formations
Csilla Balassa, Mohamed Elbalawy

P4.108 Crustal Architecture of the EMARK Oceanic Core Complex
Olivia Filson (Student), Melissa O. Anderson, Jonathan Umbsaar

P4.109 The Role of Magmas in Forming a Big Skarn System: A Case Study from La Colorada Deposit (Zacatecas, Mexico)
Pedro Francisco (Student), Zhaoshan Chang, Shiqiang Huang, Chris Emerson, Dante Juarez, Patricio Perez, Robert A. Creaser

P4.110 Geology, Mineral Alteration and Mineralization of the Santa Cecilia Porphyry Au-Cu Deposit, Maricunga Belt, Northern Chile
Jose Franco (Student), Antonio Arribas, Luciano Bocanegra, Piotr Paleczek

P4.111 Chlorite Chemistry of the New Afton Alkalic Porphyry Cu-Au-(Pd) Deposit, Canada
Andres Gonzalez (Student), Adam Simon, Devin Wade, Catherine Ryan

P4.112 Retrograde, Shear-Related Copper Mineralization at the Onganja Mining District, Namibia
Matthew P. Hales (Student), Judith A. Kinnaird

P4.113 High Grade Cu-Au-Ag Overprint at the Filo del Sol Porphyry Deposit, Argentina-Chile
Jack Halloran, Jeffrey Hedenquist, Keiko Hattori, Jose Perello

P4.114 Hitting the the Right Frequency - Not All Conductors are Created Equal
Vessela Hobson

P4.115 Nickel-Copper Sulfide Mineralization in the Maibele-North Ni Sulfide Prospect, Central-Eastern Botswana
Jacob Kaavera, Thatayaone Oletile, Elvis Mosweu, Thembiso O. Basupi, Elisha Shemang

P4.116 Unravelling Manganese Mineralization Through Intergration of Remote Sensing and Geophysical Techniques in Mansa Area of Luapula Province of Zambia
Chanda Katotobwe (Student), Kalela Banda

P4.117 Unidirectional Solidification Textures and Magmatic Silica Caps; What Do They Tell Us?
Douglas J. Kirwin, Axel Müller

P4.118 Petrogenesis of the PGE Mineralized Mafic-Ultramafic Sills on Uitloop Farm, Northern Limb
Mabatho Mapiloko (Student), Marina Yudovskaya, Paul Nex

P4.119 Controls on the Rarity of Porphyry Cu Deposits in the Archean
Xuyang Meng

P4.120 Late Cretaceous Metallogeny of the Turkish Eastern Pontides and the Georgian Bolnisi District: Evidence for a ~700 km-Long Epithermal-Porphyry and VMS Belt Along Eurasia
Robert Moritz, Şafak U. Sönmez

P4.121 Understanding of Ground Gravity Responses Over the Kiala Deposit, Western Foreland, Democratic Republic of Congo
Melanie J. Postman, Roger Salumu

P4.122 Unidirectional Solidification Textures from Two Andean Intrusion-Related Cu-Au Prospects
Misael Robles (Student), Lisard Torró, Antonio Arribas, José Franco Moraga, Douglas Kirwin

P4.123 Timing and Characteristics of Skarn-Hosted Cu-Au Mineralization at Blue Copper, Montana, USA
Erin Shervey (Student), Eric LeLacheur, Zhaoshan Chang, Shiqiang Huang, Mathias Burisch

P4.124 Eolian Organic and Pyritiferous Siltstones as Redox Traps for Sediment-Hosted Ore Deposits
Leandro Silva (Student), Peir Pufahl

P4.125 Hydrothermal Alteration and Cu-(Co) Mineralization of the Fishtie Deposit: Implication for Metal Source and Timing of Mineralization in the Zambian Copperbelt Deposits
Mukuka Simusokwe, Yasushi Watanabe, Takuya Echigo

P4.126 Physicochemical Conditions of Ore-Forming Fluids at Filo del Sol Deposit, NW Argentina: New Insights from Laser Ablation Split Stream S Isotope and Trace Element Chemistry
M. Camila Sojo Aguero, Hervé Rezeau, Michael Pribil

P4.127 The Geological Characteristics of the Newly Outlined Cu-Au Deposits at the Sharlo and Krasta Prospects, Northeast of the Chelopech Mine, Bulgaria
Anastasiya D. Stoyanova (Student), Beihan Chaush, Betina Petkova, Georgi Ivanov, Hristomir Stanev, Ilian Atanasov, Ivan Petrov, Kuncho Kunchev, Martin Dobrev, Miroslav Kavarimov, Ross Overall, Strahil Terziyski, Viktor Stoychev, Aneta Aleksieva, Mariana Trifonova, Miglena Grigorova, Stefan Metodiev, Paul Ivascanu, Istvan Marton

P4.128 Hydrothermal Alteration Footprint of the Chelopech High-Sulfidation Cu-Au Epithermal Deposit, Bulgaria
Mariana G. Trifonova, Anastasiya Stoyanova, Aneta Aleksieva, Beyhan Chaush, Betina Petkova, Georgi Ivanov, Hristomir Stanev, Ilian Atanasov, Ivan Petrov, Kuncho Kunchev, Miroslav Kavarimov, Ross Overall, Strahil Terziyski, Viktor Stoychev, Miglena Grigorova, Stefan Metodiev, Paul Ivascanu, Istvan Marton

P4.129 3D Lithological and Structural Geo-Modelling: A Case Study of The Mudilandima Deposit, Tenke Fungurume Mining District, Lualaba Province, Democratic Republic of Congo
Robert K. Tshilumba

P4.130 The Skarn to Porphyry Transition: Establishing Geochronological and Geochemical Links Between Skarn and Porphyry-Like Mineralization at Craigmont, British Columbia
Warren Wegener (Student), Shaun Barker, Farhad Bouzari, Matthew J. Manor

P4.131 An Investigation of Helium Liberation During Olivine Serpentinization as a Potential Future Experimental Method for Research into Natural Subsurface Hydrogen Production
Tori Weston

P4.132 PGE Behavior During Serpentinization of Ultramafic Sequences in the Bushveld Large Igneous Province
Marina Yudovskaya, Bruno Bartolomeu, Judith Kinnaird, Roger Key, Andy Moore, Ian McGeorge, Mabatho Mapiloko

Resource Development: ESG from Exploration to Remediation

P5.133 Embedding Good Governance in Mining Operations and Value Chains
Dennis O. Achoki (Student)

P5.134 Partitioning and Mobility of Deleterious Elements from Sulphide Oxidation in the Historical Gold Mine Tailings
Steve J. Chingwaru (Student), Bjorn Von der Heyden, Margreth Tadie

P5.135 Assessing Opportunities for Treating Acid Mine Drainages and Environmental Degradation of Kettara Mines in Morocco
Loubna Elamraoui (Student), Abdellatif Elghali, Mostafa Benzaazoua, Olutoyin A. Fashae

P5.136 Sustainable Mining Through Bioleaching in the Central African Copperbelt (CACB)
Liseli Kunda (Student)

P5.137 United Nations Resource Management Systems as a Guide for Sustainable Resource Management: Case Example from Copperbelt Province, Zambia
Michael M. Musialike (Student), Eva Marquis, Gabriel Ziwa, Mubanga Mutale, Karen Hudson-Edwards, Frances Wall

P5.138 Navigating Namibia's Energy Crossroads: Balancing Renewable Ambitions and Fossil Fuel Opportunities
Shepherd M. Nyambe (Student)

P5.139 Resilience and Governance in Artisanal Gold Mining: Addressing Challenges and Opportunities in Western Kenya
Christine A. Omuombo, Joy A. Owidhi

P5.140 Local Consideration for Blast-Induced Ground Vibration in Mining Communities
Harriet N. Tetteh (Student), Wenfeng Li, Virginia T. Mclemore, Alex Rinehart

P5.141 Minimising the Environmental and Social Impacts of Exploration: Lessons from the GREENPEG Project
Kate Smith, Frances Wall, Robert Pell

P5.142 Circular Economy Can Start from the First Stages of Exploration Geology
Frances Wall, Eva Marquis, Karen Hudson-Edwards, Markus Zils, Gawen Jenkin, Phil Bird, Carol Pettit, Vicky Smyth

Specialty Metals and Materials

P6.143 The Kuboos-Bremen Line in Namibia and South Africa: New Genetic Insights and its Potential as a Hub for REE Enrichment Research
Jorge C. Arthuzzi (Student), Robert J. Giebel, Jochen Kolb, Aratz Berañoaguirre, Michael Marks, Paul H. Macey, David L. Reid, Lorenz Kemmler, Benjamin F. Walter

P6.144 Nb-Ta Primary Oxides and U Secondary Minerals from Pegmatites in Central Portugal
Santos Barrios Sánchez, Juan G. Barreiro, Julio G. García, Vitor d. Maia, Wilck G. de Campos, Ricardo Augusto S. Cipriano

P6.145 The Control of Biotite on the Lithium, Fluorine and REE Budget of Crustal Silicate Magmas
Charles D. Beard, Vincent J. van Hinsberg, John Stix, Madeleine C. Humphreys, Owen M. Weller

P6.146 Constraining the Magmatic-Hydrothermal Evolution of a Gneiss Dome: Insights from U-Pb Dating and Trace Element Analysis of Sn-W Mineralization
Daniel Bermejo-López (Student), Lorena Ortega, Pedro Castiñeiras, Lorenzo Tavazzani, Santos Barrios, Juan Gómez-Barreiro, Elena Crespo, Cyril Chelle-Michou

P6.147 Detailed Mineralogical Study of the Bentonitic Basin of Northwestern Sardinia (Italy): the S'Aliderru and Sa Pigada Bianca Deposits
Sebastiano Coticelli (Student), Isabel Abad, Giuseppina Balassone, Piergiulio Cappelletti, Concetta Rispoli, Fabio Granitzio, Mauro Farenzena, Nicola Mondillo

P6.148 The Black Angel Zn-Pb District, West Greenland: Germanium Mineralization and Syn-metamorphic Ore Formation
Michael Eigler (Student), Benjamin F. Walter, Jochen Kolb

P6.149 Recovery Appraisal of Ag, Bi, In, Sb, Se, and Te in Processing Samples from the Ruwai Pb-Zn Skarn Deposit, Central Kalimantan, Indonesia
Shelly M. Faizy (Student), Alkiviadis Kontonikas-Charos, Mathias Burisch, Doreen Ebert, Robert Möckel, Bradley M. Guy, Joachim Krause, Valby van Schijndel, Arifudin Idrus, Ernowo Ernowo, Anggara Widyastanto, Agata V. Kindangen, Max Frenzel

P6.150 Hydrothermal Rare Earth Element (REE)-Enrichment in an Iron-Oxide-Apatite (IOA) Type Deposit at Kwyjibo, Québec, Canada: A Fluid Inclusion and Stable Isotope Study
Gary Fung (Student), Anthony E. Williams-Jones

P6.151 Remote Sensing Aided Geological Investigation of Rare Earth Element Mineralization in Sangu-Ikola Carbonatites, Southwestern Tanzania
Japhet N. Fungo (Student), Nelson Boniface, Elisante Mshiu

P6.152 Critical Metal Distribution in the Renison Bell Tin Deposit, Western Tasmania
Javier Gil-Rodriguez, David R. Cooke, Owen Missen, Wei Hong, Ivan Belousov, Colin Carter, Esther Little

P6.153 Sokli – A Multimetal Carbonatite Deposit Abounds with Critical Elements
Tuomas Gråsten, Teo Lehto, Pasi Heino, Mikko Numminen

P6.154 Cassiterite LA-ICP-MS U-Pb Geochronology of Greisen- and Vein-Hosted Sn Mineralization in the Eastern Erzgebirge (Germany/Czech Republic)
Marie Guilcher, Mathias Burisch, Richard Albert, Axel Gerdes, Jan Černý, Sam Thiele, Uwe Lehmann, Henrik Kaufmann, Jens Gutzmer

P6.155 From Eboundja Nepheline Syenite (Southern Cameroon) to Ion Adsorption REEs Deposit: Implications for REE Deposits Genesis
Kenneth K. Kwanang (Student), Licia Santoro, Louise Marie Bidjeck, Paul-Désiré Ndjigui

P6.156 Rare Earth Element Mobilization and Enrichment in Weathered Carbonatite of the Dong Pao Deposit, Vietnam
Martin Li, Martin Smith, Mei-Fu Zhou

P6.157 The Petrography and Paragenetic Sequence of U-REE-Cu-Au Skarn Mineralisation in NW Queensland
Christina Loidolt (Student), John Mavrogenes, Ross Chandler, Antony Burnham

P6.158 Insights Into U-REE-Cu-Au Skarn Occurrences in the Eastern Mount Isa Inlier from Garnet Geochemistry and Geochronology
Christina Loidolt (Student), John Mavrogenes

P6.159 Mineral Chemistry, Fluid Inclusion, and Isotopic Characteristics of the Enargite Orebodies at Lepanto, Mankayan, Philippines
Pearlyn Manalo, Ryohei Takahashi, Manuel Nopeia, Kalin Kouzmanov, Nigel Blamey, Andrea Agangi, Akira Imai, Leo Subang, Marlina Elburg, Henriette Ueckermann, Linda Iaccheri

P6.160 Hydrogeochemical Modelling of Indium in a Mine Waste Environment: Opportunities for Recovery from Aqueous Sources
Olivia Mejías (Student), Thomas Poulet, Pouria Behnoudfar, Laura Jackson, Anita Parbhakar-Fox

P6.161 Multi-stage Evolution and REE Mineralization of Wigu Hill Carbonatite, Uluguru Mountains, Tanzania
Mary C. Moshi (Student), Yasushi Watanabe, Nelson Boniface, Tatsuki Tsujimori, Carmela Tupaz, Daisuke Araoka, Shogo Aoki, Elisante E. Mshiu

P6.162 Thermal Separation of Silicate-Fluorosilicate Melts: Trace Element Partitioning and Implications for Granite-Hosted REE Deposits
Wil K. Munro (Student), Neil R. Bennett

P6.163 Rare Elements Granite-pegmatites in Kivu Region (D.R.Congo)
Rub'son H. N'nahano (Student), Moise Luemba, Huan Li, Mohammed Abdallah E. Ibrahim

P6.164 Trace Element Characterization of Scheelite in the Polyphase Intrusive W-(Sn-Ta) deposit of Santa Comba, NW Spain
Candela Pita Díaz (Student), Elena Crespo Feo, Sean McClenaghan, Fernando Tornos Arroyo

P6.165 Mineral Forms of Thallium and the Genetic Model of the Vorontsovskoe Gold Deposit (North Urals, Russia)
Fedor D. Sandalov, Mikhail A. Rassomakhin, Sergey Y. Stepanov, Elena V. Belogub

P6.166 Dolocrete as a Tool for Rare Earth Elements Exploration Undercover – A Case Study of Mt Weld Carbonatite, Western Australia
Anderson Santos (Student), Artur Deditius, Hans Oskierski, Arthur Oliveira, Denis Fougerouse, Pete Kinny, Ganesh Bhat, Malcolm Roberts

P6.167 Hunting the Magmatic Processes that Lead to Granite-Related W Mineralization
Angela I. Santos Costa (Student), Coralie Siégel, Rosa Didonna

P6.168 The World-Class Dolphin W Deposit, King Island, Tasmania – Exploration Tools and Infrared Spectroscopy
Angela I. Santos Costa (Student), David R. Cooke, Lejun Zhang, Yamila Cajal

P6.169 Features that Enhance the Development Potential of Economic, Carbonatite-Hosted REE, Nb, and P Deposits
Pete Siegfried, Frances Wall, Kathryn Moore

P6.170 Searching for Tin and Tungsten in Cornwall, UK: Insights from Geochemistry and Geological Mapping
Gabriel Smith (Student)

P6.171 The Importance of Pressure and Fractionation in Forming Compositionally-Diverse, Rare-Earth Rich Igneous Complexes: A Case Study of the Nechalacho Layered Suite, Canada
Caroline R. Soderman, Owen M. Weller, Charles D. Beard, Nicolas Riel

P6.172 Fluorite Mineralization Associated with Albitite and Skarn: An Example of Jinmu-Mihara Fluorite Deposit, Sanyo Belt, SW Japan
Masahiro Sunada (Student), Yasushi Watanabe, Takuya Echigo

P6.173 The Twihinate Carbonatite Complex, Moroccan Sahara: Genesis, REE-Nb Ore Formation, and Geochronology
Lucas Tatnell (Student), Marco Fiorentini, Mohammed Bouabdellah, Wissale Boukirou, Andrea Giuliani, Chris Fisher, Laurent Sapor

P6.174 Tracing the Metal Sources of Iberian W-Sn Deposits Through Sr-Nd Radiogenic Isotopes
Iñigo Borrajo, Fernando Tornos, John M. Hanchar, Jose M. Fuenlabrada

P6.175 The Las Herrerias Deposit: Replacive IOCG Systems in the Ossa Morena Zone (SW Iberia)
Carmen Conde, David Rodriguez, Fernando Tornos, Dolores Garcia, Gonzalo Ares

P6.176 A Geometallurgical Approach to Recovery of Critical Metals Tellurium, Bismuth, and Tungsten as By-products from a Swedish Gold Mine
Lara U. Tritton (Student), Gawen R. Jenkin, Daniel J. Smith, Helena Moosberg-Bustnes

P6.177 A Process-Based Approach to Constraining Metal Enrichment in Alkaline-Silicate Magmatic Systems Using New Thermodynamic Models
Owen Weller, Tim Holland, Caroline Soderman, Eleanor Green, Roger Powell, Charles Beard, Nicolas Riel

P6.178 Mineralogical Characterization and Location of Critical Metals in the VHMS: Reevaluating the Potential of the Tharsis Mine
Carmen Conde, Gonzalo Ares, Sergio Tenorio, Fernando Tornos

P6.179 Heemskirk Tin Project, Western Tasmania, Australia: Characterization of Severn-Queen Hill Tin Deposits
Nelao N. Naimbale (Student), David R. Cooke, Wei Hong, Robert Scott

The Energy Transition: Metals of the Future

P7.180 The 3D Distribution of Cobalt in VHMS: The Elvira Orebody, Iberian Pyrite Belt
Gonzalo Ares (Student), Carmen Conde, Guillem Gisbert, Juan C. Videira, Fernando Tornos

P7.181 The Ore Genesis of Okanjande Graphite Deposit, Central Zone, Damara Belt, Namibia
Nelao N. Naimbale (Student), Shao-Yong Jiang, Fred Kamona, Daniel Machoko

P7.182 Potential Sources of Rare Earth Elements (REEs) in Loch Borralan, Northwest Highlands, Scotland: Geochemical and Mineralogical Controls on REE Distribution in Carbonatite
Abdulkadir M. Abdulkadir (Student)

P7.183 Magmatic-Hydrothermal Alteration and Deformation of Spodumene Pegmatites from Musha-Ntunga (Rwanda)
Jolan Acke (Student), Stijn Dewaele, Justin Uwiringiyimana, Renata Barros, Anouk Borst

P7.184 Mineralogy and Phosphate Paragenesis of the Ewoyaa Lithium Pegmatites, Southern Ghana
Salaam J. Adams (Student), Marieke V. Lichtervelde, Prince O. Amponsah, Abdul-Razak S. Ballah, Iwan WIlliams, Lennard Kolff, Prosper M. Nude, Daniel K. Asiedu, Samuel B. Dampare

P7.185 Petrography and Geochemistry of Pegmatite and Leucogranite of the Ntega-Marangara Area, Burundi, in Relation to Rare Metals Mineralisation
Quesnay de Jésus Akabahinga (Student), Anthony T. Bolarinwa, Seconde Ntiharirizwa

P7.186 Petrography, Geochemistry, Tectonic Setting, and Rare Earth Elements Mineralisation Potentials of Selected Granitoids from Southwestern Nigeria
Shakirat Aminu, Akinade S. Olatunji, Maryam O. Abdus-salam, Victoria B. Omotunde, Oghenerume S. Emecho-Ted

POSTERS

P7.187 Nickel and Cobalt Enrichment Zone and Ore Potential Distribution in Sorowako Area and Surroundings, South Sulawesi, Indonesia
JJ A. Ayyala (Student), Fadlin Fadlin

P7.188 Potential of Mineral Raw Materials for Battery Production in Hungary
Mohamed Badawi (Student), Evane C. de Cuñha

P7.189 Vanadium Enrichment in Silurian Black Shales of SE Sardinia (Italy): Comparison with Other Cases of Northern Gondwana
Maria Boni, Giuseppina Balassone, Nicola Mondillo, Giovanni Di Maio

P7.190 LCT Pegmatite in the West African Craton: New Insights on the Metallogenic Context and Exploration Approach Based on the Issia District (Ivory Coast)
Christophe Bonnetti, Christophe Ballouard, Hugo Watine, Arnaud Fontaine, Anne-Sylvie André-Mayer, Karl Akueson, Rémi Bosc

P7.191 Miocene Western Balkan Lithium-Borates Metallogenic Zone
Sibila Borojevic Sostaric, Tomislav Brenko

P7.192 Monazite-Rich Carbonatite Dykes at Eureka, Namibia
Sam Broom-Fendley, Ed Loye, Pete Siegfried, Tolene Kruger, Chris Drysdale, Tim Smalley, Rainer Ellmies, Frances Wall

P7.193 Rare Earth Element Mineralization in the Dongpao Deposit, Tamduong, North Vietnam
Tinh T. Bui (Student), Yasushi Watanabe

P7.194 Hard Rock-Hosted Lithium Deposits: A Process-Centric Review
Edward Bunker, Jean-Jacques Ferguson, Chris Lambert

P7.195 Strengthening Li-Bearing Pegmatite Exploration Through Applied Geophysics
Ana M. Carvalho (Student), Ricardo Ribeiro, Alexandre Lima, Rui Moura, Klaus Brauch, Axel Müller

P7.196 Metamorphic Upgrading of Cobalt in a Volcanogenic Massive Sulfide Deposit
Tarryn K. Cawood, Jan M. Peter, Duane C. Petts, Matthew J. Polivchuk

P7.197 On the Physical Transport of Magmatic Sulfides: Insights from the Trace-Element and Isotope Signature of Sulfide Globule Halos
Maria Cherdantseva (Student), Marco L. Fiorentini, Christopher M. Fisher, Antony I. Kemp, Laure A. Martin, Matvei Aleshin, Malcolm P. Roberts

P7.198 The Origin of Stratabound Cobalt Mineralization Along the Mines-R.A.T. Contact at Kinsevere, DRC, Central African Copperbelt
Robert S. Collar (Student), Hervé Ilongo, Takalani Mandiwana, A. E. Williams-Jones

P7.199 Advances in Exploring for Ni and V Ore Deposits Using Magnetite as an Indicator Mineral
Sarah Dare

P7.200 Genesis of IOA and IOCG Deposits Along the Cordillera de la Costa, Peru: Insights from Iron (δ56Fe) and Triple Oxygen (δ18O, Δ17O) Isotopic Analysis of Magnetite
Jose C. Davila (Student), Fredrik Sahlström, Fernando Tornos, Valentin R. Troll, John M. Hanchar, Ilya Bindeman, Xin-Yuan Zheng

P7.201 Geochemical Characterization of K-Feldspar in Lithium-Rich Aplite-Pegmatites: Insights from the Barroso-Alvão Region, Northern Portugal
Filipa Dias (Student), Ricardo Ribeiro, Filipe Gonçalves, Alexandre Lima, Encarnación Roda-Robles, Tânia Martins

P7.202 Effects of Post-Mineralization Metamorphism and Deformation on Sulfides and their Remobilization in the Hotinvaara Ni Prospect, Pulju Greenstone Belt, Northern Finland
Juan C. Díaz Carreño (Student), Petri Peltonen, Jaakko Puntalo, Christoph Beier, Kathleen A. Law, Jake Williams

P7.203 Multi-Scale Characterisation of Cobaltite at Palokas, Rajapalot by X-Ray Computed Tomography and Electron Microscopy
Lily Dickson (Student), Taija Torvela, Robert Chapman, Jukka-Pekka Ranta, Alice Macente

P7.204 Lithium and Graphite Mineralisation of the Kuldjuktau Mts., Uzbekistan
Alla Dolgopolova, Reimar Seltmann, Rustam Mirkamalov, Farid Divaev, Rustam Khalmatov, Rustam Uzokov, Yulia Mun, Nigora Isokova

P7.205 Punta Corna Co-rich Vein System (Italy): Late-Alpine Hydrothermal Circulation and Metallogenesis
Giulia Domenighini (Student)

P7.206 Constraints on the Genesis of the Orbicular Rocks and Sulphide Mineralization in the Koperberg Suite, Namaqualand Metamorphic Complex, South Africa
Senamile S. Dumisa, Grant M. Bybee, Paul A. Nex, Ayesha B. Jogee, Ayesha B. Jogee

P7.207 Vanadium Distribution in Soil of Okahandja Map Sheet (2117 DA – DD) in Otjozondjupa Region, Namibia
Wilka Embula (Student)

P7.208 Characterization of Fluids Related with Alkalic Porphyry Deposits in British Columbia, Canada
María Paula Estrella (Student), Pilar Lecumberri-Sanchez, Xinyue Xu

P7.209 Sulphide Mineralization and Associated Platinum Group Elements in Mafic-Ultramafic Units of the Tantalite Valley Complex, Southern Namibia
Daniel Ferreira, Bjorn von der Heyden, Martin Klausen

P7.210 The Role of Carbonate Melts in the Transport and Concentration of Metal-Rich Magmatic Sulfides
Marco Fiorentini, Maria Cherdantseva, Isra Ezad, Michael Anenburg, John Mavrogenes, Stephen Foley

P7.211 Determining Uranium Mineral Compositions by Electron Probe Microanalysis
Guy G. Freemantle

P7.212 Future Lithium Supply: Towards a Robust Quantitative Assessment
Laura Buarque Andrade, Max Frenzel, Carolin Kresse, Britta Bookhagen, Michael Schmidt, Dirk Sandmann

P7.213 In Situ Geochemistry of Mica in Exomorphic Spodumene Pegmatite Halos: Implications for Mineral Exploration
Elena J. Geiger (Student), Julian F. Menuge, Teimoor Nazari-Dehkordi, Maurice Brodbeck

P7.214 Trace Elements in Alteration Minerals: A Case Study from the Kandilka Prospect, Eastern Rhodopes, SE Bulgaria
Yana Y. Georgieva (Student), Rossitsa Vasileva, Istvan Marton, Stefan Metodiev

P7.215 Mineralogy and Deportment of Nickel, Cobalt, and Scandium in the Laterite Deposits of Sebuku Island (SE Kalimantan, Indonesia)
Michele Giorno, Alkiviadis Kontonikas-Charos, Doreen Ebert, Joachim Krause, Robert Möckel, Dirk Sandmann, Ernowo Ernowo, Agata V. Kindangen, Yoseph Swamidharma, Yudhi Krisnanto, Max Frenzel

P7.216 Advancing Greener Economy: Insights from Studies on Bauxite Deposits in Bosnia-Herzegovina (Jajce and Posušje) and Porphyry Copper System in Bulgaria (Assarel)
Marko Holma, Šime Bilić, Ivica Pavičić, Franjo Šumanovac, Gordana Deljak, Ivan Grubišić, Irena Peytcheva, Stoyan Georgiev, Elitsa Stefanova, Dimitrina Dimitrova, Atanas Hikov, Desislav Ivanov, Ivaylo Vasilev, Pasi Kuusiniemi, Timo Enqvist, Gergely Surányi, László Bálazs, Sandór Demők, Jarmo Korteniemi, Giulio Casini, Eduard Saura, David Marti, Helena Seivane, Martin Schimmel, Markku Pirttijärvi

P7.217 Temporal Evolution of the Majuba Hill Cu-(Mo)-(Sn) Deposit, Pershing County, Nevada
Alexander P. Holmwood, Matthieu Harlaux, Philipp Ruprecht

P7.218 A Fresh Look at the World-Class "Five-Element" Vein Deposits, Cobalt Embayment, ON (Canada): Fluid and Metal Sources and Systematics, and Mineralization Ages
Joshua Jackman (Student), Jacob Hanley, Daniel Kontak, Robert Creaser, Richard Stern, Jonathon Toma

P7.219 Lithium-Caesium-Tantalum-Mineralised Granitic Pegmatite Systems in the Palaeoproterozoic Bedrock of Sweden
Erik Jonsson

P7.220 The Structural Setting and Relative Timing of the Bergby LCT-Pegmatite Field, Central Sweden
Karin Högdahl, Erik Jonsson, Magnus Leijd, Anders Zetterqvist, Pablo Petri

P7.221 High Magma Differentiation Controls Rare Metal Mineralization: Traces by Tourmaline Major and Trace Elements, and Boron Isotopes
Haiquan Li, Youye Zheng, Xin Chen

P7.222 Copper and Molybdenum Isotope Variation Along a Vertical Transect Through the Epithermal-Porphyry Valeriano Cu-Au System, El Indio-Maricunga Belt, Northern Chile
Stephanie Lohmeier, Bernd Lehmann, Jan Pašava, Lukas Ackerman, Ondrej Šebek, Alexandre Andronikov, Andrew Hodgkin, Francisco Bravo

P7.223 Different Sn-W Vein Textures: The San Finx Deposit, NW Spain
Ivan Losada, Iñigo Borrajo, Fernando Tornos

P7.224 Magmatic Evolution and Sulfide Saturation Mechanisms Forming the Ultramafic-Hosted Stormi Ni-Cu-(Co) Deposit of the Vammala Belt, SW Finland
Jakob Machleidt (Student), Karsten Haase, Manuel Keith, Reiner Klemd, Christoph Beier, Patrick Hehn

P7.225 Geochemical Fingerprinting of Lithium Pegmatites: A Whole-Rock Perspective
Ryan J. Manton, Rebecca M. Montsion, Anusuriya Devaraju, Jessica M. Walsh

P7.226 Overview of the Potential of Strategic Minerals in Bahia State - Brazil
Hevelyn E. Martins (Student), Maria J. Mesquita

P7.227 Carbon Isotopic Studies of Calcite Veins Associated with U Mineralization in the Jahaz Deposit, North Delhi Fold Belt, Northwestern India
Priyanka Mishra (Student), Rajagopal Krishnamurthi, Baskaran Saravanan

P7.228 Estimation of Lithium Content in Brines of La Union Salt Flat, Located in Ahumada, Chihuahua, Mexico
Héctor Montes Carreón (Student), Perla Rodríguez Contreras

P7.229 Application of Apatite Chemistry in LCT Pegmatite Exploration in the Late Archean Yellowknife Pegmatite Province, NWT Canada: A Preliminary Analysis
Gilles N. Ngoran (Student), David R. Lentz, Gideon Lambiv Dzemua

P7.230 Raw Materials from Geothermal Fluids
Simona Regenspurg, Lydia Olaka, Franziska Wilke, Katrin Kieling, CRM-geothermal Team

P7.231 Survey for Critical Raw Materials in Tanzanian East African Rift Geothermal Areas
Franziska Wilke, Simona Regenspurg, Anna Jentsch, Timmu Kreitsmann, Martin Zimmer, Michael Bau, Samuel Niedermann, Lydia Olaka, Albano Mahecho

P7.232 The Magmatic-Hydrothermal Nature of Granite-Hosted Uranium Mineralization on the Example of the Omahola Project in the Damara Orogen, Namibia
Alex Otto

P7.233 Insight from Field and Petrographic Studies of Saltpond and Winneba Pegmatites and Other Associated Rocks: Implications to Ghana's Lithium Occurrence
Theophilus Owusu Agyapong (Student), Mary N. Selby, Samuel Nunoo, Prince O. Amponsah

P7.234 A Conceptual Mineral Systems Model for the LCT Pegmatites
Shanta Panja (Student), Alok Porwal, Majid Keykhay-Hosseinpoor

P7.235 Lithium Clays in Volcano-Sedimentary Deposits
Francesco Putzolu, Robin Armstrong, Tom Benson, Chris Emproto, Rebecca Ball, Isabel Abad, Adrian Boyce, Laura Hepburn, Nicolas Bompard, Richard Herrington

P7.236 A Mineral Systems Approach for Li Volcano-Sedimentary Deposits: Commonalities with Other Li Deposit Models
Robin N. Armstrong, Francesco Putzolu, Tom Benson, Richard Herrington

P7.237 Unveiling the Geological Potential of Bauxite Deposits in the Democratic Republic of Congo (DRC) for Sustainable Aluminium Production and the Green Transition
Derick U. Ramazani (Student), Dominique W. Osomba

FREEPORT-MCMORAN

Foremost in Copper
Foremost in Copper Exploration
Please talk with one of our reps at the conference:

Wolf Schuh
+1-520-780-3640
Darija Pajic
+381 60 4234 721
Paul Albers
+1-928-925-3289
Delcio Teixeira
+1-520-237-3294
Alex Holmwood
+1-520-697-4819

P7.238 Palladium Enrichment and Mobilization through Alteration and Replacement of Base Metal Sulfides in the Parisien Lake Zone, East Bull Lake Intrusion, Ontario, Canada
Marc Rochette (Student), Daniel Layton-Matthews, Dave Peck, Reid Keays

P7.239 Overview of Uranium Activities in Namibia and the African Continent
Gabi Schneider

P7.240 Exploring Lithium Metallogenesis Within the Groundwaters of the Arid-Hyper Arid Environments of the Northern Cape, South Africa
Mary N. Sithole (Student), Khuliso Masindi

P7.241 Mineralogical and Geochemical Characteristics of the Carbonatite Laterite at Sokli, Finland
Martin P. Smith, Martin Y. Li, Sam Broom-Fendley, Tuomas Gråsten, Frances Wall

P7.242 Compositional Features of Lithium-Bearing Pegmatites of the Sinsikourou and Kola Areas Around the Bougouni Region, Southwestern Mali
Elhadji Mory Traore (Student), Akinade S. Olatunji, Mory Sidibe, Seko Sanogo

Vital High-Volume Base Metals

P8.243 Proterozoic Sediment-Hosted Mineralisation in Southeastern Australia and Links to the Belt-Purcell Basin of North America
Sheree Armistead, Sebastien Meffre, Ralph Bottrill, Andrew Cross, David Huston, Grace Cumming

P8.244 Assessing Proximity to ISCG and IOCG Mineralisation Using REE-Bearing Phosphate Chemistry
Travis Batch (Student), Caroline Tiddy, Adrienne Brotodewo, David Giles, Michael Taylor, Vaclav Metelka

P8.245 Is the Tennant Creek Mineral Field an IOCG Province? Insights from Chlorite Chemistry
Damian Braize (Student), Jeffrey A. Steadman, Sebastien Meffre, Ana Liza Cuison

P8.246 Hydrothermal System Evolution of the Santa Maria Pb-Zn-Cu-(Au, Ag) Deposit, Camaquã Basin, Southern Brazil
Lucas F. Brito (Student), Caetano Juliani, Jamie J. Wilkinson

P8.247 Geochemistry and Origin of the Magnetite-Vonsenite Mineralization of the Monchi Mine (SW Iberia)
Alberto Campanero (Student), Carmen Conde, Gonzalo Ares, Fernando Tornos

P8.248 Vectoring in Metamorphosed VMS Deposit: Insights from Chlorite, White Mica, and Garnet
Cendi Dana (Student), Steven Hollis, Darryl Podmore, Megan James, Ian Butler, Florian Fusseis, Adrian Boyce, Riquan Azri

P8.249 Constraints on the Pre-Eruptive Metal and Volatile Content of Magmas Associated with Archean Volcanogenic Massive Sulfide Deposits from Zircon-Hosted Melt Inclusions
Priyal Daya (Student), Jacob J. Hanley, Kevin Neyedley, Sarah Speight, Stephen J. Piercey, Alexandra Tsay, Zoltan Zajacz

P8.250 Combined Micron-Scale EBSD Study and LA-ICP-MS Analysis Reveals the Evolution of Pyrite and Cu-Fe-Sulphides in the Olympic Dam Cu-U-Au-Ag Deposit, South Australia
Samuel A. King, Kathy Ehrig, Nigel J. Cook, Cristiana L. Ciobanu

P8.251 Geology and Mineralogy of Pb-Ag-Cu Veins of the Central Plateau Deposit of the Jbel Addana District, (Western Anti-Atlas, Morocco)
Zoubair El Ouad (Student), Mustapha Souhassou, Moha Ikenne, Hibti Mohamed, Ilmen Said, Abia El Hassan, Younes Moundi, Atmane Madi, Abdelhak Karfal

P8.252 Mineralogy and Trace Element Composition of Sulfides from Porphyry-Stage and Minor Late Polymetallic Mineralization in the Gaby District, Northern Chile
Alexandre Cugerone, Kalin Kouzmanov, Lorenzo Tavazzani, Lluís Fontboté, Sergio P. Henríquez, Juan C. Toro, Rubén Pardo, Carolina Rodriguez

P8.253 White Mica Alteration in Porphyry Systems: Insights from the Gaby Porphyry Cu Deposit, Northern Chile
Alexandre Cugerone, Kalin Kouzmanov, Lluís Fontboté, Luis Krampert, Alexandra Tsay, Zoltan Zajacz, Sergio P. Henríquez, Juan C. Toro, Rubén Pardo, Carolina Rodriguez

P8.254 Alteration Assemblages and Quantitative Mineralogy of the Boundary Zone Zn-Pb ± Ag Deposit, Yukon, Canada
Haruna M. Grema (Student), Joseph M. Magnall, Sarah A. Gleeson, Jack E. Milton, Anja M. Schleicher, Alicja Wudarska, Hans-Martin Schulz

P8.255 Factors for Classification of the Precious and Base-Metal Mineralization in Epithermal Deposits: Evidence from Apatite Chemistry in Ore-Related Igneous Activity
Masaya Koseki (Student), Shota Satori, Yasushi Watanabe, Takuya Echigo, Shogo Aoki

P8.256 Fluid History of Zambian Copper Deposits Based on Metal Zonation in Pyrite
Ross Large

P8.257 Origin of the Magnetite-(Apatite) and IOCG Deposits of the Coastal Cordillera of the Andes
Ignacio Ledesma, Fernando Tornos

P8.258 Understanding Overprinting Mineralizing Systems of the Ruby Hill Deposit, Eureka, NV: Implications for Future Exploration
Alyssa Lindsey (Student), Simon Jowitt

P8.259 Facies-Constrained Lithogeochemistry in the Irish Midlands: A Comparison Between the Collinstown Prospect and the Stonepark Zn Deposit
Joseph Magnall, Sarah Gleeson, Philip Rieger, Mark Holdstock, Kacie Wellington, David Furlong, Murray Hitzman

P8.260 Mineral Prospectivity Mapping for Porphyry Copper-Gold Mineralization Style Using Fuzzy Logic Method in the Arthurs Seat Project, Jamaica
William Miguel Ortiz

P8.261 The Characteristics of Chromite in the Lateritic Nickel Deposit from East Sulawesi Ophiolite, Indonesia
Ibnu Munzir (Student), Marek Michalik

P8.262 The Essential Role of High-Volume Base Metals: Catalysts of Modern Industry
Chawanangwa S. Mwansa (Student)

P8.263 Allanite U-Th-Pb Geochronology of IOCG and Zn-Ag-Pb Deposits in the Mount Isa Inlier, Australia
Luís Portela (Student), Coralie Siégel, Charlotte M. Allen, Jim R. Austin

P8.264 Copper Bio-Enrichment: Results from Isotopic Analysis at the Las Cruces Deposit, Spain
J. Javier Rey-Samper (Student), Ryan Mathur, Fernando Tornos

P8.265 Lithostratigraphic Controls on the Mineralizing System at the Stonepark Zn-Pb Deposit, SW Ireland
Philip Rieger, Murray W. Hitzman, Victoria Figueira Susin, Mark Holdstock, Eoin Dunlevy, Aline Melo

P8.266 Hematite Orebodies Formed by Multiple Geological Events: A Case Study from the Akatani Deposit, Niigata Prefecture, Japan
Kotaro Seno (Student), Yasushi Watanabe, Takuya Echigo, Keita Itano, Shogo Aoki, Yoshiaki Kon, Shota Satori

P8.267 Geological-Mining Characterization of the Sierra Rancheria Using Remote Sensing in the Northern Region of Chihuahua, Mexico
Carlos A. Silva (Student), Luis A. Jiménez, Perla Rodríguez

P8.268 Geophysical, Geochemical, and Isotope Studies of the Pb-Zn-Cu-(Ag-Au) Santa Maria deposit, Dom Feliciano Belt, Brazil
Nadine Silva (Student), Farid Chemale Jr., Lucy T. Chemale, Glaucia Queiroga, Marly Babinski, Samuel Lago, Claúdia Teixeira, Tiago Girelli, Monique Rizzi

P8.269 Lithogeochemistry and Magnetic Susceptibility for Stratigraphy and Alteration Identification in the Western Foreland Shelf of the DRC
Amanda M. Stoltze, Tim Dunnett

P8.270 Gorno and Piani Resinelli: Genesis and Correlations of Two Alpine-type Pb-Zn Districts
Luca Summino (Student), Michele Andrea Giorno, Luca Barale, Carlo Bertok, Max Frenzel, Marta Gasparrini, Luca Martire

P8.271 Unveiling Geometallurgical Potential: Insights from the Soledad Cu-Au-Ag Mineralized Tourmaline Breccias, Central Peru
Victor J. Torres, David R. Cooke, Lejun Zhang, Julie Hunt, David L. Kelley

P8.272 Mineral Chemistry as a Potential Exploration Tool Applied to the Lusale Basin and Fishtie Deposit, Central African Copperbelt (Zambia)
Veronica G. Trevisan, Jamie J. Wilkinson, John Spratt

P8.273 Highway to Grade: The Importance of an Early Quartz Vein Stockwork for Channeling Late Ore-Bearing Fluids While Favoring Sulfide Precipitation
Martin G. Vallejos (Student), Laura Maydagán, David Selby, Kevin B. Heather, Mariano Poodts

P8.274 Analysis of Visual Grading of the Hydrothermal Alteration of the Vazante Zn-Pb Deposit, Minas Gerais, Brazil: A New Vector for Exploration?
Poliana Vidal Salgado (Student), Luiz G. S. Correia, Bruno A. A. C. Arruda, Rafael F. de Freitas, Samuel Boucas do Lago

P8.275 Influence of Mafic Units on the Mineralising System Around the Redmoor Project, Cornwall
Georgina C. Woodward (Student)

P8.276 Texture, Trace Elements, and Sulfur Isotopes of Pyrite from the Zhao'anzhuang Iron Deposit, Southern Margin of North China Craton: Implications for Ore Genesis
Hengfeng Zhang (Student), Hao Hu, Jianwei Li

OSINO
RESOURCES

BUILDING VALUE FOR ALL

Osino will soon begin constructing Namibia's third gold mine, the Twin Hills Gold Project. To build value for all stakeholders, Osino established the Trust as a development vehicle for those financially benefiting from the Project. Many vendors and shareholders now support this model to collectively share value with the host communities and the Namibian nation.

www.twinhillstrust.org

TWIN HILLS
TRUST | NAMIBIA

Student Travel Support

Travel support for 94 SEG Student presenters was made possible through the generous contributions of our conference sponsors and the SEG New Century Fund. Thank you!

Fabrizzio Abello, University of British Columbia
Jolan Acke, KU Leuven
Aaron Adsit, Colorado School of Mines
Imad Aflla, Ibn Zohr University
Marcus Angus, University of Nevada, Reno
Edwin Animah Obeng, University of Mines and Technology
JJ Ayasze Gifari Ayyala, Jenderal Soedirman University
Mohamed Badawi, Miskolci Egyetem
Alessia Barbieri, University College London
Travis Batch, University of South Australia
Juan Bello-Rodriguez, University of Toronto
Daniel Bermejo Lopez, Universidad Complutense of Madrid
Damian Braize, University of Tasmania
Catriona Breasley, University of British Columbia
Kiara Brooksby, University of Exeter
Evelyn Caiza, Central University of Ecuador
Ian Campos, Laurentian University
Hugo Carrasco, University of Geneve
Ramesh Chandra Behera, Indian Institute of Technology Dhanbad
Rory Changleng, Penn State University
Axel Cima, University of Tasmania
Takeshy Coaquira, University of Tasmania
Robert Collar, McGill University
Angela Isaura Santos Costa, University of Tasmania
Deniro Gonçalves Costa, Universidade Federal de Minas Gerais
Cendi Dana, University of Edinburgh
Jose Carlos Davila Zegarra, Pontifical Catholic University of Peru
Juan Diaz Carreno, University of Helsinki
Lily Dickson, University of Leeds
Giulia Domenighini, Università degli Studi di Torino
Kelvin dos Santos Alves, Universidad de Salamanca
Michael Eigler, Karlsruhe Institute of Technology
Halleluya Ekandjo, iCRAG
Mohamed Ayed Ibrahim Fahd Elbalawy, University of Miskolc
Danis Ionut Filimon, Università di Pisa
Olivia Filson, University of Toronto
Jose Franco Moraga, University of Texas at El Paso
Sze Yui Fung, McGill University
Elena Geiger, University College Dublin
Silvia Gioiello, Università degli Studi di Torino
Silas Goetz, University of Nevada, Reno
Martin Gonzalo Vallejos, Universidad Nacional del Sur - INGEOSUR
Andres Felipe Gonzazlez-Duran, University of Michigan
Marie Guilcher, TU Bergakademie Freiberg
Matthew Hales, University of Witwatersrand
Charlie Hamblet, Camborne School of Mines
Joshua Jackman, Saint Mary's University

Michael Kirschbaum, Colorado School of Mines
Lot Koopmans, University of Oxford
Kenneth Kathy Kwanang, University of Yaounde I
Ignacio Ledesma, IGEO-CSIC-UCM
Liu Jialong, China University of Geosciences (Wuhan)
Christina Loidolt, Australian National University
Jakob Machleidt, Friedrich-Alexander University Erlangen-Nuremberg
Daniel Alejandro Castano Madrigal, University of Texas at El Paso
Shelly Mardhia Faizy, TU Bergakademie Freiberg
Lucy Mathieson, Curtin University
Duncan McLeish, McGill University
Olivia Mejías, University of Queensland
Julian Melo-Gómez, Laurentian University
Lebogang Mokane, Akita University
William Munro, University of Toronto
Allyson Murray, University of Michigan
Nelao Naimbale, University of Tasmania
Nonhlanhla Ndlovu, University of the Witwatersrand
Kevin Neyedley, St. Mary's University
Terence Cho Ngang, RWTH Aachen University
Abner Nghoongloka, Akita University
Zahra Nourizenouz, Helmholtz-Institut Freiberg für Ressourcen-technologie
Valentin Oge, Helmholtz-Institut Freiberg für Ressourcentechnologie
Victor Torres Pacheco, University of Tasmania
Shanta Panja, Indian Institute of Technology Bombay
Kelebogile Phili, Akita University
Candela Pita, IGEO
Maxwell Porter, University of British Columbia
Daniel Ferreira, Stellenbosch University
J. Javier Rey-Samper, Universidad Complutense of Madrid
Federico Saponara, National University of La Plata
Erin Shervey, Colorado School of Mines
Felix Sihombing, University of Oxford
Leandro Silva, Queen's University
Koray Tasbicen, Colorado School of Mines
Lucas Tatnell, University of Western Australia
Lauren Terry, Colorado School of Mines
Erin Thompson, University of Leicester
Helen Twigg, University College Dublin
Arthur Vicentini de Oliveira, Curtin University
Victor Vincent, University College Dublin
Lisa Wasitschek, Universität Würzburg
Warren Wegener, University of British Columbia
Hengfeng Zhang, China University of Geosciences (Wuhan)
Pengcong Zhang, China University of Geosciences (Beijing)
Aoyue Zheng, China University of Geosciences (Beijing)
Peace Zowa, University of the Witwatersrand

Conference Accessibility Program (CAP)

SEG's new Conference Accessibility Program (CAP) empowered 100 local and regional students, post-doctoral researchers, academics, and early-career professionals to participate in the SEG conference. Generous support from our CAP sponsors (+CAP) assisted in their travel, conference registration, and 2024 SEG membership costs.

Abdulkadir Mohamed Abdulkadir, UCL
Rahidat Abubakar, Nassarawa State University
Salaam Jansbaka Adams, University of Ghana
Quesnay de Jésus Akabahinga, Pan African University, Life and Earth Sciences (including Health and Agriculture), University of Ibadan, Ibadan, Nigeria
Shakirat Aminu, Kwara State University, Malete, Nigeria
Maxwell Amponsah-Dacosta, University of the Western Cape
Abia Amukongo, University of Namibia
Martha Amwaama, University of the Free State
Tuutaleni Angula, University of Namibia
Nana Yaw Asamoah, Kwame Nkrumah University of Science and Technology
Kamogelo Baloyi, University of the Witwatersrand
Kesia Nikita Bezuidenhout, UNAM
Joseph Boateng, Radford University College
Steve Chingwaru, Stellenbosch University
Emmanuel Christopher, University of Portharcourt
Abdullahi Dauda, University of Jos
Akina Denk, University of Namibia (UNAM)
Victor Diyelmak, University of Jos
Michael Eino, University of Namibia
Zoubair El Ouad, Ibn Zohr University
Loubna Elamraoui, Mohammed VI Polytechnic University (UM6P)
Munghatihe Emmanuel, Makerere University
Kaducu Emmanuella Lagen, Makerere University
Daniel Reuben Ferreira, Stellenbosch University
Abubakarr Fonti-kanu, Fourah Bay College
Japhet Fungo, University of Dar es Salaam
Borbor A.K.K. Gibson, University of the Witwatersrand
Sarah Gyamera-Amoako, University of Mines and Technology
George Ishabairu, University of Dar es Salaam
Selma Jeremia, University of Namibia
Gaone Joseph, University of the Witwatersrand
Junias H. David Junias, University of Namibia
Rebecca G. L. Kallon, Fourahbay College
Eunice Kamosho, University of Namibia
Sarafina Kashihalwa, University of Namibia
Neto Kasuto, University of Namibia
Chanda Katotobwe, University of Zambia
Tebogo Kelepile, Botswana International University of Science and Technology
Theophilus Kolobeki, University of Zambia
Samuel Asare Konadu, University of Mines and Technology
Yusuf Lahai, Fourah Bay College, University of Sierra Leone
Luise Leonard, University Of Namibia (UNAM)
Filippus Tangeni-Omwene Lukas, University of Namibia
Movan Mahanya, University of Zimbabwe
Luti Manjimela, University of Zambia
Fred Manyando, University of Namibia
Mabatho Mapiloko, University of the Witwatersrand

Rowin Marais, Stellenbosch University
Mumba Mbasela, University of Zambia
Tamba Mbayo, Fourah Bay College
Gosaye Melesse, Pan African University of Life and Earth Science Institute
Lemogang Mokhosi, University of the Witwatersrand
Matebello Petunia Mokoena, University of the Witwatersrand
Innocent Mufungizi, University of Kinshasa
Theophile Mugerwa, University of Rwanda
Abdullah Bello Muhammed, University of Mines and Technology
Joshua Munkinyi, University of Zambia
John Mutale, University of Zambia
Takudzwa Mwagura, University of Zimbabwe
Chawanangwa Mwansa, University of Zambia
Stella Nankinga, Makerere University
Mhleli Ncube, Stellenbosch University
Siphesihle Natatia Ndwandwe, Stellenbosch University
Julie Neidel, University of Namibia
Lemewihbwen Ngiamte George, Pan African University Life and Earth Science Institute, University of Ibadan
Manuel Nopeia, Akita University
Cedrick Nsengiyumva, Botswana International University of Science and Technology
Henry Nule, University of Port Harcourt
Maryce Obindah, The University of Nairobi
Newman Odonkor, University of Ghana
Christine Omuombo, The Technical University of Kenya
Theophilus Owusu Agyapong, University of Ghana
Michael Kwame Peters, University of Ghana
Faith Magdalenè Rooi, University of Namibia
Tevin Sherura Rumanzi, Makerere University
Johannes Sales, University of the Witwatersrand, Johannesburg
Ali Seiba, CK Tedam University of Technology and Applied Sciences
Mikidadi Shaha, Akita university
Nailoke Shali, Osino Gold Exploration and Mining Pty Ltd
Benson Shikulunde, UNAM
Josia Shilunga, Stellenbosch University
Sidy Sidibe, Insititute of Applied Technologies of Africa
Shafiishuna Silvanus, University of Namibia
Rejoice Simon, University of Namibia
Mukuka Simusokwe, University of Zambia
Mary Sithole, University of the Witwatersrand
Quinter Wenzi Tanni, Pan African University of Life and Earth Sciences Ibadan Nigeria
Mussie Tesfai Menghisteab, University of Free State
Elhadji Mory Traore, Pan African University Life and Earth Science Institute, University of Ibadan, Ibadan Nigeria
Robert Kapajika Tshilumba, The University of Zambia
Mbili Tshiningayamwe, University of Namibia
Derick Ungu Ramazani, University of Kinshasa
Elkan Utoni, University of Namibia
Louis van Eeden, Stellenbosch University
Paul Visaa, University of Port Harcourt
Dickson Wachira, Taita Taveta University
Yann Waku Mpaka, Stellenbosch University
Rowan Walker, Stellenbosch University
Diallo Zoumana, École Nationale D'Ingénieurs Abderhamane Baba Touré

SEG 2024 Affiliates

Thank you to our SEG 2024 Affiliates—professional societies and organizations that share our commitment and vision for the future of geology. Affiliate organizations assist SEG with promoting the conference to their members.

- Association of Applied Geochemists
- European Federation of Geologists
- Geological Society of Africa
- Geological Society of South Africa
- International Association on the Genesis of Ore Deposits
- Mineralogical Society of America
- Society for Geology Applied to Mineral Deposits
- South African Geophysical Association

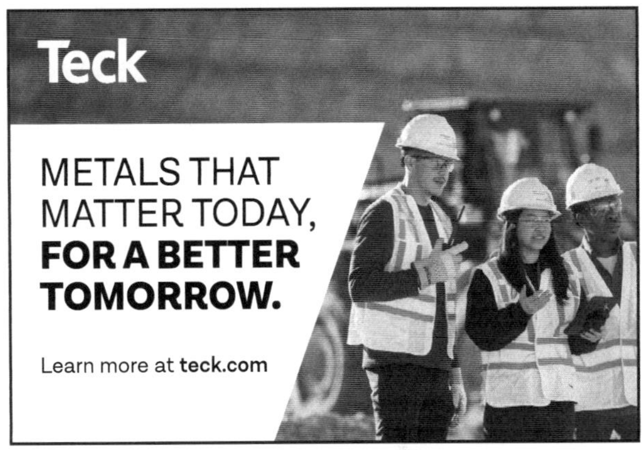

Newmont™

Exhibitor Booths

#	Exhibitor	Website
1, 2	RioTinto	www.riotinto.com
3	GeologicAI	www.geologicai.com
4	Geoscience Council of Namibia	www.geocouncil.org.na
5	Axiom Group	www.axiomex.com
6, 11	Society of Economic Geologists	www.segweb.org
7	iCRAG	www.icrag-centre.org
8	Actlabs	www.actlabs.com
9	Veracio	www.veracio.com
10	Gemstone Society of Namibia	
12	Bell Geospace	www.bellgeo.com
13	IRIS Instruments	www.iris-instruments.com
14	Ivanhoe Mines	www.ivanhoemines.com
15	Kodo Drilling	www.kododrilling.com
16, 21	KoBold Metals	www.koboldmetals.com
17	Innov-X Africa	www.innovxafrica.com
18	Osino Resources	www.osinoresources.com
19	Teck	www.teck.com
20	Exige	www.exigesa.com
22	TerraCore	www.terracoregeo.com
23	Mount Sopris Instruments	www.mountsopris.com
24	Viridien	www.viridiengroup.com
25	Critical Productions	www.critical-productions.com
26, 27	B2Gold	www.b2gold.com
28	Terratec Geophysical Services	www.terratec-geoservices.com
29	Barrick	www.barrick.com
30	Alaska Critical Minerals Collaborative	www.segweb.org/ACMC
31	Geological Society of Namibia	geolsocnamibia.org

Floor Plan Map

Mövenpick Hotel Windhoek
Corner of Auas and Aviaton Road
Windhoek, 9000 | Namibia
+264 61 296 8000

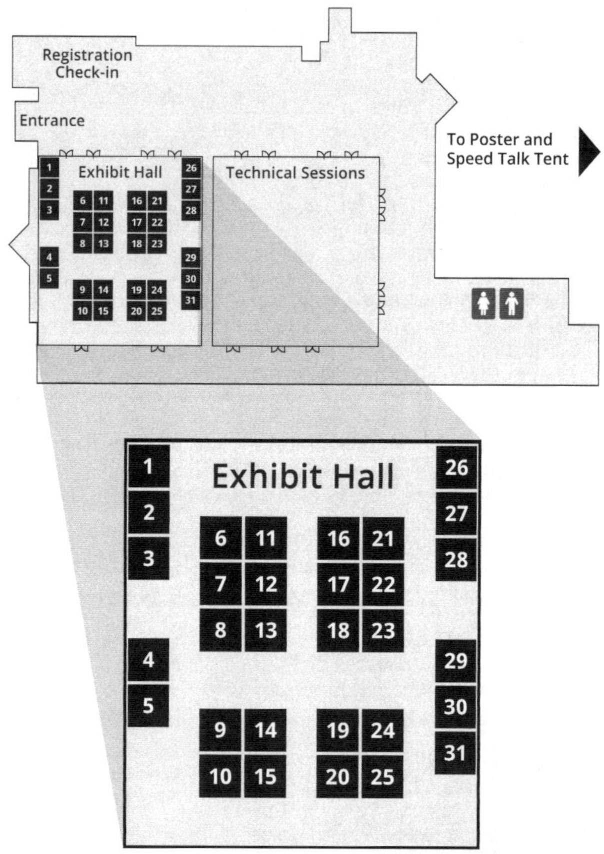

Visit www.seg2024.org/venue for hotel information.

The information in this program is current through September 1, 2024. Please refer to the online platform for updated information:

seg2024.org/virtual

Thank You, Exhibitors

Actlabs	INNOV-X AFRICA
ACMC Alaska Critical Minerals Collaborative	IRIS INSTRUMENTS
AXIOM GROUP	IVANHOE MINES
B2GOLD	KoBold Metals
BARRICK	KODO DRILLING
BellGeospace	MOUNT SOPRIS INSTRUMENTS
CRITICAL PRODUCTIONS	OSINO RESOURCES
EXIGE	RioTinto
GEMSTONES SOCIETY OF NAMIBIA	SOCIETY OF ECONOMIC GEOLOGISTS FOUNDED 1920
GeologicAI	Teck
GEOSCIENCE COUNCIL OF NAMIBIA	TERRACORE GEOSPECTRAL IMAGING
iCRAG SFI RESEARCH CENTRE IN APPLIED GEOSCIENCES	terratec Geophysical Services
	VERACIO™
	VIRIDIEN

Thank You to Our SEG 2024 Conference Sponsors

PATRON

B2GOLD

PREMIER

BHP

GeologicAI

GOLD

BARRICK

EQUINOX GOLD

KoBold Metals

Newmont

SILVER

Critical Productions

Teck

OSINO RESOURCES

VERACIO

COPPER

Sandfire

BRONZE

AB GM (A&B Global Mining)

Dawn Zhou

iCRAG (SFI Research Centre in Applied Geosciences)

NAMIBIA CRITICAL METALS INC.

QKR NAMIBIA (Navachab Gold Mine)

CNNC Rössing Uranium Working for Namibia

Threesixty Emergency Services CC

SUPPORTER

cimera

MUNDORO

Thank You to Our 2024 Conference Accessibility Program (CAP) Sponsors

Thank you to the sponsors of our SEG 2024 Conference Accessibility Program. The additional CAP sponsorship (+CAP) empowered **100** local and regional students, postdoctoral researchers, academics, and early-career professionals to participate in the SEG conference. Generous support from our CAP sponsors assisted in their travel, conference registration, and 2024 SEG membership costs.

PATRON

Rio Tinto

GOLD

KoBold Metals

SILVER

FREEPORT-McMoRAN

IVANHOE MINES

COPPER

Richard Bedell

unesco

BRONZE

Anonymous Donor

We also sincerely thank the 30 individuals whose generous sponsorship of the Conference Accessibility Program helps drive our vision for the future of geology.

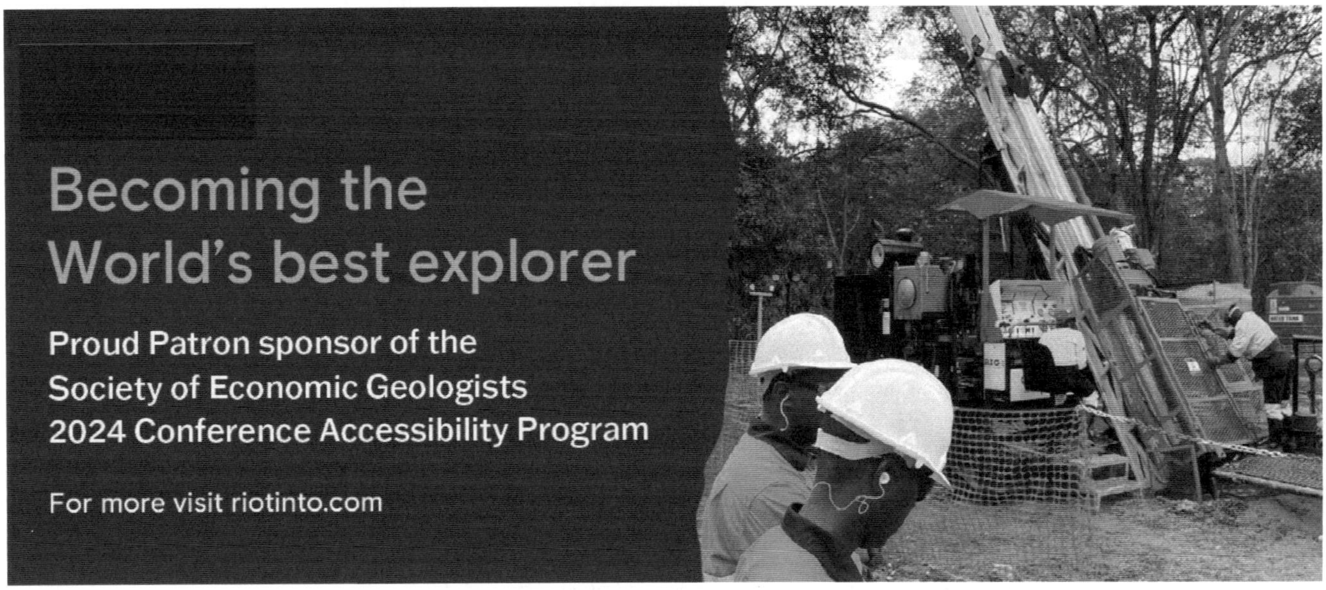
Becoming the
World's best explorer
Proud Patron sponsor of the
Society of Economic Geologists
2024 Conference Accessibility Program
For more visit riotinto.com

SEG 2025 BRISBANE, AUSTRALIA

September 26–29, 2025

The skills, knowledge, and capabilities of economic geologists are more important than ever before to the world's population, as we seek to source and help produce the mineral supply that will feed the increasing metal demand of the energy transition. SEG 2025 attendees will learn about the most inspiring and impactful developments that address many facets of this challenge. Resources and developments in the Asia Pacific region will be in focus, but economic geology innovation is global, and attendees can expect to experience a global breadth in the conference program. The conference content will range from cutting-edge science to practical discovery case studies, delivered through a mix of field trips, workshops, posters, and an innovative combination of shorter- and longer-format technical presentations. In addition, the conference will present a great opportunity to build networks and collaborations across an international group of attendees that will span industry, government, and academia and cater to everyone from students and early career geoscientists to industry and discipline leaders.

Rick Valenta, Director - Sustainable Minerals Institute, The University of Queensland, SEG 2025 Chair

Mark Noppé, Director – WH Bryan Mining Geology Research Centre, The University of Queensland, SEG 2025 Co-Chair

Jennifer Craig, SEG Executive Director

SEG 2024 Conference: Sustainable Mineral Exploration and Development

---

The Impact of the Pan-African Reporting Code on the Future of Mining in Africa

Olugbenga Okunlola[1], Theophile Ndougsa Mbarga[2], Prosper M. Nude[3], Felix Bob Ocitti[5], Deng Ngang Deng[6], Tunde Arisekola[5]
1. Geological Society of Africa, Ibadan, Nigeria, 2. University of Yaoundé, Yaoundé, Cameroon, 3. University of Ghana, Accra, Ghana, 4. United Nations Economic Commission for Europe (UNECE), Geneva, Switzerland, 5. African Minerals Development Centre, Addis Ababa, Ethiopia, 6. Target Mining Resources Pty Ltd, Kalgoorlie, WA, Australia

The Pan-African Resource Reporting Code (PARC) is a comprehensive framework for the public reporting of mineral and energy resources in Africa. It aims to bolster investor and stakeholder confidence by aligning resource reporting with international standards and the Africa Mining Vision (AMV) and development goals. PARC serves a wide range of stakeholders, including investors, local communities, governments, mining companies, and professional bodies.

The PARC public report uses African Mineral and Energy Resources Classification and Management System (AMREC) classifications codes specific to each resource type and offers guidelines and templates for reporting on various resources such as metallic and development minerals, construction raw materials, petroleum, coal, renewable energy, nuclear fuel, environmental impacts, and social benefits. Notably, it is the only code with guidelines for artisanal and small-scale mining reporting.

The code is built on four key principles: transparency, materiality, competency, and the social, environmental, and economic benefits of resource projects. These principles are defined as follows:

1. Social, Environmental, and Economic Benefits: Reports must detail how projects address social and environmental impacts and contribute to ecosystem benefits.
2. Transparency: Reports must provide clear, unambiguous information to avoid misleading the public.
3. Materiality: Reports must include all relevant information necessary for investors and stakeholders to make informed decisions.
4. Competency: Reports must be prepared by qualified individuals subject to a professional code of ethics.

Supporting these principles are ethics and codes of conduct, which mandate adherence to professional standards, confidentiality, and avoidance of conflicts of interest. The compliance is enforced through an Ethics and Standards sub-committee, which addresses complaints and recommends disciplinary actions. In essence, PARC aims to enhance the credibility and quality of resource reporting in Africa, supporting sustainable development and responsible resource management. The presentation will focus on PARC (2023) and its relevance to the African mining landscapes.

Fostering Sustainable Development—Opportunities for the Extractive Industry

Thabisile Phumo
Sibanye Stillwater Ltd, Johannesburg, South Africa

The extractive industry has a historical reputation of destruction, loss, and exploitation. What often goes unnoticed is that behind that legacy lies advancement of humankind and opportunities to do right for humans and the planet.

The global south, particularly Africa, has borne the brunt of the legacy with countries well endowed with minerals not being able to change the lived realities of their citizens and protect them from the negative impacts of the industry. Responsible stewardship of minerals and a sustainable approach provides for a new reality of resource blessing and a positive legacy of exploration to extraction.

Imagining that new legacy and the opportunities of the industry of the future provides for a responsible reality of co-created sustainable solutions, shared value, and protection of the ecosystems in which the industry operates, with the primary objective of leaving them in an equal or better position at end of life of the operations.

The divide between the source of minerals and the industry is unsustainable. Africa remains on the backfoot in a context where global sustainability depends on the social, ecological, political, and economic development of those impacted by and those who benefit from the value created by a responsible extractive industry.

Ultimately, it is about creating capacity for responsible harvesting of minerals and processing them into commodities that benefit the rest of the world. Therein lies the opportunity to explore the possibility of turning the curse into a blessing in the interest of people and the planet, thereby leaving a positive legacy for generations to come.

More than Mining – B2Gold's Legacy in Namibia

John Roos
B2Gold, Windhoek, Namibia

B2Gold's Otjikoto gold mine is home to world-class production standards and high-quality output. Apart from consistently achieving its production targets, the Otjikoto mine is a shining example of how to incorporate exploration and production targets with sustainability goals.

B2Gold's visionary management team is keen to lead in sustainability and has invested heavily in projects that are not directly linked to the bottom line. In Namibia, the company is building a legacy through implementing industry-leading initiatives on progressive rehabilitation, investment into Corporate Social Responsibility (CSR), and assessment of post closure land use opportunities.

B2Gold's postclosure vision is a top priority for the local management team and is driven by an overarching strategy at international level. B2Gold's CSR activities follow the company's wider commitments that fall into four categories: health, education, environment, and livelihoods.

On the energy transition front, B2Gold Corp., in 2023, set a target to reduce the group's Scope 1 and 2 greenhouse gas (GHG) emissions by 30% in 2030 against a 2021 baseline. As a company, B2Gold recognises that climate change is a critical global challenge and understands that it must act accordingly. This commitment to GHG emissions reduction forms a key part of B2Gold's climate strategy, incorporating climate management as a part of its planning process.

The Otjikoto Mine is leading the charge when it comes to B2Gold's transition to green energy. With 7mWdc installed solar capacity on the mine and another 10mWdc capacity being installed, a significant portion of the operation's energy requirements will be delivered through renewable solar energy.

B2Gold's ultimate goal is to leave the area in which we extracted gold in a better state than we found it, both environmentally and socially.

First Major Ore-Grade Gold Concentration on Earth's Surface: Why, When, and Where?

Hartwig E. Frimmel[1], Ryan Mathur[2]
1. Department of Geodynamics and Geomaterials Research, Institute of Geography and Geology, University of Würzburg, Würzburg, Germany, 2. Geology Department, Juniata College, Huntingdon, PA, USA

The first major gold concentration in the Earth's Crust is recorded by largely conglomeratic rocks of the c. 2.9 to 2.8 Ga Central Rand Group in the Mesoarchean Witwatersrand Basin, Kaapvaal Craton, South Africa, which accounts for almost one third of all known gold worldwide. Here we present new Cu isotope data on pyrite, carbon seams and gold from conglomerates at the base of the Central Rand Group (c. 2.90 Ga), the base of the Ventersdorp Supergroup (c. 2.78 Ga), and the base of the Transvaal Supergroup (c. 2.66 Ga). Our results indicate an overall increase in oxidation state of post-depositional fluids from 2.9 to 2.6 Ga but also synsedimentary gold precipitation by biological activity following an oxidative gradient. The highest $\delta 65Cu$ were obtained from carbon seams (+2.41 ± 0.24‰) and are explained by Cu isotope fractionation due to acidophilic bacteria, thus confirming the hypothesis of first large-scale gold concentration having been triggered by oxidative microbial activity under an overall reducing atmosphere. This process probably occurred on all Archean cratons at around 2.9 Ga, but unique preservation of the Witwatersrand strata accounts for their exceptional gold endowment. Paleoplacers significantly younger than 2.9 Ga received their detrital gold load largely from the mechanical reworking of older, 2.9 – 2.8 Ga placers and, after the first peak in endogeneous gold deposit formation at around 2.75 – 2.50 Ga, more and more from the latter kind of deposits. This explains the spatial and temporal distribution of gold in Neocharchean and Paleoproterozoic conglomerates in all cratons and can help in the assessment of exploration targets within such conglomeratic successions.

Beyond the State-of-the-Art Geophysical Surveys for Mineral Exploration: Where Do We Go Next?

Musa Manzi
Wits Seismic Research Centre, School of Geosciences, University of the Witwatersrand, Johannesburg, South Africa

As the energy transition accelerates, the mining industry faces an increasing push to produce more raw materials in more difficult and challenging mining environments prone to seismicity and characterized by complex geologic structures. Geophysical methods, seismic methods in particular, are getting more attention than ever to address these challenges because they provide high-resolution images of key subsurface geological structures and relatively retain their resolution at target depths. This paper addresses these challenges and provides several solutions. Beyond the state-of-the-art seismic solutions for deep-seated mineral deposit targeting are presented through case studies from several deep mine sites in the Witwatersrand Basin (South Africa). Firstly, we discuss the value of reprocessing legacy data using new processing algorithms for gold exploration, particularly focusing in the mining regions where no other data are available or acquisition of new data is difficult and expensive. Secondly, we showcase the state-of-the-art surface-tunnel-borehole solutions developed under the EU-funded Eramin Future project for shallow and deep-seated mineral deposit targeting. Finally, we discuss the current status and limitations of geophysical methods in difficult underground mining environments (e.g., lack of GPS signal) and present our newly developed solutions for in-mine (~ 3-km mine tunnel) seismic surveys that can be utilised to image the gold deposits and complex geologic structures several metres ahead of the mine face and metres-to-kilometres below and above the tunnel floor. Essentially, these new in-mine seismic solutions take advantage of the distributed acoustic sensing (DAS) technology that incorporates fiber-optic cables and wireless seismic recorders, which are then GPS-time synchronised using a newly developed GPS-time transmitter system developed under the Smart Exploration project. The GPS time transmitter opens up new opportunities in exploration projects as it allows the recording of seismic data in GPS-denied environments such as in-mine infrastructure (e.g., mine tunnels and underground boreholes).

Gold in Africa – History, Endowment, and Discovery

Mike Robertson
The MSA Group, Johannesburg, South Africa

History

Gold has been mined in Africa for centuries with gold artefacts and ancient workings known from many regions including Northeast Africa, West Africa, South Africa and Zimbabwe. Africa was the world's second largest gold producer by continent between 2010 and 2022. Over this period Africa produced on average one quarter of the world's gold and showed a steady proportional increase in production from 22% to 27% over this period.

Endowment

Gold endowment in Africa is associated with a variety of deposits including orogenic, placer, intrusion-related, skarn, IOCG, porphyry, epithermal and alluvial types. The Archean Witwatersrand basin is the world's greatest goldfield and was the world's largest producer for 100 years. On a continent-scale however, gold endowment is largely related to orogenic deposits within greenstone belts on Archean Cratons in South Africa, Zimbabwe, Tanzania, the DRC and West Africa.

Discovery

In recent times, mining-friendly legislation has been a catalyst for gold exploration. Following a long history of ancient workings and small colonial mines, new legislation in Tanzania prompted extensive exploration activity in the Lake Victoria Goldfields resulting in the development of five mines within five years and an overall increase in production from 10,000 ounces in 1997 to 1.68Moz in 2005. A similar situation in West Africa over the past 30 years has led to numerous new discoveries and a reported Mineral Resource inventory of about 300 Moz, the majority of which represents new ounces. Over 40 new mines have come on stream during this period. In many instances, new African mines have been established on ancient workings. However, success has also been driven by a mineral systems approach to exploration, using process-based predictive models. Improved insights, particularly on structural controls on mineralization, have led to exploration success.

Source of fluids in orogenic and intrusion-related gold deposits: insights from trace elements in pyrite

Dominique Genna[1], Sarah Dare[2], Damien Gaboury[2], Michel Jébrak[3]
1. CONSOREM-UQAC, Chicoutimi, QC, Canada, 2. UQAC, Chicoutimi, QC, Canada, 3. UQAM, Montréal, QC, Canada

A diverse range of gold mineralization has been observed in greenstone belts worldwide, spanning from synvolcanic to syntectonic deposits. Whereas most syntectonic deposits are considered orogenic in origin, formed by metamorphic fluids, the contribution of magmatic-hydrothermal fluids remains a debate. This study examines the trace element composition of over 1500 pyrite grains from 50 deposits within the Archean greenstone belt of Abitibi, Canada, using LA-ICP-MS. Our case studies are representative of various types of syntectonic mineralization, including 1) As-rich and As-poor orogenic deposits 2) intrusion-related deposits.

Using our novel multielement discrimination diagram, normalized to Archean sedimentary pyrite, we demonstrate that the variability in the trace element pattern of pyrite reflects the source of the ore-forming fluids. Pyrite from As-rich deposits (both sediment-hosted and volcanic-hosted) exhibit a flat signature, similar to Archean sedimentary pyrite, but with positive anomalies in Au-As-Sb-W. Compilation of pyrite from major As-rich gold districts worldwide, typically sediment-hosted have the same signature. In these cases, devolatilization of underlying sediments release Au-As-Sb from pyrite and W from rutile into the metamorphic fluids. The pyrite signature of these gold deposits thus reflects the composition of these fluids. In contrast, pyrite from As-poor orogenic deposits are depleted in most trace elements. This pattern is similar to pyrite from altered magmatic sulfide droplets in basaltic rocks. We propose that in this case, the source of the ore-forming fluids may result from the devolatilization of mafic crust. Finally, pyrite in intrusion-related deposits, associated with sanukitoid intrusions, display a flat to moderately depleted pattern with distinctive positive anomalies in Au-Te-Ag-W. This signature closely resembles that of modern alkalic porphyry deposits, where the magmatic contribution of gold and fluids has been irrevocably established. Detection of a magmatic input, using pyrite, could influence exploration strategies, early-on, for future gold deposit discoveries.

Large Gold Systems of Central Asia: Geodynamics and Metallogeny Revisited

Reimar Seltmann
CERCAMS NHM London, London, United Kingdom

Major gold deposits occur across the Central Asian Orogenic Belt (CAOB), also referred to as the Altaids orogenic collage. They are attributed to diverse deposit types of which the most important ones are porphyry deposits. For example, in Uzbekistan the Kalmakyr-Yoshlik porphyry system (Middle Tienshan) hosts more gold than Muruntau (Kyzylkum accretionary zone).

In the western Altaids, the Paleozoic accretionary belts of Kazakhstan and of the Northern, Middle and Southern Tienshan as well as suture/shear zones and sedimentary basins establish central Asia as an emerging commodity basket including undiscovered deposits under cover or undeveloped mineral potential still to be unlocked.

As a result of post-collisional extension and orogenic collapse, each accretionary cycle (early, mid, late Paleozoic) is concluded by the formation of major lode Au deposits representing "orogenic-style" (and intrusion-related gold systems, IRG) as well as "Carlin-style" formations (Au associated with Sb-Hg mineralisation). These are often associated with alkaline, dominantly shoshonitic intrusions and are related to major shear zones, often of trans-crustal nature, tapping deep reservoirs through tectonic pumping related to rapid uplift and decompression, that aid focused fluid flow driven by elevated thermal gradient. The peak gold mineralisation in the CAOB (comprising the vast majority of the known gold endowment) took place around the Carboniferous-Permian boundary at ~305-290 Ma, referred to as "jackpot" event. It is represented by famous mineral systems including Muruntau, Amantaytau, Kumtor, Bakyrchik orogenic gold deposits, Zharmitan (IRG), Kochbulak (mesothermal gold), Dzhezkazgan (sedimentary copper), and Khaidarkan, Kadamjay (auriferous Sb-Hg deposits).

In addition to the tectonic, geologic and metallogenic setting and distribution of the main Au-bearing mineral systems (porphyry Cu–Au/Mo, meso- and epithermal Au, orogenic Au, VHMS, Carlin) within the CAOB, the setting, geology, alteration and mineralisation of these deposits is re-assessed based on new geochronological and isotope data (U-Pb zircon ages, Re-Os ages, Sr-Nd-Hf-Pb isotope data).

Insights into the Metallogenic Evolution of the Timmins Gold Camp Using Quartz-Hosted Melt Inclusions from the Porcupine Intrusive Suite

Sarah Speight[1, 2], Jacob Hanley[1], Stephen Piercey[2], Jim Crowley[3]
1. Saint Mary's University, Halifax, NS, Canada, 2. Memorial University of Newfoundland, St. John's, NL, Canada, 3. Boise State University, Boise, ID, USA

The Abitibi greenstone belt in northeastern Ontario, Canada is home to the Timmins gold camp, a world-class gold district with >60 Moz of high-grade gold (past production and future reserves). Deposits are spatially associated with the synorogenic Porcupine-Destor deformation zone (PDDZ). The formation of these deposits is attributed to a few key processes, including large-scale CO_2-rich fluid flow and clastic sediment deposition along the PDDZ, the presence of ultramafic rocks, and the occurrence of competent intrusions that pre-date metamorphism. These intrusions are believed to have prepared the host rocks for enhanced fluid flow during mineralization. Despite the coincident timing of early magmatism with these events, the contribution of metals from these intrusions is often disregarded.

However, quartz-hosted melt inclusions (MI) from porphyry intrusions within the main Timmins gold camp show evidence of variably elevated, arc-related metals (e.g., Cu, Mo, Bi, W, As, Sb, Ag, Sn). The Paymaster and Crown Lake porphyries, along with the Krist fragmental horizon, also have MI with similar primary bulk compositions. SEM analysis of the MI shows co-precipitation of Cu and Au, suggesting a genetic link and shared metallogenic history. Assessing the compositional evolution of an Archean porphyry system through primary MI, and comparing with modern environments, will help us understand magmatic-hydrothermal processes in the Timmins district. By examining MI from the Porcupine Intrusive Suite and applying time constraints with accompanying U-Pb geochronology, early magmatic degassing events can be evaluated with respect to the overall metal budget of the Timmins gold camp.

Orogenic Gold Grain Coarsening from Colloidal Precursor by Ostwald Ripening

Bjorn von der Heyden, Rutger La Cock
Stellenbosch University, Stellenbosch, South Africa

The growth of large gold grains within a hydrothermal ore forming system is difficult to reconcile with the relatively low solubility of gold in most crustal fluids. The presence of gold colloids within the hydrothermal fluid can help to explain this dilemma. Several workers have demonstrated the presence of colloid-sized gold in vein material from both epithermal and orogenic gold forming systems. However, the mechanisms by which these colloids overcome their electrostatic repulsion to coalesce and form large grains remain incompletely understood. A recent sampling campaign in a tectonically late and highly auriferous black quartz vein in the Archean Barberton greenstone belt has unveiled unique gold textures comprising a dispersion of gold-silver colloids and nanoparticles in close association with coarse gold grains. These coarse gold grains were liberated using a non-destructive SELFRAG sample preparation technique, and the colloidal gold was imaged using high-resolution scanning electron microscopy. The observed individual gold spheroids range in diameter between < 5 nm and > 1 μm, and some of the best-preserved spheroids are found in association with micrometer-scale botryoids and conical protrusions growing from the adjacent coarse gold grain surface. Raman spectroscopy revealed the presence of both graphitic carbon and amorphous shungite carbon in association with these gold textures. Based on these results, we suggest that a viscous hydrocarbon phase was instrumental during the transient stages of gold grain growth, which occurred by an Ostwald ripening mechanism. Arrestation of the Ostwald ripening process, due to increasing viscosity of the hydrocarbon phase, resulted in a ‘freezing in’ and fortuitous preservation of the observed spheroids and protrusions. This novel mechanism for gold grain growth may serve as one explanation for the formation of anomalously gold-enriched ore shoots in some hypogene gold deposits (i.e., those running at 100s of ppm Au).

New genetic type of gold deposit for Timok, Serbia: the high-grade gold skarn mineralization at Coka Rakita project

Mladen Zdravkovic[1], Dragana Davidovic[1], Jelena Zivanovic[1], Ivan Strmbanovic[1], Nutu Groza[1], Dragan Milosevic[1], Olivera Dukic-mijucic[1], Snezana Janjic[1], Aleksandar Sestak[1], Vuk Milic[1], Dragana Drobnjak[1], Bojana Kotlajic[1], Vaso Kitanovic[1], Janja Mihajlovic[1], Nikola Novicic[1], Tijana Mitic[1], Ivana Suzic[1], Paul Ivascanu[2], Kuncho Kunchev[2], Stefan Metodiev[2], Ivan Krumov[2], Istvan Marton[2]
1. DPM Avala doo, Bor, Serbia, 2. Dundee Precious Metals Chelopech EAD, Chelopech, Bulgaria

The Čoka Rakita deposit is located in the NW part of the Timok Magmatic Complex (TMC), in eastern Serbia, which is part of the Late Cretaceous Apuseni-Banat-Timok-Srednogorie magmatic-metallogenic belt. The spatial relationships between mineralization styles suggest that the TMC is generated by a long lasting (from ~89 Ma to 76 Ma) westward migrating arc magmatism beginning with the Majdanpek – Bor – Čukaru Peki Cu-Au belt, succeeded by the younger Kuruga high-sulfidation Cu-Au belt, the Timok Diorite Cu-Au porphyry belt, and the still younger NW Timok Au belt.

The various gold mineralization styles in the NW Timok are part of a large zoned magmatic-hydrothermal system; they are hosted by different lithological units and formed in different hydrothermal environments (from proximal porphyry-skarn mineralization toward distal epithermal). The distal sedimentary rock-hosted gold deposits have been the focus of intensive exploration of the company during the past decade. The Čoka Rakita deposit was discovered recently at the northeastern flank of a large monzonite intrusion, which is part of a distinct alkalic magmatic event in TMC and is associated to syn- and post-monzonite fertile hornblende-biotite-plagioclase±pyroxene-phyric monzodiorite sills and dikes. Low-grade disseminated gold mineralization has been intersected during historic exploration programs within the epiclastic and the diorite intrusive units. During recent years, an intensive drilling program has targeted the deeper levels of this system, which revealed the potential for new manto-type skarn mineralization, defined primarily as stratigraphically controlled stratabound gold-rich garnet skarn lenses within calcareous sandstones in the hanging-wall of the sill-like monzodiorite bodies.

The Čoka Rakita project is in the initial development stage with intensive exploration drilling and a Maiden Mineral Resource estimate completed in December 2023, reporting 1.78 million ounces of gold. The unique mineralogical and grade characteristics are new not only for the Serbian TMC, but for the entire Western Tethyan Belt.

Permeability Evolution Within an Orogenic Disseminated Gold Deposit: Evidence from Liba Gold Deposit, Central China

Pengcong Zhang, Kunfeng Qiu
China University of Geosciences, Beijing, Beijing, China

The nature of permeability associated with pervasive hydrothermal alteration within disseminated orogenic gold deposits is under-researched. The Liba orogenic gold deposit, located in West Qinling, Central China, hosts 112t of gold resources, predominantly in the form of disseminated gold. The slate wall rock mainly consists of quartz and sheeted silicate minerals such as sericite, biotite, and chlorite. Interparticle pores, found between sheeted minerals and/or quartz, have straight boundaries and appear as relatively regular polygons ranging in size from 5μm to 20μm. They are isolated and not interconnected, representing the intrinsically low permeability of the wall rock. In slate ore, iron-bearing minerals biotite and chlorite have been completely altered to pyrite and sericite. To be receptive to pervasive hydrothermal alteration, wall rock permeability needs to be enhanced. Notably, there are no significant differences in the types, sizes, and densities of pores within the ore compared to those in the wall rock, but these pores in the ore present distinct tortuous boundaries and irregular shapes. Significant pore differences imply that interconnected pores provide channels for pervasive fluid flow. Moreover, disseminated pyrite within the ores indicates the absence of macroscopic fractures, implicating the infiltration of hydrothermal fluid into the rocks through grain-scale microcracks. Mass balance and theoretical calculations suggest that after hydrothermal alteration, the solid volume of the rock increases (maximum up to 38%), leading to a decrease in rock permeability. Grain-scale microcracks thus should be induced by tectonic loading and hydraulic fracturing, which connect the pre-existing pores and serve as channels for fluid flow. The isolated pores within the ores represent destroyed permeability, which should be caused by the precipitation of hydrothermal minerals. Therefore, the evolution of permeability is controlled by a competition between permeability enhancement processes and permeability destruction processes.

Synergies of Drone-Borne and Satellite Data for Noninvasive Mineral Exploration in Namibia

René Booysen, Rupsa Chakraborty, Samuel T. Thiele, Richard Gloaguen
Helmholtz Institute Freiberg for Resource Technology, Freiberg, Germany

The transition toward a net zero energy economy has led to an increased demand for the critical raw materials required for green technologies. Recycling alone is not capable of compensating the requirements for the foreseeable future. At the same time, the extractive sector is facing increasing difficulties in getting stakeholder support to develop new projects. Therefore, our goal is to adopt exploration techniques that minimize environmental impact and prioritize non-invasive methods. For this, we use innovative remote sensing methods to not only improve mineral detection and mapping, but also foster social acceptability for the mining and exploration industry. Hyperspectral imaging (HSI) is a rapidly developing technology that allows for fast and systematic identification of key minerals at the Earth's surface and provides information about mineral abundances and associations. Several recently launched satellites and the rapid rise of NewSpace (commercial providers) are also opening new opportunities. In this contribution, we illustrate a process including joint drone-borne HSI and satellite-based HSI. We leverage the potential of the different platforms and imaging systems, taking into account their respective advantages and disadvantages. We suggest a vertical integration of drone-borne high spatial and spectral resolution but with limited coverage, and large-scale imaging with 30 m ground sampling provided by satellites such as EnMAP, PRISMA, and the soon to be launched Planet (Tanager) and CHIME. We argue that a combination of machine learning and spectroscopy, accompanied by a structural analysis provides an ideal solution to map potential targets accurately. The processing chain includes radiometric and geometric corrections, coregistration, and spectral and structural mapping. We showcase this approach at two study sites in Namibia: The Marinkas Quellen Carbonatite Complex and the Uis pegmatite-hosted tin mine. Both of these deposits host CRMs used in today's green technology, i.e., REEs and lithium, respectively.

Influence of Mineral Microinclusions on Vectoring and Fertility Assessment Tools in the Green Rock Environment: A Case Study from Batu Hijau Porphyry Cu-Au Deposit, Indonesia

Axel Cima[1, 2], Ivan Belousov[1, 2, 3], David R. Cooke[1, 2], Matthew J. Cracknell[1, 2]
1. University of Tasmania, Hobart, TAS, Australia, 2. Centre for Ore Deposits and Earth Sciences (CODES), Hobart, TAS, Australia, 3. CODES Analytical Laboratories (CAL), Hobart, TAS, Australia

Valuable chemical information is contained within alteration minerals such as chlorite and epidote, providing useful tools for vectoring and fertility assessment in porphyry exploration. Although previous studies have focused on these alteration minerals, it is common for them to contain microinclusions of other minerals that are also analyzed when acquiring LAICPMS data. Addressing this gap in current knowledge, this study reevaluates over 3000 legacy LA-ICPMS analyses from the Batu Hijau Porphyry Cu-Au deposit, focusing on the distribution and composition of microinclusions in chlorite and epidote.

Microinclusions in chlorite and epidote increase in frequency toward the center at Batu Hijau, reflecting intensity of hydrothermal activity and elevated trace elements in hydrothermal fluids. Through new LA-ICPMS and EPMA analyses on selected samples, this research confirms that mineral microinclusions, particularly titanite and apatite within chlorite and epidote associated with propylitic alteration, can serve as effective vectors for porphyry exploration and aid in fertility assessment.

In a magmatic environment, apatite co-crystallization has been demonstrated to affect the extents of Ce and Eu anomalies in zircon thereby impacting estimates of magma oxidation based on zircon compositions as well as influencing other trace elements. Ti-bearing microinclusions (rutile, titanite) are also prevalent in chlorite and epidote. Titanium contents can serve as temperature indicators in chlorite, vectoring toward the heat center of the system. Thus, the presence of microinclusions may artificially increase the measured Ti, thereby influencing vectoring outcomes.

Preliminary results indicate an increase in Zr, LREE, U, and As in titanite and LREE/HREE ratios in apatite microinclusions toward the deposit center. Sulphur contents in apatites potentially highlight the extents of the pyritic zone. Although further research is necessary, these findings reveal promising paths for new exploration tools within the porphyry-epithermal environment.

Do Mineral Systems Have Fingerprints? (Yes, and They Can Be Used to Find World-Class Orebodies!)

Paul Hodkiewicz, Bruce Hobbs, Alison Ord
University of Western Australia, Nedlands, WA, Australia

Recent advances in drill core scanning and downhole sensing technologies used in the mining industry provide continuous, high-resolution mineralogical, chemical, textural and petrophysical drill hole data. These objective data sets are inputs to a range of novel analytical and rock mass characterisation methods that allow mineral systems and orebodies to be 'fingerprinted' in ways that were not possible with more traditional, manual, subjective, and interval-based core logging codes. The novel analytical methods include recurrence analysis, probability distributions and entropy analysis, which can be used to characterise the distinct downhole signatures of mineral assemblages and chemical abundances associated with mineralisation. These non-linear analytical methods have been used widely in the medical and material sciences to distinguish between, for example, healthy and sick (cancerous) cells and organs, and high- and poor-quality manufactured materials. In economic geology, these methods highlight that mineral systems are non-linear and dynamic, open geochemical systems. More important, the methods can be used to distinguish well-endowed mineral systems from others. This presentation will demonstrate that these new, continuous downhole data sets and analytical methods reveal fingerprints and spatial patterns that can improve exploration targeting and orebody knowledge. Results can be used in mine planning and processing to deliver better inputs to integrated mining operations for the downstream customers of the orebody. This includes a better understanding of spatial geometallurgical characteristics of the orebody that will improve predictions of rock mass fragmentation, comminution, crushing, grinding, mineral processing and metal recoveries. The opportunity for the mining industry is to take advantage of this new orebody knowledge to discover new resources and to develop more efficient and lower footprint mining operations with lower energy and water requirements. Examples from orogenic gold, porphyry copper-gold, and magmatic-hydrothermal polymetallic mineral systems will be presented.

Development of New Exploration Vectors by Quantitative Mineralogical Analysis at the Spremberg-Graustein-Schleife Cu-Ag Kupferschiefer Deposit, Lusatia, Germany

Yonghwi Kim[1], Max Frenzel[1], Bradley M. Guy[1], Samuel T. Thiele[1], Ralph Braumann[2], Jens Gutzmer[1]
1. Helmholtz-Zentrum Dresden-Rossendorf, Freiberg, Germany, 2. Kupferschiefer Lausitz GmbH, Spremberg, Germany

The Spremberg-Graustein-Schleife deposit is part of the Kupferschiefer district in the southern Permian Basin and comprises stratabound copper, silver, and other valuable and critical metals in mineralized rocks hosted by pre-Zechstein sandstones, Kupferschiefer, and Zechstein carbonates. Due to the significant exploration potential across the southern Permian Basin, the goal of the present study is to utilize mineralogical signatures as exploration vectors. The specific focus of this study is on gangue mineralogy, especially solid solutions such as carbonates and dioctahedral micas since their quantitative and compositional changes can potentially provide a larger detectable footprint in the mineralizing system. All samples and data from the last drilling campaigns were provided and discussed by KSL GmbH.

Multiple analytical methods were combined to investigate samples from three mineralized and nonmineralized drill cores. Scanning electron microscope (SEM)-based image analysis (MLA) was carried out to obtain quantitative data on mineralogy as well as major-element carbonate chemistry of each stratigraphic unit. Quantitative X-ray powder diffractometry was performed as an external validation. The mineral chemistry of carbonates and dioctahedral micas was studied by EPMA and LA-ICP-MS.

The Zechstein carbonates in the well-mineralized core are dominantly composed of dolomite, whereas in the weakly-mineralized core, calcite is the more abundant carbonate phase. Three successive generations of carbonate in the mineralized drill-hole were identified from backscatter electron (BSE) imaging, comprising (i) dolomite to (ii) Mn-Fe dolomite and (iii) ankerite rims. A quantitative comparison between dolomite and ankerite shows a correlated trend with the Cu-sulfide contents. Dioctahedral micas in the footwall sandstones were compositionally subdivided into muscovite and phengite, and the mineralized drill hole tends to contain more muscovite than the weakly mineralized drill hole. The marked mineralogical differences between well-mineralized and weakly-mineralized cores may indicate overprinting by mineralizing fluids and may thus be useable vectors toward mineralization.

Geochemical Signatures of Magmatic Hydrothermal Alteration in Carbonate Rocks Around the Basin Porphyry in the Drum Mountains of West-Central Utah

Michael J. Kirschbaum[1], Zhaoshan Chang[1], Jeff Nicholes[2], Adam Pacey[3]
1. Colorado School of Mines, Golden, CO, USA, 2. Jeff Nicholes LLC, Elko, NV, USA, 3. Rio Tinto Exploration, Bundoora, VIC, Australia

The Basin porphyry prospect in Utah, USA, is subeconomic with a historic noncompliant supergene resource of 4 Mt at 0.22% Cu. Mineralization styles in the district include minor carbonate-hosted vein/replacement/jasperoid Cu-Au-Ag, carbonate-hosted Mn replacement, pebble dike-hosted Au, and sediment-hosted disseminated Au. We investigate the whole-rock geochemical and C-O isotope signatures in the carbonate wall rocks up to 7 km from the intrusion to compare them with that of the giant Bingham porphyry deposit, Utah.

The stratigraphy is composed by a series of gently W-dipping early Phanerozoic platform carbonates overlying a thick sequence of siliciclastic units (Prospect Mountain quartzite). These have been intruded by composite diorite and monzonite stocks extending into the base of the carbonate sequence. We selectively sampled vein and wall rock from three lower carbonate units (the Tatow Member of the Pioche Formation, the Howell Limestone and the Dome Limestone) along a transect extending 2 km north and 7 km south of the Basin porphyry.

Multielement bulk-rock geochemistry displays good trends in many elements, increasing closer to the Basin porphyry. The strongest element trends are for Be, Ce, Co, Ni, S, Se, Ta, and Te, which show highest values in the immediate vicinity of the Basin porphyry and decrease to background levels about 1 km outboard. $\delta^{13}C$ isotope compositions of bulk rock show depletions of ~5‰ around the Basin porphyry relative to typical early Phanerozoic marine carbonates, and return to the typical values 1 km outboard. $\delta^{18}O$ values of the bulk samples show strong depletions extending out to ~3 km from the Basin porphyry, whereas all the rest show weak depletions relative to typical early phanerozoic marine carbonate values. Carbon and oxygen isotopes from carbonate veins reveal similar trends to the bulk isotopes, except that the higher stratigraphic units display less noise.

The Transformative Impact of Enhanced Geoscience Data Collection and AI-Powered Modeling on the Mining and Exploration Industry

Patrick B. Redmond[1], Grant Sanden[1], Ryan Barnett[2], John Manchuk[1]
1. GeologicAI, Calgary, AB, Canada, 2. Resource Modeling Solutions, Edmonton, AB, Canada

Delivering on the UN Sustainable Development Goals and transitioning to a low-carbon global economy requires technologies that utilize a wide range of minerals in vast quantities. Unfortunately, as the world's need for minerals is greater than ever, discovery rates have dropped and discovery costs have risen. In addition, the timeline from discovery to mine development has grown longer and most new mines fail to achieve design production in the early years, resulting in a delayed return on investment and a global shortfall in the minerals needed for the green energy transition. Multiple factors contribute to these trends, however, the issue of inadequate geoscience data collection and analysis stands out as a major contributor, since it leads to poor characterization of mineral deposit properties and their associated uncertainty. This creates a poor understanding of downside risk and upside potential, suboptimal mine and mill design, and generally degraded decision making.

Fortunately, more accurate, consistent, high-resolution and near real-time geoscience data collection has become available using advanced multi-sensor scanning technology that captures RGB, XRF, LIBS, hyperspectral, petrophysical, and LiDAR data, among others. Simultaneously, rapid advances in AI-powered modeling solutions permit effective integration of these data in geology, grade, geometallurgical, geotechnical, and recovery modeling of a deposit, significantly improving the holistic characterization of a mineral resource to facilitate improved decision making throughout its life cycle. Our industry is rapidly adopting these technologies, which are transforming how we collect and interact with geoscience data across the entire exploration and mining value chain.

This presentation will summarize the referenced technologies, presenting examples and case studies that illustrate their application across a range of deposit types. It will conclude with a discussion of how we as geoscientists can combine foundational geoscience knowledge with the data from these technologies to improve exploration, deposit evaluation, and mining outcomes.

Valley – A High-Grade Reduced Intrusion Related Gold System in Yukon, Canada

Thomas K. Branson
Snowline Gold Corp., Vancouver, BC, Canada

Valley is a significant new Reduced Intrusion-Related Gold System (RIRGS) discovery in the Selwyn Basin of the Yukon Territory (Canada). Since 2021, Snowline has delineated a 700 m x 500 m mineralization footprint from surface up to 450 m vertical depth with drill hole intercepts up to 2.5 g/t Au over 553.8 m from surface, including 132 m of 5.0 g/t Au also from surface.

The Valley stock is an elongate 1 km x 800 m mid-Cretaceous Mayo-suite multi-phase granodiorite hosted in siltstones and shales of the Ordovician-Silurian Road River group. The granodiorite is surrounded by a pyrrhotite-rich contact-aureole and hosts a high density of gold-bearing sheeted quartz+carbonate-scheelite veins with up to 30 veins per metre. The sheeted veins are sulphide poor (<5%) with minor pyrite, pyrrhotite, chalcopyrite and arsenopyrite. Texturally, visible gold is related to bismuthinite, scheelite and telluride minerals consistent with Bi-Te pathfinder element associations present in surface geochemistry. Veins are dominantly controlled by steep northwest-striking faults, though veins are known to occur in at least three structurally-controlled orientations. Locally, high grade mineralization appears to be spatially associated with a fine-grained porphyritic phase.

Valley is part of the Tombstone Gold Belt (TGB), a trend of RIRGS deposits that include Kinross' Fort Knox and Victoria Gold's Eagle mines. Compared to these two active sites characterized by average production and reserve grades well below 1 g/t Au, Valley's intercepts are multiples above what is currently being profitably extracted in the TGB. Situated in the underexplored eastern part of the TGB, Valley is the focus of exploration activities within Snowline's Rogue Project, which covers a 30 km by 60 km magnetic geophysical anomaly associated with mid-Cretaceous intrusions of the Rogue Plutonic Complex (RPC). In addition to Valley, the RPC hosts numerous underexplored intrusions with high potential for further RIRGS discoveries.

U-Pb Geochronology of the Archaen Tati Greenstone Belt gold deposits, NE Botswana: Evidence of two post-Archean and post-metamorphic mineralization events.

Charles Geraud E. Kono, Thierry B. Bineli, Peter N. Eze
Botswana International University of Science and Technology, Palapye, Botswana

The Tati Greenstone Belt (TGB) of northeastern Botswana, in the southwestern edge of the Zimbabwe Craton, hosts numerous gold deposits/occurrences including, the Golden Eagle and Mupane gold deposits. Despite being critical to understanding gold mineralization in the greenstone belt, the geochronology of the TGB gold deposits has been remarkedly under studied.. Here, we report on the U-Pb ages obtained from apatite, titanite, and garnet, assessed to be in equilibrium, and therefore, cogenetic with gold-bearing arsenopyrite from the Golden Eagle and Mupane gold deposits to constrain the time of gold introduction in these deposits. In the Mupane gold deposit, garnet in sample 1 yields a Tera-Wasserburg lower intercept $^{206}Pb/^{238}U$ age of 2189.3 ± 22.3 Ma (MSDW = 1.54; n = 66/92), which overlaps with the garnet concordia age 2154 ± 13.3 Ma (MSDW = 1.94; n = 80/177) from sample 2. In the Golden Eagle deposit, titanite yields a Tera-Wasserburg lower intercept $^{206}Pb/^{238}U$ age of 1927 ± 18.7 Ma (MSDW = 1.54; n = 21/21), whereas apatite yields a Tera-Wasserburg concordia age of 1960.5 ± 112.5 Ma. This suggests two episodes of post-Archean gold mineralization events within the TGB. The U/Pb age from apatite overlaps with the second cycle of the Limpopo-Liberian tectonic event of 2050 ± 1.5 Ma, which is here considered as precursor event to mineralization in the area. On the other hand, the ages obtained on garnet do not correspond to any recorded geo tectonic event in the TGB, and may be related to either partial resetting of these metamorphic garnets by the second Limpopo tectonic event or a concealed intrusion. Considering the age of the metamorphism (2630 ± 70–2570 ± 70 Ma) in TGB, the Mupane and Golden Eagle deposits are therefore here regarded as post-metamorphic orogenic-type gold deposit.

Intrusion-Associated Au Mineralization in the Archean Mooshla Intrusive Complex, Doyon-Bousquet-LaRonde Mining Camp, Abitibi Greenstone Belt, Québec, Canada

Kevin Neyedley[1], Jacob J. Hanley[1], Patrick Mercier-Langevin[2, 3], Richard A. Stern[4]

1. Department of Geology, Saint Mary's University, Halifax, NS, Canada, 2. Geological Survey of Canada, Quebec, QC, Canada, 3. Agnico Eagle Mines Limited, Toronto, ON, Canada, 4. Department of Earth and Atmospheric Sciences, University of Alberta, Edmonton, AB, Canada

The Mooshla Intrusive Complex (MIC) is an Archean polyphase magmatic body located in the Doyon-Bousquet-LaRonde mining camp of the Abitibi greenstone belt, Québec, Canada, that is spatially associated with numerous Au-rich VMS, epizonal 'intrusion-related' (or porphyry-epithermal-*style*) Au(-Cu) vein systems, and shear zone-hosted Au deposits. To elucidate any genetic links between deposits and the MIC, mineralized samples from two epizonal 'intrusion-related' Au-Cu vein deposits (Doyon and Grand Duc Au(-Cu)) have been characterized using a combination of detailed petrography and trace element and sulfur isotope analyses of sulfide phases.

Based on a wide variety of geochemical data, it is clear that numerous processes were occurring to produce the different ore assemblages at Doyon and Grand Duc. However, the main gold precipitation mechanism is interpreted to be a reduction in the fluid composition that was influenced by wall-rock sulfidation for both the Doyon high-grade and Grand Duc lower-grade vein systems.

Multiple fluid processes were operating in the mineralization at Grand Duc as the texture, trace element signatures, and $\delta^{34}S$ composition of pyrite all are indicative of fluid boiling leading to an increase in fO_2 of the hydrothermal fluid and gold mineralization is late-pyrite to syn-pyrrhotite + chalcopyrite + sphalerite. The common occurrence of tellurobismuthite in the high-grade veins suggests that gold precipitation may be related to a reduced fluid event as Te is more soluble in oxidized fluids and precipitates as fluid conditions become more reducing.

Importantly, the $\Delta^{33}S$ composition of pyrite and pyrrhotite is similar in both deposits and does not vary within individual pyrite grains, implying the source of sulfur was from a uniform reservoir. Additionally, there is no evidence of a sedimentary $\Delta^{33}S$ signal for a source of sulfur, and therefore it is likely to be magmatic, indicating the MIC could have supplied volatiles to these Au-rich deposits.

Petrography, Ore Mineral Paragenesis, and Gold Deportment of the Twin Hills Gold Deposit, Central Namibia

Johannes Sales[1, 2], Paul Nex[2]
1. Osino Resources, Omaruru, Namibia, 2. School of Geosciences, University of the Witwatersrand, Johannesburg, South Africa

The Twin Hills Gold Deposit is an orogenic gold deposit hosted within the metamorphic turbidites of the Kuiseb Formation in the Southern Central Zone of the Damara Belt. This study aimed to understand the petrography of the host rocks, the formation of the ore minerals and assess the deportment of the Twin Hills Gold Deposit. The host rocks are characterized by graded bedding ranging from pelitic mica-rich interbeds to psammitic quartz-rich interbeds, forming an interbedded metagreywacke package. The Interbedded metagreywacke is made up of quartz, biotite, muscovite, plagioclase, orthoclase and cordierite. The ore mineral assemblage is characterized by gold, pyrrhotite, arsenopyrite, and pyrite. Gold mineralization at the Twin Hills Gold Deposit is associated with two main hydrothermal alteration styles, namely, potassic alteration and silicification. This research suggests at least three stages of mineralization within the Twin Hills Gold Deposit. The first stage is the diagenetic stage, which involved the enrichment of diagenetic arsenian pyrite (Py_0) in the turbiditic sediments with gold and other trace elements via the exhalation of reduced, deep-seated Au-As enriched H_2S-rich basinal brines onto the sea floor. The second stage involved the release of gold initially locked up within the arsenian pyrite (Py_0) crystal lattice into metamorphic fluids and the subsequent precipitation of free gold (Au_1) from gold-bearing fluids. The third stage is the quartz veining stage, which was responsible for the remobilization of fine gold grains (Au_1) from the groundmass of the metagreywacke and concentrating them into the biotite selvages. This led to further upgrading of the gold grades to higher economic levels. The gold deportment analyses revealed three modes of occurrences of gold within the Twin Hills deposit. This includes gold disseminated within the groundmass, gold grains disseminated within biotite selvages associated with quartz veinlets and gold grains included within arsenopyrite crystals.

Distribution and Partitioning of Trace Elements During Hydrothermal Sulfidation of Magnetite by Pyrite, KCD Gold Deposit, DRC

Yann Waku Mpaka[1, 2, 3], Bjorn P. Von der Heyden[1]
1. Stellenbosch University, Stellenbosch, South Africa, 2. Barrick Gold Corporation, Toronto, ON, Canada, 3. University of Kinshasa, Kinshasa, Democratic Republic of the Congo (Kinshasa)

Magnetite plays a crucial role during the sulfidation process within iron-rich host rocks like banded iron formation (BIF). During this process, fluid-mineral interaction between magnetite and H_2S-rich hydrothermal fluids results in precipitation of pyrite, which can be auriferous. Research on the partitioning behavior of trace elements during this process is notably scarce. This study investigated ten samples ranging from unaltered to strongly altered BIF units in the Karagba-Chauffer-Durba (KCD) gold deposit. We employed detailed micro-analytical LA-ICP-MS analysis on pyrite-magnetite pairs along with whole-rock geochemistry to evaluate the distribution and partition coefficients of various elements, including Au, As, Ag, Cu, Co, Ni, Mo, Pb, Sb, S, Cd, Hg, Se, Gd, Zn, Mn, Cr, V, and Ti. Petrographic assessments revealed alteration assemblages characterized by Fe-carbonate, sericite, quartz, pyrite, chlorite, and magnetite, accompanied by major and trace element changes. The decrease in SiO_2 content from unaltered to altered samples suggests desilicification during carbonate-pyrite alteration, with the formation of secondary magnetite, under near-neutral to acidic conditions. Mineral trace elements such as V, Co, Ni, Zn, Ga, Ge, and Cr displayed distinct behavior during alteration. Vanadium, primarily occurring as V^{5+} at high oxygen fugacity and V^{3+}/V^{4+} at low oxygen fugacity, exhibited sensitivity to reducing conditions, influencing its solubility and mobility in magnetite. Co and Zn, considered compatible elements in pyrite, showed depletion in magnetite, indicating sequestration in pyrite. Ni showed variability between unaltered and altered BIFs, highlighting the influence of alteration on its distribution. Ga, Ge, and Cr remained relatively immobile during alteration, suggesting their compatibility with magnetite and potential as tracers of magnetite origin. Furthermore, the Co/Ni ratios in magnetite and pyrite served as proxies for assessing hydrothermal versus metamorphic minerals. This study contributes to understanding the trace element evolution during sulfidation hydrothermal alteration and partitioning of trace elements in both magnetite and pyrite.

Hydrothermal Alteration and Volcanic Stratigraphy at the Paleoproterozoic Kay Mine Volcanogenic Massive Sulfide Deposit, Black Canyon City, Arizona USA

Aaron J. Adsit[1], Thomas Monecke[1], Mark D. Hannington[2], Miranda R. Lehman[1], David Smith[3]
1. Center to Advance the Science of Exploration to Reclamation in Mining, Department of Geology and Geological Engineering, Colorado School of Mines, Golden, CO, USA, 2. Department of Earth and Environmental Sciences, University of Ottawa, Ottowa, ON, Canada, 3. Arizona Metals Corp., Toronto, ON, Canada

The Kay mine is a past-producing volcanogenic massive sulfide deposit located 70 km north of Phoenix Arizona. A historic estimate of resources and reserves totals 5.8 million tonnes grading 2.2% Cu, 3.03% Zn, 54.9 g/t Ag, and 2.8 g/t Au at a cut-off grade of 2% Cu equivalent. Kay is hosted in a bimodal volcanic and volcaniclastic succession of the 1750 Ma Black Canyon Creek Group. The massive sulfides are located along a favorable stratigraphic position that is marked by a change from mafic to felsic-intermediate volcanic rocks. A mudstone-matrix monomict rhyolitic to rhyodacitic breccia interpreted to be a peperite marks the paleoseafloor position. Massive sulfides at Kay formed through subseafloor infiltration and replacement of permeable volcaniclastic rocks beneath this breccia unit. Utilization of continuous core scanning X-ray fluorescence with a Minalyzer CS and hyperspectral core scanning with a HySpex SWIR-384 camera revealed the chemostratigraphy at Kay and uncovered the mineralogy of the hydrothermal alteration halo. Downhole geochemistry shows Na depletion surrounding the mineralization, a halo of K enrichment in the proximal hanging wall and distal footwall, and a zone of Mg enrichment beneath the mineralization. Hyperspectral imaging demonstrated that the enrichments in K and Mg correspond to strong muscovite and chlorite alteration. This was confirmed by thin section petrography, SEM-EDS analyses, and automated mineralogy scans. The observed alteration zonation reflects variations in the temperature of alteration with the high-temperature hydrothermal feeder zone being characterized by chlorite alteration and the surrounding lower-temperature alteration being dominated by muscovite. In addition, carbonate alteration dominated by ankerite is widespread at Kay. The combination of geochemistry and mineralogy determined by coregistered XRF and hyperspectral core scanning provides a means of precisely mapping the alteration zonation where deformation and metamorphism have obscured the primary textures.

DrillDown: An Open-Source Python Package for "Drilling Down" Into Ore Deposits Datasets

Robert S. Collar, A. E. Williams-Jones
McGill University, Montréal, QC, Canada

Even as the metal demands of the Global Energy Transition have increased, the rate at which new ore deposits are discovered has fallen. In response, members of industry and the research community are collecting datasets of increasing size and diversity along drillholes, the only means by which deposits can be sampled at depth. At the same time, the tools by which we integrate, visualize, and analyze such drillhole datasets have not significantly evolved. Existing tools exhibit shortfalls, not least of which are their high license fees. They are also generally closed-source, OS-limited, and siloed, particularly from the diverse collection of tools that make up the broader Python Scientific Ecosystem.

DrillDown is a python toolkit for visualization and analysis involving drillhole datasets. It provides functionality for constructing drillholes from collar, survey, and one or more downhole datasets, which may include interval data (e.g., assays or lithologies) or point data (e.g., structural measurements or sample locations). Drillholes and associated downhole datasets can be visualized in 3-D along with other meshes (e.g., ore grade shells or DEMs). DrillDown weaves together both cursor-based and programmatic ways of interrogating datasets, including cursor-based selection in 3-D, boolean filtering, and cross-filtering between the 3-D scene and a variety of 2-D plots. This rich functionality is best leveraged when DrillDown is paired with jupyter notebooks, resulting in interactive and shareable analyses. For example, data selected in 3-D can be accessed in a jupyter notebook directly for further programmatic analysis.

DrillDown is free, open-source, and hosted on GitHub, promoting community contribution, transparency, and extensibility. As an open-source python package, DrillDown can be readily paired with other python packages. In this way, it both serves as a jumping off point for bringing the Python Scienific Ecosystem to bear on ore deposits problems and as a springboard for the geoscientist-turned-novice-programmer.

Perkins Discontinuities: Structurally Controlled Grade Patterns Diagnostic of Epigenetic Mineralization at the Deposit-Scale

Jun Cowan[1], Bruce Hobbs[2]
1. Monash University, Melbourne, VIC, Australia, 2. University of Western Australia, Perth, WA, Australia

Our research seeks to rejuvenate the investigation into mineralization timing by examining deposit-scale grade patterns that are discernible solely through the analysis of assay data obtained from drilling, without the need for any specialized analytical equipment other than a 3D viewing software. It builds on the observation and significance of mineralization asymmetry made by pioneering economic geologists as far back as the nineteenth century. The asymmetrical mineralization patterns identified by these pioneers at small scales have largely been overlooked by contemporary economic geologists since the 1960s. However, they are crucial for determining mineralization timing, and Perkins employed this technique to determine the epigenetic timing of sphalerite mineralization at Mt Isa. Despite the unequivocal conclusion reached by Perkins, his observations and conclusions are actively ignored by modern economic geology researchers who continue to assert that Pb-Zn mineralization at Mt Isa is syngenetic.

These asymmetrical mineralization patterns, which terminate laterally against planar boundaries, are termed Perkins Discontinuities in recognition of the significant scientific contribution made by Perkins. These discontinuities are observed across a wide range of commodities at the deposit-scale from hundreds of meters to many kilometers leading us to conclude that the processes responsible for these geometries are independent of commodity types. In fact, they are readily modelled as subtle contrasts in permeability during ore deposition. Perkins Discontinuities form at the late stages of orogenesis, cross-cutting earlier formed folds, therefore diagnostic of epigenetic deposits, yet they are also observed in deposits traditionally interpreted as syngenetic, such as volcanic massive sulfide (VMS), sedimentary exhalative (SEDEX), and nickel sulfide deposits. Previous interpretations of these later deposits as syngenetic are challenged based on the observable presence of Perkins Discontinuities.

Mineral Systems: A Discussion on the Continuing Challenges and Emerging Opportunities

Joel Hrominchuk, Caroline Perring, Thompson C. McCuaig
BHP, Perth, WA, Australia

Application of mineral systems in the exploration industry continues to expand, improving the understanding of how mineral deposits are formed, however, after years of application their full adoption and value have been hampered by some continuing challenges. Mineral systems are often only applied at the earliest stages of exploration, the systems approach though is most effective across scale, as conceptual assumptions are replaced with knowledge and proxies to critical processes are verified with direct evidence. Throughout exploration mineral systems provides the framework for understanding knowledge gaps and what data will improve this knowledge. The acquisition of knowledge identified in this gap analysis defines the exploration program needed to confirm critical processes and vector towards areas of highest prospectivity.

Continuing challenges in identifying and understanding of the critical processes, specifically which are truly essential for the formation of a mineral deposit while others may be absent or simply modifying factors. In addition to criticality, the interdependence, timing, and efficacy of processes must be evaluated to confirm their relevance. Although challenging to constrain, inclusion of these factors into systems thinking represents an emerging opportunity for understanding how and why ore deposits form. Further opportunities include improving our understanding what data is effective at what resolution and scale, improving our ability to predict and detect mineral systems, also promoting targeted research and development in deliberately designed geoscience research.

Our ability to compare geological or political regions and prioritise effort including opportunities at all scales from global prospectivity to potential acquisitions is greatly enhanced by mineral systems. The most important opportunity in the application of mineral systems to the mining industry is as the foundation for commodity supply evaluation and forecasting future production. The systems approach to supply, costs and products produced can then be integrated with other market fundamentals to formulate growth strategy.

One Person's Trash is Another's Treasure; Extracting Additional Information on Fluid Rock interaction from Disturbed Isotopic Systems

Lucy M. Mathieson[1], Christopher L. Kirkland[1], Matthew L. Daggitt[2], Simon Bodorkos[3]
1. Curtin University, Bentley, WA, Australia, 2. University of Western Australia, Perth, WA, Australia, 3. Geoscience Australia, Canberra, ACT, Australia

Crustal fluids can be tracked using traditionally dismissed discordant uranium (U)-lead (Pb) data, offering a novel tool for understanding mineralization processes. This study confronts traditional limitations in geochronology posed by Pb-loss in zircon, which has typically led to the dismissal of discordant U-Pb data, which is data where the isotope systems do not yield similar apparent ages. By reinterpreting these data as critical indicators of past geological processes, such as mineralization and hydrothermal activity, we challenge the conventional perception that views Pb-loss as an analytical hindrance. Employing a comprehensive geochronological database from Geoscience Australia and state geological surveys, our research utilizes Bayesian modelling through Monte Carlo simulations to investigate radiogenic-Pb mobility in the Arunta region of Central Australia. Our findings reveal significant Pb-loss events that correspond to major tectonothermal episodes in the region's history. Specifically, we resolve Pb-loss events at approximately 850 Ma, likely corresponding to the opening of the Centralian Superbasin; 550 Ma, an age associated with the Petermann Orogeny; 475 Ma, potentially linked to the extensional Larapinta Event; and 320 Ma, which corresponds to the Alice Springs Orogeny. These insights underscore the utility of discordant U-Pb data for reconstructing detailed geological histories and promote the use of these data as a valuable tool for understanding fluid mobility in the crust, mapping fluid pathways, and pinpointing potential sites of mineralization within complex geological terrains.

Mineral Prospectivity Mapping on Point Data using Graph Deep Learning Regression: Case Study of Copper Occurrence in Northern Ireland

Felix M. Sihombing[1, 3], Richard M. Palin[1], Hannah S. Hughes[2], Laurence Robb[1]
1. Department of Earth Sciences, University of Oxford, Oxford, United Kingdom, 2. Camborne School of Mines, University of Exeter, Penryn, Cornwall, United Kingdom, 3. Department of Geosciences, Universitas Indonesia, Jakarta, Indonesia

Over the past decade, machine learning (ML) has become an increasingly important tool within geoscience that can support conventional mineral exploration approaches. Here, we present a simple approach for graph deep learning-based mineral prospectivity mapping (MPM) that uses geochemical data collected during standard exploration practices, requires minimum preprocessing, and generates better results than the conventional method. Critically, we structured data into a graphical format and defined the problem as a regression problem instead of a commonly used binary classification problem. We used two deep graph learning techniques: graph isomorphic network (GIN) and graph convolutional network (GCN). Both GIN and GCN performance were benchmarked against results obtained using a random forest (RF) approach, a widely used machine learning algorithm for MPM. Our new method requires two types of point-format information: exploration data and mineral occurrences. As a case study, we consider prospectivity mapping for copper mineralization using stream sediment geochemistry point data from Northern Ireland. Both GIN and GCN graph deep learning approaches perform better than the RF benchmark method, with GIN giving the best performance (R2 = 97.9%), GCN being intermediate (R2 = 89.6%), and RF showing the relatively poorest performance (R2 = 61.7%). Compared to RMSE values for the benchmark method (RF), GIN reduces prediction error by 77%, and GCN reduces this error by 45%. Alongside a significant improvement in identifying known deposits, we also show that GIN can 'virtually' identify new prospective regions in the study area by comparing two sets of mineral occurrence data collected during different periods of time, which may be subsequently confirmed via on-site investigation. Given the minimal amount of conventional data and preprocessing required, these graph deep learning algorithms show notable potential to replace conventional techniques (e.g., RF) and can be easily adopted to a broad range of exploration scenarios.

SEG 2024 Conference: Sustainable Mineral Exploration and Development

Using Mg-Carbonates, Kyanite, and Immobile Elements as Pathfinders for Technology Metals in the Central African Copperbelt

Anna Bidgood[1, 2], Paul Slezak[2, 3], Murray Hitzman[2, 3]
1. KoBold Metals, Dublin, Ireland, 2. Irish Centre for Research in Applied Geoscience, Dublin, Ireland, 3. University College Dublin, Dublin, Ireland

Fluids are an important component of a mineralizing system, leaching, transporting and concentrating the metals that ultimately make up an ore deposit. The extensive geochemical and mineralogical footprint of a fluid is often used to vector towards potential localized mineralization. The Central African Copperbelt (CACB) records a >250-m.y. history of basin extension and inversion in addition to ~500 m.y. of potential supergene fluid interaction, where multiple generations of fluids have altered the surrounding rocks, changing their geochemical signatures. Owing to the longevity and pervasiveness of fluid flow in the basin, alteration footprints can be difficult to unravel, and potential links to mineralization may remain cryptic. Nevertheless, alteration footprints contain a wealth of information relating to the physiochemistry of the fluids that passed through the district's rock packages. Information such as pressure, temperature, salinity, oxygen fugacity, and sulfur fugacity may be recorded in the mineralogy of the alteration footprints which, if deciphered, can provide important clues as to the location of prospective fluid flow pathways.

For example, hydrothermal kyanite-bearing assemblages have been locally observed to relate to Mg metasomatism, which appears to be a basin-wide occurrence. These assemblages act as an indicator for extremely saline (>30 wt %) fluids that have increased capacity to transport metals such as Co, Cu and Ni, as well as traditionally immobile elements such as Al, Ti, and rare earth elements (REEs). The footprint of kyanite alteration is smaller than the pervasive Mg alteration, but broader than that resulting from sulfide mineralization, providing a potentially useful pathfinder mineral for a mineralizing system.

By leveraging the totality of information recorded in the minerals preserved in the basin, we will begin to construct a spatial and temporal picture of alteration and mineralization within such systems improving both our understanding and potentially exploration success.

Mitigating the Streetlight Effect: Stepping Into the Darkness is an Important Step in Understanding the Copperbelt

David Selley
CODES Centre of Ore Deposit and Earth Sciences, Hobart, TAS, Australia

Our capacity to explore successfully on the Copperbelt, and moreover to transfer exploration strategies to other basins, is dependent upon the robustness of our ore deposit models. In recent years, there have been important advances in constraining the nature and origin of ore fluids, as well as the scale and intensity of alteration systems. An appreciation of the vast time scales at which the mineralising process operated and the dynamic, evolving nature of hydrologic framework, have been demonstrated by discovery of "non-classical" ores hosted at stratigraphic levels previously considered nonprospective. Despite our progressively expanding knowledge of the basin's form, facilitated partly through the assembly of modern geophysical data sets, deep drilling, and reinterrogation of geologic maps, notable exploration successes remain restricted to (or incrementally extending along) an arcuate "line of lode" defined some 100 years ago, or a subsidiary belt of high-tonnage, low-grade ores: the classical Cu arc and Domes Region, respectively. While we continue to explore outboard of these long-established mineralised domains, it seems our lack of success points towards limitations in our understanding of the mineral system. Even within this most spectacularly endowed, relatively data-rich basin, we are still not at a position to determine the prospectivity of a particular compartment.

Within the northern compartment, metal endowment is neither uniform nor ubiquitous, with the bifurcating mineralised corridors or the classical Cu arc and Domes Region enclosing an apparently barren lozenge-shaped domain that I refer to herein as the "empty quarter."

It this paper, I present what we believe we understand about the links between basin compartmentalisation, hydrologic evolution, and ore distribution, but focus on what limited material we have on the empty quarter in an attempt to build a more complete picture of the mineral system.

Provenance of the Katangan Basin in Zambia and the Implications for Basin Evolution and Exploration Search Space

Munshya Zimba[1], Michael C. Daly[2]
1. First Quantum Minerals Limited, Solwezi, Zambia, 2. Department of Earth Sciences, Oxford University, Oxford, United Kingdom

Sedimentary provenance is a crucial element of understanding the stratigraphy, sediment routing, compartmentalization, and tectonics of basins. The regional provenance history of the Neoproterozoic Katangan basin, host to world-class Cu-Co-(Ni) deposits, is poorly understood. A study utilising Kernel Density Estimating and Multidimensional Scaling (MDS) of over 3,000 detrital U-Pb zircon data from regionally spread samples and regional geology provides insights into the provenance of the Katangan Supergroup, basin trends, sediment routing, and major tectonic discontinuities. This information has potential to influence mineral exploration ideas. MDS plots revealed three distinct clusters. MDS1 is characterised by frequent neo-Archean ages in the northwest part of the basin (Mwinilunga) and ends abruptly on the northern margin of the Kabompo Dome. These peaks are interpreted as being sourced from NW in the Archean Congo craton of the Kasai area of the Democratic Republic of the Congo (DRC). This abrupt termination is a strong indication of a major tectonic boundary and Mwinilunga basin being isolated from an eastern sediment source. MDS2 is characterised by dominantly Paleoproterozoic ages of the Bangweulu craton and spatially covers the Domes Region and extends to the eastern margin of the Bangweulu basin. This is interpreted to indicate a local provenance predominantly from underlying Bangweulu aged granites in the basement dome inliers. MDS3 spatially covers central Zambia and is characterised by both strong Paleoproterozoic and Mesoproterozoic frequencies. The two high frequencies result from sediments derived from the Bangweulu, Irumide terranes to the northeast and the Munali Hills type granitoid basement in the south. Distribution of orebodies in the defined basin compartment margins is coincident with MDS2 cluster. Although many local controls are responsible for mineralisation, the extension of MDS2 in central and southern Zambia may imply continuation of prospective sequences at depth, and further investigations might open up new exploration search space.

Effective Bottom-Up Social and Environmental Strategies for Exploration Teams: Lessons from an Orphaned Mine

Robin N. Armstrong, Richard J. Herrington, Bio+Mine Team
The Natural History Museum, London, United Kingdom

Future-proofed Environmental and Social Governance (ESG) will demand that the risks associated with a potential mine's closure are better evaluated and considered at the exploration phase. Arguably, social implications and associated risks are amongst the most difficult to evaluate and without clear methods to integrate these with potential environmental impacts. Past operations with legacy issues demonstrate that during the exploration / discovery phase key strands of data were not collected or effectively integrated to 'predict' emerging hazards. This is in part due to a historical dependence solely on hard engineered solutions. In the tradition of the "present is the key to the past," we will present the lessons learnt from a multidisciplinary team working with the indigenous community at the orphaned Sto. Niño copper mine (Benguet, Philippines). This work demonstrates the power of early and transparent community engagement and the need to acknowledge the uniqueness of each site at both the local and regional scale. The Sto. Niño example will illustrate the advantages of effectively integrating geological, environmental (including biodiversity), social data along with indigenous peoples' knowledge early in a project's development. We will present a potential methodology adopted from the social sciences for the analysis of contrasting data types. Our experiences demonstrate that effective Environmental and Social governance must reflect shared values and aspirations between the resident communities and the visiting exploration teams and not be a top-down dictate.

Exploration, Mining and Biodiversity in Angola: A New Business Model

Luis Chambel, Sandra Amaro, Luis Caetano
Sinese, Sardoal, Portugal

Geopolitical tension is heightened and increasing, with war, threats to free trade, and an increasing fragmentation of the global economy into conflicting blocks. There is a markedly increased supply risk, especially in energy, high-tech, energy transition, and defence commodities.

The world needs a sustainable economic model and a new energy paradigm, for which we need an expanded set and higher tonnage of minerals, with a smaller negative impact on the planet and human societies. Yet, mineral deposits are increasingly deeper, of lower grade, and with higher complexity, in more remot areas, with stricter mining regulation, with a harder-to-get social license to operate, and with competing land uses; with decreasing grades, the energy and water consumption and land use per ton of mined metal increase, with biodiversity and ecosystems under increasing pressure -- a trend that can only be countered with innovation.

Alongside the use of innovative technologies – including machine learning, drones, multispectral imagery and eDNA kits, we've adopted a new business model focused on core stakeholders aiming for a sustainable, nature-positive exploration and exploitation of natural resources.

This work describes the application of biodiversity assessment – identifying hotspots, endangered species, and vulnerable ecosystems -- in multiple mineral projects in Angola as part of the exploration-up-to-closure economic geology toolkit. Rather than just finding, evaluating, and mining mineral resources, the focus of our approach is on social and natural systems – simultaneously exploring for, e.g., superdeep Type II diamond mineralization in kimberlite and alluvial deposits, copper in the Angolan side of the Copperbelt, or REE in carbonatites, sources of renewable energy, biological resources, and archaeological sites.

This approach allows mineral companies to plan their activities with a better knowledge of the local social and natural systems, define impact metrics, and report to society under ESG-compliant standards, namely, PERC, TFND and others.

Redefining Viability: Integrating Social and Environmental Dimensions into Mineral Exploration Success

Matthew Grimshaw, Sarah Gordon
Satarla, London, United Kingdom

In the field of mineral exploration, a standard narrative suggests that only one in a thousand projects culminates in success, highlighting the industry's inherent risks and the significant uncertainty involved in identifying viable projects. This prevailing statistic reflects the technical and geological challenges the industry faces and underscores a potential oversight in the traditional metrics used to define project viability.

In this presentation we propose a re-evaluation of what constitutes a 'viable' mineral exploration project, advocating for a paradigm shift beyond the conventional geological and technical parameters. We introduce an integrated framework that considers social and environmental factors from the project's inception, suggesting that these elements are essential to the project's overall success and sustainability.

Our approach involves a detailed methodology that intertwines social responsibility and environmental stewardship with technical exploration strategies. This includes engaging with local communities to understand and address their concerns, conducting comprehensive environmental impact assessments, and developing strategies that minimise ecological footprints while exploring.

The potential benefits of this integrated approach are multifaceted. By considering social and environmental aspects early in the exploration process, companies can build stronger relationships with local communities, enhance their social license to operate, and potentially identify and mitigate environmental and social risks before they escalate. Moreover, this holistic view of project viability could lead to more sustainable mining practices, aligning the industry with global sustainable development goals and potentially improving the public perception of mining activities.

Through case studies and theoretical models, this presentation will explore how redefining viability in mineral exploration can lead to more responsible and successful exploration projects. We will discuss the implications of this shift for industry practices, policy-making, and the broader goal of sustainable development in the context of mineral exploration.

Charting Sustainable Resource Development: ESG Perspectives in the DRC Mining Sector

Innocent Mufungizi[1], Yann M. Waku[2]
1. University of Kinshasa, Kinshasa, Democratic Republic of the Congo (Kinshasa), 2. 3Department of Earth Sciences, Stellenbosch University, Private Bag X1, Matieland 7602, South Africa 2, Stellenbosch, South Africa

In the face of escalating climate change concerns and the global push for greener technology, the Democratic Republic of Congo (DRC) emerges as a pivotal player in the transition to a sustainable future. Renowned as a leading producer of cobalt and the second-largest producer of copper globally, the DRC boasts a wealth of mineral resources. With partially estimated unexploited resources valued at $24 billion, the country continues to attract investors. However, alongside the promise of economic prosperity, the mining sector in the DRC also confronts significant environmental and social challenges, underscoring the importance of integrating Environmental, Social, and Governance (ESG) principles throughout the entire mining lifecycle. This review delves into the state of resource development in the DRC, examining the process of acquiring mining licenses, assessing the social and environmental impacts of mining activities, exploring recent advancements in remediation engineering, and outlining some recommendations. The DRC Mining Cadaster data reveals a significant number of 567 exploitation permits, 1729 research permits, and 631 other mining titles valid as of 2022, reflecting the country's substantial mining activity. With copper, cobalt, and zinc dominating exports, the mining sector plays a crucial role in the country's economy, contributing over 70% to Gross Domestic Product (GDP). However, this economic prosperity comes with environmental and social costs, including soil degradation, CO_2 emissions, and social issues such as city relocation and conflicts. It is indicated a notable absence of robust ESG policies applicable across the mining sector in the DRC. While some major mining operations adhere to comprehensive ESG standards, the overall landscape suggests that significant gaps and challenges persist. Therefore, innovative approaches such as advanced remote sensing or genomic approaches for heavy metal remediation are highly recommended for vigilant monitoring. This study highlighting the urgent need for concerted efforts towards sustainable resource development in the DRC.

An Effective Formalization of Artisanal Mining Will Improve the Public Acceptability of the Mineral Extraction Industry in Mozambique

Manuel Nopeia[1], Ivan Baptista[2], Simão Diquissone[3]
1. Akita University, Akita, Japan, 2. Lurio University, Pemba, Mozambique, 3. District Services of Economic Activities of Montepuez, Montepuez, Mozambique

The exponentially growing extractive industry in Mozambique is exposed to different risks; one of those is the lack of public acceptability, which has led to the interruption of mining activities, destruction and vandalization of mining company's properties, and uprising against the mining operators and central government. Based on an extensive literature review and assessment of companies' performances from reports and news, combined with field surveys in artisanal mining areas in northeastern Mozambique, we evaluated the key factors controlling the level of public acceptance of mineral extraction in Mozambique. Among the key factors identified is the ineffective artisanal mining formalization scheme. Mozambique's current approach for formalization of artisanal mining entails delimitating designated areas and issuing mining passes for artisanal mining within those areas. However, artisanal gold mining activities in Mozambique are more prevalent in undesignated areas, which overlap with large-scale mining (LSM) licenses. Interviews with artisanal gold mining (AGM) operators and local officials in Namuno district, one of the main mining districts in Mozambique, revealed that most of the designated areas for AGM were delimitated based on discoveries of gold by local farmers, rather than an extensive geological mapping. Furthermore, the compilation of LSM licenses and the designated areas for AGM in northeastern Mozambique showed an overlap of most of the LSM licenses and designated areas. Effective formalization of artisanal mining may lead to a better coexistence between both industries, which are equally indispensable. Considering the current formalization approach, we recommend the prioritization of detailed geological characterization of designated areas, outside the LSM licenses. In addition, providing AGM operators with easily understandable geological information about those areas will encourage them to apply for mining passes for those areas, rather than invading the LSM licenses; consequently, improving the acceptability of the mineral extractive industry.

The Benign Nature of Tailings from the Tumas Uranium Project, Namibia: Aligning Environmental Safeguards with Economic Outcomes

Alex Otto
Deep Yellow Ltd., Subiaco, WA, Australia

The Tumas project has the potential to produce 3.6 Mlbs of U_3O_8 with a potential Life of Mine (LoM) exceeding 30 years. Uranium mineralization is found within the calcretised sediment fill of the Tumas paleochannel, with carnotite being the primary ore mineral. Geologically, the Tumas deposits exhibit many similarities to the nearby Langer Heinrich mine. Following the Langer Heinrich mine's commencement in 2009, the Tumas project will become the world's second conventional alkaline leach operation. The design of the Tumas processing plant and its economic performance is aligned with protection of the surrounding environment. Alkaline leach solutions offer several advantages for tailing management. The high pH of the leach solution selectively dissolves carnotite while leaving over 99% of the rock mass unaffected. Moreover, concerns regarding mine acid drainage are alleviated since sulphides are not soluble. Advanced membrane technology enables the recycling of a significant portion of the leach agent, thereby minimizing costs. Deep Yellow collaborated with CSIRO to model the interaction between tailing liquor and groundwater. Key findings indicate that when mixed with groundwater, any residual uranium will precipitate as newly formed carnotite in close proximity to the tailing storage facility. Furthermore, the uranium content in the groundwater is projected to decrease below natural levels. Numerical groundwater flow models demonstrate that uranium migration beyond the mining footprint is not anticipated for at least 100 years. Both studies independently confirm the benign nature of Tumas tailings and the safe storage capability within in-pit tailing storage facilities.

Geological Characterisation, Stratigraphic Correlation and 3D Modelling of the Musinga Copper Exploration Play, Northeast Kabompo Dome, Democratic Republic of Congo

George Aldiss
Ivanhoe Mines, Kolwezi, Democratic Republic of the Congo (Kinshasa)

Musinga is a greenfield exploration licence located in the Democratic Republic of Congo, adjacent to the border with Zambia. The area is masked by laterite, saprolite, and Kalahari sand with limited rock exposure, hampering mapping. Ivanhoe mines explored the area between 2012 and 2019 with a combination of geophysical surveys and diamond drilling. Eight of fourteen holes were re-logged and sampled for polished thin sections to characterize the lithologies and build a lithostratigraphic framework. The lithostratigraphy comprises Pre-Katangan basement gneissic schists unconformably overlain by Proterozoic Mwashya of the Katanga Supergroup. Mwashya is locally characterised by pale grey phyllites and altered micaceous siltstones, fine-grained sandstones, and some thin, strongly pyritic black shales. The black shales could provide excellent sulfur reservoirs for copper reduction. Intense alteration, metamorphism, deformation, and gradational lithology changes complicate hole correlation. Identification of distinct pyritic black shales that appear in three of the deep Musinga holes acted as key marker horizons for correlation. The Musinga rocks have been moderately deformed by folding, faulting, and zones of intense quartz, calcite, and dolomite veining which sometimes contain strong pyrite, trace chalcopyrite, and pyrrhotite.

The Sentinel Copper and Enterprise Nickel mines located on the Zambian side of the Kabompo Dome provide useful geological analogues to Musinga. Both deposits show that structural deformation and host-rock facies are key factors controlling mineralisation along the edge of the Kabompo Dome. This could be due to structures enhancing permeability in the naturally low-permeability phyllites. Brittle deformation such as dilational strike-slip fault jogs and related fracturing could have allowed copper brines to migrate to reductant pyritic shales around the Kabompo Dome, inducing deposition of copper sulfides. Exploration in the Musinga area should target pyritic black shales proximal to steeply dipping faults identified from magnetic discontinuities, which could act as higher-permeability conduits for mineralising brine migration.

Beyond Kamoa and Kakula: Exploration in the Western Foreland

Yves Ilayao[1], Kennedy Lukengu[1], Theodore Mukanda[1], Amanda M. Stoltze[1], Tim Dunnett[2]
1. Ivanhoe Mines Exploration, Kolwezi, Democratic Republic of the Congo (Kinshasa), 2. Ivanhoe, San, South Africa

The Western Forelands (WF) area of the Democratic Republic of Congo is host to the Kamoa-Kakula complex, which contain 37.91 Mt of Cu at an average grade of 2.72% (measured and indicated). After the discovery of this world-class complex in the 2000s, the WF has been the focus of extensive exploration. The discovery of the Makoko Deposit (2017), the Kiala Deposit (2018), and more recently, high-grade mineralisation at Kitoko at over 1,000-m depth, demonstrates the continued prospectivity of the region.

The mineralisation model has continued to evolve and benefit from the sustained exploration effort. One key feature of the mineralisation model is the pre-sedimentary basement architecture which informs the deposition of sediment facies. The understanding of basement architecture is ongoing and geophysical tools are being trialled to improve pre-drilling models. Mineralisation discovered in the western portion of the shelf is located where pyritic units, generally siltstones, are forming at onlap positions close to basement. These pyritic units supply the sulphur required to generate Cu sulphides, acting as the reductant.

At Kamoa-Kakula, a thick sequence of Roan underlies the deposits and was thought to be essential to generate high-grade mineralisation. Recent high-grade mineralisation in Nguba sediments directly overlaying basement has shown that a proximal "red-bed" aquifer is not needed. Lack of an aquifer necessitates a different mechanism for introducing mineralising fluids to the host lithologies. Two working hypotheses are that fluids could be introduced by regionally significant faults joining the WF to the greater copper belt, or alternatively that fracture networks in the basement have allowed ingress of fluids. Another area where the mineralisation model is evolving is the alteration assemblages that are forming associated with ore. These alteration packages are forming distinct haloes where grades are low, but at the edge of a large mineralising fluid system.

Camp Creek Porphyry Alteration Mineralogy and Geochemistry: Exploring for Blind Copper Mineralization in Northern British Columbia, Canada

Maxwell L. Porter[1], Shaun Barker[1], Farhad Bouzari[1], Nichole Moerhuis[1], Maria A. Rodriguez-Mustafa[1], Daniel Guestrin[2], Christina Antsey[2], Robert G. Lee[3], Brock Riedell[4]
1. University of British Columbia, Vancouver, BC, Canada, 2. Brixton Metals, Vancouver, BC, Canada, 3. BHP, Vancouver, BC, Canada, 4. Riedell Exploration Ltd., Vancouver, BC, Canada

Camp Creek is a Late Cretaceous calc-alkalic Cu-Au-(Mo) porphyry deposit, located within the Stikine terrane in northern British Columbia, Canada. Late Triassic mafic volcanic and clastic sedimentary rocks of the Stuhini Group host numerous Late Cretaceous granodiorite to monzonite stocks and dykes. This research aims to characterize the mineralogy, composition, timing, and shape of various porphyry units at Camp Creek, and to link these features to mineralization. Drilling programs by Brixton Metals have discovered Cu-Au-(Mo) mineralization approximately 400m below the surface. Multiple phases of porphyry units have previously been classified at Camp Creek based on detailed trace element analysis, particularly Zr, Ti, and Sc. In this study, host rock and alteration types were characterized by petrography, SWIR, microXRF element mapping, and whole-rock geochemistry. U-Pb zircon dating of the porphyry phases and post-mineral dykes has constrained temporal relationships. Mineralization is typically hosted within porphyry X (PX), a hornblende-quartz-biotite-plagioclase porphyry, dated (U-Pb zircon) at 86.60 ± 0.36 Ma, cut by A-type quartz-vein stockwork, occurring together with K-silicate and biotite alteration. In addition to PX, several other texturally similar porphyry bodies occur within the deposit. Both PZ and PY have apparent sill-like geometry, with PZ being a shallow, relatively unmineralized porphyry phase that transitions to PY with increasing depth, correlating with increased grade and quartz veining density. However, the contact relationships of these porphyries are obscured by intense alteration. SWIR data indicates a vertical zonation of alteration from pyrophyllite, transitioning into illite, then muscovite and ending with K-silicate dominant minerals with depth. Sericite type generally transitions from white to pale green and then dark green with depth. Initial observations indicate multiple generations of biotite-altered volcanic rocks proximal to the PX stock. The results of this study will identify key vectors to locate buried mineralization in porphyry copper deposits, enhancing regional exploration approaches.

Framboids, Fluids, and Field Mapping: Understanding Sedimentary-Rock-Hosted Metal Prospectivity Across 11 Orders of Magnitude

Elizabeth C. Turner
Harquail School of Earth Sciences, Sudbury, ON, Canada

Research and mineral exploration in sedimentary-rock-hosted ore systems (e.g., Cu, Zn, Ni, Pb, Au), commonly focus on deposit-specific (10^{-1} to 10^{3} m) grade distribution, alteration trends, and structural controls. However, the phenomena that control the nature and geographic distribution of sedimentary-rock-hosted ore systems and ore deposits span a range of scales from (for example) micron-scale (10^{-6}) porosity-permeability-mineralogy distribution to district-scale (10^{5}) tectonic events, stratigraphy, and fluid-migration patterns—scales that are seldom studied or meaningfully incorporated into exploration philosophies. Discovering new critical mineral deposits requires an expanded, more holistic approach that incorporates the full range of controlling factors. This talk explores the crucial importance of multiple scales of underappreciated controlling factors in several critical-mineral districts: the seemingly familiar Central African copperbelt and the more remote, underexplored Polaris Zn-Pb-Cu and Borden Basin (Nanisivik) Zn districts in Canada's high Arctic islands.

Evolution of the Neoproterozoic Katangan Basin in the Kipushi Mine Area, Southern DRC and Northern Zambia: The Importance of Halokinesis, Global Glacial Events, and Tectonics

Helen Twigg[1], Murray W. Hitzman[1], Ghislain Kabunda[2], Munshya Zimba[2]
1. iCRAG, University College Dublin, Dublin, Ireland, 2. First Quantum Minerals, Solwezi, Zambia

The Neoproterozoic Katangan basin in the southern Democratic Republic of Congo (DRC) and northwestern Zambia is host to the Central African Copperbelt (CACB), a major source of copper and cobalt. Recent regional studies highlighted the impact of salt on base metal mineralisation through hydrological compartmentalisation, halokinetic traps, and its effects on deformation styles during the Lufilian orogeny. This study investigates district-scale basin development in the Kipushi mine area and identifies structures and unconformities associated with extension, salt evacuation, global glacial events, and inversion in sedimentary rocks of the Roan, overlying Nguba, and top Kundelungu groups. The study uses lithofacies analysis combined with stratigraphic and structural reconstructions to investigate basin controls around the diapiric Kipushi Anticline. Sediment thickness variations in the district display evidence of extensional normal faults with throws of ~600 m active during deposition of Mwale Formation glacially influenced diamictites at the base of the Nguba Group that resulted in the formation of a basement high at Kipushi. Halokinesis in the area appears to have been initiated during deposition of the Roan Group Mwashya Subgroup sediments and continued into the period of deposition of the lower Kundelungu Group sediments. Isolated minibasins between diapiric salt structures were developed during deposition of the Katete Formation. The minibasins widened and joined during deposition of the upper Monwezi Formation. Salt withdrawal continued through the Marinoan glaciation at the start of deposition of the Kundelungu Group sediments and appears to have waned during deposition of the Kanianga Formation. A salt canopy domain is interpreted, which could have contributed to early diapir collapse and welding along sub-basin margins, thus affecting hydraulic connectivity within the basin. The Lufilian inversion reactivated and reversed salt and suprasalt structures. The non-cylindrical folding style present in the area was likely related to original minibasin geometries.

Exploring Geological Complexities of the Kakula Cu Deposit, Democratic Republic of Congo

Franck B. Twite[1], George Gilchrist[2]

1. Kamoa Copper SA, Kolwezi, Democratic Republic of the Congo (Kinshasa), 2. KoBold Metals, Berkeley, CA, USA

The Kakula Cu deposit, situated west of the Kolwezi Cu-Co district in the Democratic Republic of Congo, is a rift basin that hosts exceptionally high copper grades. Kakula is understood to be a redox-controlled sedimentary rock-hosted stratiform copper deposit that developed at a redox boundary between the Mwashya Subgroup and the overlying Grand Conglomérat. Following the discovery of the Kakula deposit in 2016, resource drilling indicated that the deposit includes well-defined higher- and lower-grade zones that trend in a WNW orientation. The resource drilling identified areas of steeper dip, but was not tight enough to model smaller-scale features. Mine development (at February 2024) includes approximately 130 kilometers of linear development that has allowed detailed morphology and grade trends to emerge. Detailed underground geological mapping on 5-meter spacing was conducted to better understand structures and their impact on the Kakula mine design. Syn-sedimentary faults formed during the deposition of the Grand Conglomérat were recognized by steeply dipping bedding, soft-sediment deformation, and thickness variation in siltstones towards the base of the Grand Conglomérat. These structures, which create an offset varying between 5 and 40 m, occur over short distances of <25 m, impacting the mining production. Asymmetric folds possibly linked to basin inversion were identified in the Kakula area; they have a wavelength of 30 to 60 m and create an uplift of 4 to 12 meters. Understanding these structures and how to predict them is crucial for effective resource management and mine planning, but the short distance over which they develop precludes them from being properly characterized by surface drilling alone. The presentation will provide a comprehensive overview of the Kakula deposit, focusing on its basin architecture and controls on mineralization.

Geological and Metallogenic Characterisation of Manganocopper Mineralisation in the Tizi N'Isdid Deposits (Ounein Massif, High Atlas, Morocco)

Imad Aflla, Mustapha Souhassou
Egerne, Polydisciplinary Faculty of Taroudant, Ibn Zohr University, Agadir, Morocco., Taroudant, Morocco

The Tizi-N-Isdid region, renowned for its significant mineral resources, particularly manganocopper, is located in the Ounien massif, on the border between the Anti-Atlas and the High Atlas, on the western edge of the Precambrian Ouzellarh promontory. The Tizi N'Isdid deposit is hosted by a detrital series and overlain by Adoudounian dolomitic bars. A precise lithostratigraphic description is provided, highlighting the presence of volcanic and volcano-sedimentary formations of the Ouarzazate group, as well as formations of the Taroudant azoic group. Mineralization occurs mainly in reddish-brown claystone, with braunite as the main manganese ore. The gangue is made up of elements of claystone, calcite, quartz in mosaic or fibroradial clusters and, secondarily, barite. The paleogeography of the site shows stratification of mineralization on a shallow subhorizontal platform, with minor variations in the thickness of basaltic flows and coarse conglomerates. Manganese mineralization extends over 6 km in a north-south direction, in the form of lenticular layers and discontinuities. Mineralization styles are examined in detail, showing a variety of morphologies ranging from massive mineralization to bands in claystones. Geochemical analyses suggest a Sedex-type hydrothermal origin for the mineralization, with braunite as the dominant mineralogical indicator. In conclusion, the geological study of the Tizi-N'Isdid deposit provides a detailed overview of its geology, the horizontal and vertical distribution of its mineralization and its genetic typology. This information makes a major contribution to the characterization and development of manganese and copper mineralization in the Ounien High Atlas region.

Bringing 20 Years of Research and 500 Reef Intersections Together on Ivanhoe's Tier 1 Platreef Deposit: A Synthesis from the Northern Limb of the Bushveld Igneous Complex

Tim Dunnett
Ivanhoe Mines, Johannesburg, South Africa

Ivanhoe Mines' exploration commenced on the Northern Limb of the Bushveld Igneous Complex in 1998. Work attempted to unpack the enigmatically thick polymetallic Platreef developed on the Northern Limb. Initially exploration focussed on the near surface expression defining a resource amenable to open pit exploitation, before switching to a deeper target in 2010. A flattening of the Main Zone – Critical Zone contact, ultimately resulting in the "Flatreef" resource of 346Mt of 3PGE+Au at 3.77g/t as well as 0.16% Cu and 0.32% Ni using a 2g/t cutoff, defined using 482 drillholes.

In the past 20 years a large body of scientific work has been completed. A total of 12 drillholes from the open pit and 36 from the Flatreef area have been studied, in a total of 4 PhD, 10 MSc, and 4 Honours studies, producing 17 academic papers. The drillholes studied can be separated into three groups based on the part of the resource drilled; the open pit; the Flatreef and the deep reef. The footwall to the open pit and Flatreef is a zone of sediment-magma assimilation, divided into areas assimilated with shales and dolomites, whereas in the footwall to the deep reef horizon, cyclical norite (NC2) stratigraphy is developed. This insight into less contaminated lithologies that has been used to define stratigraphy and compare the Flatreef with the Merensky Reef of the Eastern and Western Limbs. All three drillhole groups have been combined to understand the Flatreef formation. The highest grade and thickest mineralised intervals are associated with the areas of sediment assimilation and not in areas with NC2 developed, underscoring the importance of sediment-magma interaction in the development of the Platreef.

The detailed, focussed scientific information and the holistic knowledge from core and underground exposures is synthesized in this review paper.

Primary-Hosted Uranium Resources, Geology, and Mineralisation in the Damara Orogen, Namibia

Guy G. Freemantle
University of the Witwatersrand, Randburg, South Africa

The bulk of Namibian uranium deposits occur in the Erongo Region, where two categories of deposits host the nation's >380 kt indicated U resources: 77 % in primary-hosted deposits of the Rössing type; and 23 % in carnotite-dominant surficial deposits of the Langer Heinrich type. The primary-hosted deposits are those in which U occurs in the sheeted leucogranites (SLG) that are commonly referred to as alaskite. Presently ˜130 kt U indicated resources are hosted in at least four unexploited alaskite deposits within a 50-km radius of Rössing and Husab. The major primary deposits are products of the Pan African Damara Orogeny, in which the high-T°/low-P southern Central Zone (sCZ) hosts the highest-grade metamorphism, and consequently the greatest density of granitic intrusions occur. All of the sCZ primary deposits are similar, having formed in structurally prepared sites in granulite facies Neoproterozoic metasediments of the lower Damara Supergroup; spatially proximal to basement-cored domes and inliers, and temporally proximal to the regional peak metamorphism. Economic deposits occur on the southwest edge of the Khan Inlier at Etango, on the southeastern margins of the Ida Dome at Omahola, and on the eastern edge of a basement inlier at Valencia. Contemporary interpretations for the formation of these large, low-grade deposits favour models of low-percentage partial melt of weakly uraniferous source material that interact with reactive metasediments that encourage the precipitation of U(IV) oxides and silicates. The uranium repositories in the source rocks include common accessory minerals, such as zircon, monazite, biotite, and apatite, that retain or release U, Th, lanthanides, and other incompatible elements into the silicate melt and magmatic fluids. The bulk of U occurs in uraninite [$(U^{4+},U^{6+})O_2$], coffinite [$U(SiO_4)_{1-x}(OH)_{4x}$], and secondary hydrate minerals, predominantly uranophane [$Ca(UO_2)_2SiO_3(OH)_2 \cdot 5(H_2O)$]. A relatively low proportion of U is hosted in the refractory minerals betafite [$(Ca,U)_2(Ti,Nb,Ta)_2O_6(OH)$] and/or brannerite [$UTi_2O_6$].

Petrogenesis and Structure of the UIS Pegmatites, Namibia: A New Source of Battery Metals in Africa

Laurence J. Robb[1], Lewis Lloyd[1], William Maloney[1], Frederick Wright-Morris[1], Simon Tapster[2]
1. University of Oxford, Oxford, United Kingdom, 2. British Geological Survey, Keyworth, United Kingdom

The Neoproterozoic Damara Orogen (~900-500 Ma) hosts several pegmatite belts containing economic concentrations of minerals such as cassiterite (Sn), columbite group minerals (Nb-Ta), lepidolite, petalite, spodumene (Li) and U-bearing phases. This study focuses on the Uis pegmatites which form part of the Cape Cross-Uis Pegmatite Belt (CUPB) in central Namibia. Several hundred Sn-Ta-Li rich pegmatites occur in the Uis district but no parental granite source is observed and the question whether such a body exists at depth, or the pegmatites are derived by local anatexis of a metasedimentary source, is addressed.

The Uis pegmatites intrude the Zerrisene Group, a thick sequence of marine turbiditic sediments, and are likely to have been derived by melting of this dominantly metapelitic precursor. During peak orogenesis the area experienced tight to isoclinal folding and thrusting as a consequence of the transpressional collision between the Congo and Rio de la Plata Cratons. Several stages of granitic plutonism occurred between 540-510Ma, creating contact related cordierite-schists over wide areas. Orogenic collapse and cooling caused extension and multiple stages of syn- to post-tectonic pegmatite emplacement – cassiterites from the Uis pegmatites have been dated at 520 Ma, an age that is identical to the youngest concordant zircon also dated from these intrusions.

Barren pegmatites are different in mineralogy and composition to the mineralized Uis pegmatites and these differences, evident in tourmaline compositions, suggest that such features could be useful in identifying other prospective pegmatite belts in Namibia. Current ore reserves, from just two of several historical open pits at Uis, stand at more than 70 Mt of ore at 0.13% Sn, 0.63% Li_2O and 85ppm Ta. Excellent infrastructure and a ready global market, coupled with ambitious development plans, will ensure that this region becomes an important battery metal producer of the future.

The Formation of High-Grade Iron Formation-Hosted Iron Ores of the Northern Cape Province of South Africa: Insights from Hematite Ages

Albertus Smith[1], Nicolas Beukes[1], Conrad De Kock[1], Martha Masangane[1], Hayden Miller[2], Boitumelo Ntlhoro[1], Kenneth Farley[2]
1. University of Johannesburg, Johannesburg, South Africa, 2. California Institute of Technology, Pasadena, CA, USA

Most of South Africa's high-grade iron ores occur on the western margin of the Kaapvaal Craton in the Northern Cape Province. The ores are hosted within the Paleoproterozoic Transvaal and Keis Supergroups, respectively occurring directly below and above the pre-Gamagara unconformity. Advances in hematite age dating utilizing the combined (U-Th)/Ne and (U/Th)/He technique provide an opportunity to better understand the formation of these ores. Samples were taken from the northern and southern Maremane Dome, Wolhaarkop Dome, Kalahari Manganese Field (KMF) and Griqualand West region. Iron formation-hosted laminated, massive and brecciated ores occurring in the Asbesheuwels, Koegas and Voëlwater Subgroups of the Transvaal Supegroup and conglomeratic ores and iron-rich sandstones occurring in the basal Gamagara Formation of the Keis Supergroup were analyzed.

Three hematite (U-Th)/Ne age populations occur in the analyzed samples: ~2100-1800 Ma; ~1700-1400 Ma; and ~1200-1000 Ma. Hematite (U-Th)/He (cooling) ages were more variable and were always either similar or younger than their respective hematite (U-Th)/Ne ages.

The oldest hematite (U-Th)/Ne age population is prevalent in the northern Maremane Dome and overlaps with the age of the pre-Gamagara unconformity (~2.2-2 Ga) as well as the lower Keis Supergroup (~2.0-1.9 Ga). This hematite likely formed by supergene enrichment below the pre-Gamagara unconformity and endoclastic deposition above the unconformity. The middle age population has no overlap with any known major geological events in the region, occurs across most of the study area, and falls within the depositional age range of the Keis Supergroup (~2.0-1.3 Ga). The youngest age population appears more commonly towards the south and southwest of the Maremane Dome and in the KMF and corresponds well to the age of the Kheis-Namaqua orogeny (~1.2-1.0 Ga). This implies that there was either further ore formation or hematite age resetting due to orogeny-driven hydrothermal fluid flow along the pre-Gamagara unconformity.

Magmatic Stratigraphy of the Deep Platreef at Sandsloot, Northern Bushveld Complex: Carbonate Contamination and Controls on Ni-Cu-PGE Mineralisation

Erin S. Thompson[1], David A. Holwell[1], Marc Reichow[1], Iain McDonald[2], Thomas Blenkinsop[2], Hannah Hughes[3], Katie McFall[4], Kate R. Canham[1], Matthew A. Loader[5], Lara Du Preez[6], Kofi Acheamponga[6], Andy Lloyd[6]

1. Centre for Sustainable Resource Extraction, University of Leicester, Leicester, United Kingdom, 2. School of Earth and Environmental Sciences, Cardiff University, Cardiff, United Kingdom, 3. Camborne School of Mines, University of Exeter, Penryn, United Kingdom, 4. Department of Earth Sciences, University College London, London, United Kingdom, 5. Anglo American UK, 17 Charterhouse Street, London, United Kingdom, 6. Anglo American South Africa, 144 Oxford Road, Rosebank, Johannesburg, South Africa

The Platreef, northern limb of the Bushveld Complex, South Africa, is widely regarded as one of the world's largest resources of platinum-group elements (PGE). These PGE, as well as significant amounts of base metals (Ni, Cu, Co) also found in the northern limb, are essential for the growth of many sustainable and environmentally friendly technologies. The Platreef is part of a complex package of Critical Zone (CZ) rocks which differ from the wider Bushveld in that high grade PGE mineralization is spread over a much greater thicknesses compared to the Merensky Reef. Additionally, it lies north of the Thabazimbi-Murchison lineament, and rests directly on the metasedimentary Transvaal Supergroup. The resulting footwall interactions are known to have had a significant impact on the fluctuating geochemical and mineralogical signatures of the northern limb magmas, as well as their Ni-Cu-PGE grades.

In this study, the petrology, bulk geochemistry, including Nd-Sr-Hf-Pb isotopic compositions, and silicate mineral chemistry of samples from the Platreef at Sandsloot are examined, with the aim of establishing the magmatic stratigraphy of the deeper Platreef, down-dip from the Mogalakwena open pit. Here, the Platreef is composed of up to five pyroxenitic packages, with significant PGE and base metal mineralisation observed in two. Varying CaO/Al_2O_3, and Eu/Eu* ratios, Pb and Sr anomalies are associated with different units and allude to varying degrees of contamination. Variations in the parental magma compositions for each of the identified units are more difficult to distinguish. Hybrid skarn textures are observed within the Base Metal Zone (BMZ), indicating the importance of interactions with the underlying Malmani Dolomite in forming economic Ni-Cu-PGE mineralisation at this location. Significant PGE mineralisation is also commonly accompanied by extensive hydrothermal alteration, which is not observed in barren units, and suggests that rock-fluid interactions may be key to concentrating PGE grades at Sandsloot.

Discovery of the Oak Dam West IOCG Deposit, South Australia

Thiago Oliveira[1], Jamie King[1], Kathy Ehrig[2], Marcel Menicheli[2]
1. BHP Exploration, Adelaide, SA, Australia, 2. BHP Copper SA - Olympic Dam, Adelaide, SA, Australia

Oak Dam West is a breccia-hosted iron oxide-copper-gold (IOCG) deposit located approximately 65 km southeast of the iconic Olympic Dam IOCG deposit. The Oak Dam district consists of two distinctive gravity anomalies separated by ~3 km, i.e., Oak Dam East (coincident magnetic anomaly) and Oak Dam West (no magnetic anomaly). The former Western Mining Corporation (WMC) drilled the first hole into Oak Dam East (OKD-E) in 1976, which tested the coincident gravity and magnetic anomalies, intercepting Fe oxide-rich rocks with low copper and high uranium grades under approximately 500 m of post-mineral sedimentary cover. Drilling was completed on 13 holes intermittently until 2010. The first hole into the center of the de-magnetized gravity anomaly at Oak Dam West (OKD-W) was drilled in 1981, and intercepted sediment- and hematite-rich, copper-barren breccias at ~730-m depth. Two further holes were completed, with one reaching basement at ~875 m.

Based on the mineral systems approach and multi-disciplinary data integration, a regional prospectivity map and ranking tools were developed to prioritize BHP Metals Exploration 2018 exploration campaign. Oak Dam West was highlighted as the best prospect, based on the strong gravity anomaly (geophysics) associated with hematite-rich breccias (geology), and consistent geochemical footprints with a central zone of a hematite-dominant IOCG hydrothermal mineral system (geochemistry).

A conceptual model of barren hematite-rich core surrounded by mineralized hematite-rich breccias was generated to guide the drilling plan, supported by 3D inversion and forward modeling of gravity. Drill holes were designed to test the margins of the central barren breccia hoping to find the highest copper-grade zones. AD22 successfully intersected chalcopyrite-pyrite mineralized breccias at Oak Dam West. AD23 intersected high-copper-grade, chalcocite-rich hematite breccias, at the northern contact with the central Cu-barren breccia, with 425.7 m at 3.04% Cu and 0.59 g/t Au, becoming the discovery hole of Oak Dam West.

Ore-Forming Processes at the Rosh Pinah Zn-Pb-Ag Deposit, Southern Namibia: Evidence from Mineralogy, Geochemistry, and Metal Distribution

Halleluya Ekandjo, Koen Torremans, Murray Hitzman, Philip Rieger, Simon Jones
SFI Research Centre in Applied Geosciences (iCRAG), School of Earth Sciences, University College Dublin, Dublin, Ireland

The Rosh Pinah deposit (19.94 Mt @ 7.38 wt % Zn, 1.83% Pb and 27.7 g/t Ag) in Namibia is hosted in carbonaceous argillites, siltstones, carbonate rocks, and volumetrically minor volcaniclastic rocks of the Neoproterozoic Rosh Pinah Formation. To better understand the metallogenesis of the deposit, we present an integrated analysis of petrography, metal distributions, C-O-Sr isotope and in-situ and whole-rock geochemistry. Despite the simple sulfide mineralogy (sphalerite, galena, pyrite, and chalcopyrite), the paragenetic sequence of alteration and mineralization was complex, strongly controlled by sedimentary lithofacies. The highest Zn, Cu, and Fe contents are located adjacent to structures that appear to have served as conduits for metal bearing fluids. The richest orebodies occur where these are juxtaposed to pre-existing carbonate horizons. Replacement of pre-existing marine carbonate cements by hydrothermal carbonates with concomitant sulfides indicates that carbonate buffering played a key role in metal precipitation. This is corroborated by C-O-Sr isotope data and REEY patterns from LA-ICP-MS trace element data on subsequent carbonate phases. These observations combined with textural relationships indicate that the Zn-Pb-Ag ore formed via sub-seafloor replacement by hydrothermal fluids and was overprinted by later barite during and after the main ore stage. Petrography and whole-rock geochemistry show increasing enrichment of Ba (as barite, barian K-feldspar, and barian carbonate minerals) and depletion of K and Na in argillites and siltstones towards well-mineralized zones in carbonate bodies provide potential vectors to Zn-Pb mineralization. Although the ore at Rosh Pinah is associated with subtle alteration haloes, exploration for additional orebodies in the district should focus on identifying favorable sedimentary facies and structural architecture for hydrothermal fluid flow and metal precipitation.

Triple-Halogen (Cl-Br-I) Signatures of Salt Melt Inclusions in Kiruna-Type Magnetite-Apatite Ore Systems – Proof of Formation Via Evaporite Anatexis

Tobias Fusswinkel[1], Xinyue Xu[2], Matthew Steele-MacInnis[2], Peter Koděra[3], Fernando Tornos[4], John M. Hanchar[5], Wyatt M. Bain[6]
1. Institute of Applied Mineralogy and Economic Geology, RWTH Aachen University, Aachen, Germany, 2. Department of Earth and Atmospheric Sciences, University of Alberta, Edmonton, AB, Canada, 3. Department of Geology of Mineral Deposits, Comenius University, Bratislava, Slovakia, 4. Instituto de Geociencias, Consejo Superior de Investigaciones Científicas, Madrid, Spain, 5. Department of Earth Sciences, Memorial University of Newfoundland, St. John's, NL, Canada, 6. British Columbia Geological Survey, Victoria, BC, Canada

The origin of Kiruna-type magnetite-(apatite) (MtAp) deposits is widely debated, with proposed formation models ranging from purely orthomagmatic to entirely hydrothermal processes. The debate largely centers on the nature of the fluids forming in these deposits, especially the occurrence of iron-rich ore-stage melt inclusions that are composed of chlorides, sulfates, and carbonates. Inclusions of this type are ubiquitous in many MtAp deposits and they have been controversially interpreted to either represent partial melts of evaporite-bearing carbonate rocks in response to magma intrusions, or to be hydrosaline liquids forming halite condensates during magmatic fluid exsolution in shallow crustal settings.

We present the first-ever triple-halogen (Cl-Br-I) in situ LA-Q-ICP-MS melt inclusion results from MtAp deposits at El Laco and Laco-Salar (Chile). Additionally, we determined halogen signatures in chloride-rich Fe-K-Na salt melt inclusions from the subvolcanic Biely Vrch porphyry-Au deposit. These inclusions have previously been interpreted to have formed by magmatic exsolution of a hypersaline brine followed by halite condensation and provide a comparative basis for the magmatic-hydrothermal formation mechanisms proposed for salt inclusions in MtAp deposits.

These data show that the MtAp melt inclusions from both sites exhibit exceptionally low Br/Cl and I/Cl ratios, closely resembling those in surficial evaporitic halite and differing markedly from those found in typical magmatic-hydrothermal settings. In contrast, the fluids from Biely Vrch display significantly higher Br/Cl and I/Cl ratios, aligning with those observed in other porphyry-type and arc-related magmatic-hydrothermal systems globally. These results strongly support the anatectic model of salt melt generation and demonstrate that halogen signatures are a valuable diagnostic tool for identifying and understanding the role of evaporite anatexis in the ore genesis of MtAp deposits worldwide and for the further development of MtAp ore system models.

Linking the Nature and Timing of Diagenesis of Source Rocks with Hydrocarbon Migration and Copper Mineralisation Events in the Neuquén Super Basin, Argentina

Richard J. Herrington[1], Josefina Pons[2], Ana L. Rainoldi[3], Jamie Kelly[4], Huw Griffiths[5], Nick Roberts[6], Veronica Trevisan[1], Julia Woitischek[5]
1. Natural History Museum, London, United Kingdom, 2. IIPG, Consejo Nacional de Investigaciones Científicas y Técnicas, Universidad Nacional de Rio Negro, Neuquen, Argentina, 3. Instituto Geológico del Sur (INGEOSUR), Universidad Nacional del Sur (UNS)-CONICET, Bahía Blanca (Argentina), Bahia Blanca, Argentina, 4. University of Southampton, Southampton, United Kingdom, 5. Imperial College, London, United Kingdom, 6. British Geological Survey, Keyworth, United Kingdom

The Mesozoic Neuquén Super Basin in Argentina, with up to 6km thickness of clastic sediments, contains more than 800 million m^3 of hydrocarbon reserves as well as sediment-hosted copper occurrences in altered sandstones. Known metal deposits each contain between 10^3 and 10^6 tonnes of metal at grades ~0.3% Cu with V credits. Previous work has established that early diagenesis of synorogenic sedimentary rocks (Cenomanian Santonian age) is followed by a protracted hydrocarbon migration event (upper Cretaceous to Middle Tertiary) that is closely followed and partially overlaps with mineralization events (Middle-Late Tertiary) resulting in deposition of sulfides dominated by chalcocite and minor bornite and chalcopyrite.

In this new study, novel U-Pb dating methods have been applied to three calcite generations from synorogenic red beds that indicate early diagenesis between 100-94 Ma, hydrocarbon migration 76-55 Ma, and mineralization at around 28 Ma. This supports the previous model of hydrocarbon migration promoting redbed bleaching and secondary porosity formation followed by a second event bringing copper-rich brines into sandstones impregnated with bitumen. To integrate basin development, metallogenic, and hydrocarbon histories, a range of potential metal source rocks were collected across the basin and studied using a range of analytical techniques, complimented by preliminary laboratory leaching experiments applied to bulk samples. A programme of Pb isotope analysis is now underway on a range of sediments, leachates, and sulfides to confirm the likely source rocks for the metals in the deposits. Preliminary mass balance work measuring the volumes of potential source rocks versus the known basin occurrences will be presented in an assessment of the potential for further undiscovered deposits.

We will discuss the likely origins of the copper, probable pathways, and depositional history for the copper mineralization with reference to the primary metal sources for the coatings and matrix cements from the sedimentary source regions.

Geometallurgical Characterization of Potential By-Products in the Spremberg-Graustein-Schleife Kupferschiefer Deposit

Zahra Nourizenouz[1], Max Frenzel[1], Bradley M. Guy[1], Joachim Krause[1], Doreen Ebert[1], Aratz Beranoaguirre[2, 3], Jens Gutzmer[1]
1. Helmholtz-Zentrum Dresden Rossendorf, Helmholtz Institute Freiberg for Resource Technology, Freiberg, Germany, 2. Karlsruhe Institute for Technology, Chair of Economic Geology and Geochemistry, Karlsruhe, Germany, 3. Goethe University-Frankfurt, Institute of Geosciences, Frankfurt am Main, Germany

The Kupferschiefer ore deposits that occur along the southern margin of the Central European Permian Basin constitute one of the largest and highest-grade sediment-hosted copper provinces in the world. Mineralization formed when oxidizing and metal-rich basinal fluids interacted with a thin but laterally persistent carbonaceous black shale/marl unit at the base of the Permian Zechstein succession. In addition to copper, Kupferschiefer ores also contain elevated concentrations of silver, lead, zinc, cobalt, nickel, rhenium, and gold. While some of these metals (especially Ag) already contribute as by-products to the overall revenue generated, quantitative geometallurgical studies exploring the deportment and spatial variability of all these metals are still lacking.

In this study, we investigate multi-metal deportment (i.e., Ag, Pb, Zn, Co, Ni, Re, Au) at the Spremberg-Graustein-Schleife deposit, located in Lusatia in SE Germany. This investigation is carried out through a comprehensive approach that integrates data from laser ablation ICP-MS (LA-ICP-MS), electron microprobe analysis (EPMA), and mineral liberation analysis (MLA) on samples from the mineralized intervals of three exploration drill cores. The results illustrate that the deportment of Ag is particularly complex, as it occurs as i) substitution in the crystal lattices of ore-forming Cu and Cu-Fe sulfides, ii) micro-inclusions within Cu-enriched zones in pyrite/marcasite, and iii) isolated grains of electrum. Galena and sphalerite are the main hosts of Pb and Zn, respectively. Cobalt and Ni, in contrast, are found as Co-Ni-arsenides and sulfides, with chalcopyrite and pyrite exhibiting enrichment in cobalt. Such quantitative information can not only facilitate the integration of processing of the by-products into the geometallurgical models, but also enable the prediction of the mineralogical distribution of these elements within drill cores, using bulk geochemical data.

Using Geochemical and Petrophysical Properties to Target Sediment-Hosted Copper Deposits

J K. Porter[1], D E. Kelsey[2], C Adams[3]
1. University of South Australia/Geological Survey of Western Australia, Perth, WA, Australia, 2. Geological Survey of Western Australia, Perth, WA, Australia, 3. CSIRO, Perth, WA, Australia

Airborne geophysical surveys, such as airborne magnetics and gravity, are important techniques employed in mineral exploration, particularly for unexposed, inaccessible regions under cover. Characterising the petrophysical and chemical properties of the rock types that occur within and near a deposit can improve geophysical models and identify targets based on relevant anomalies. These data can be utilised to prioritise geophysical surveying programs in order to improve exploration efficiency and costs.

The Neoproterozoic Yeneena Basin in the Paterson Province of Western Australia contains several known Cu deposits, including the Nifty Cu deposit. Nifty is a base metal (Cu ±Zn ±Pb) deposit hosted primarily in carbonaceous shale with lesser carbonate beds in stratiform lenses within a syncline structure. Nifty has been well studied and thus provides an opportunity to characterise the mineralised host rock and related alteration zones in order to define the stratigraphy using only petrophysical and chemical properties.

Whole-rock geochemistry and petrophysical measurements were taken at ~10-m intervals in seven diamond drill cores across the Nifty deposit. Using principal component analysis and machine learning techniques, five units with distinctive chemical and physical properties were identified, including those identifying the main Cu orebody, silicic alteration, carbonate alteration, pyrite-rich shales, and less-altered black shales. These units are laterally extensive across the mine site mineralisation and provide targets for further site-scale geophysical surveying. Based on these classified lithological groups, an induced polarisation/resistivity survey would identify the conductive Cu ore, surrounding resistive silica-rich alteration and the overlying, chargeable pyrite-rich shale layer. This technique could be used to trace these layers across the extent of the mineralisation footprint and could be further applied to areas of less well-explored sediment-hosted Cu prospects.

Constraining Mineralization and Alteration Through New Geologic Mapping at Spring Peak Low-Sulfidation Epithermal System, Mineral County, NV

Marcus J. Angus, Simon Jowitt
University of Nevada, Reno, Reno, NV, USA

Spring Peak is a newly discovered low-sulfidation epithermal system with a well-preserved epithermal footprint at surface. The high-grade gold mineralization at Spring Peak is located within the boiling horizons of thermal upwelling zones, one of which is located directly beneath the outcropping sinter, which preserves high-temperature lithofacies. Sinters are surficial expressions of low-sulfidation epithermal systems and form where silica oversaturated alkali-chloride fluids are expelled in hot spring deposits, forming a predictable gradient in lithofacies with distance from the vent source. This project involves detailed 1:2000 geologic mapping, which aims to identify alteration trends and the location of favorable structural controls in order to delineate additional upwelling zones with potential to host previously unrecognized gold mineralization. Geologic mapping to date has identified a window of Mesozoic units exposed by faulting and Tertiary bi-modal volcanism. Epithermal alteration is most intense around NNE-NE striking high-angle normal faults and includes sinter, hydrothermal breccia, quartz veins with preserved boiling textures, adularia and argillic alteration. This detailed mapping will be complemented by research into the Spring Peak sinter and preserved lithofacies to identify the locations of high temperature vent conduits and low temperature, distal-apron fluid pathways and to identify the spatial and temporal relationships of sinter deposition with gold mineralization. The age of mineralization at Spring Peak has yet to be constrained and $^{40}Ar/^{39}Ar$ dating of adularia in mineralized drill core, coupled with rhyolite domes collected at surface, will determine whether Spring Peak has similar Miocene mineralization analogous to the nearby Aurora system, or if Spring Peak is a potentially younger sinter-vein system. Geologic mapping of the sinter-vein system at Spring Peak offers a unique opportunity to study epithermal alteration and associated sinter deposition with proximate gold mineralization at depth.

B2Gold's Antelope Discovery – Fresh Insights Lead to Exploration Success at Otjikoto

James Bursey
B2Gold Namibia, Windhoek, Namibia

The Otjikoto Gold Mine is located in the north-central part of Namibia and is the country's biggest gold producer. A new discovery, the Antelope deposit, approximately 3km south of Otjikoto, has the potential to supplement mine production starting in 2026, subject to additional engineering studies of the initial mineral resource estimate.

Discovered in 2022, following deep drill testing on three-dimensional models of magnetic inversion data, the Antelope deposit may consist of as many as three separate mineralized structures, of which the southernmost Springbok zone has been defined by 100 x 50m spaced drilling, over a strike length of approximately 800m. Mineralization has a dip of extent of approximately 150m and ranges from 7 to 10m thick. Northeast, along strike of Springbok, the Oryx zone appears to represent a second high-grade shoot, stacked stratigraphically above Springbok, for an overall combined plunge extent of approximately 1,500m. Shoots remain open to the northeast and southwest along the plunge azimuth. Mineralization is characterized by sheeted quartz–pyrrhotite veins, overprinted by deformation focused along two main marble beds that serve as major stratigraphic markers in the Otjikoto stratigraphy. The shoot-like geometry of the Antelope deposit mineralization derives, in part, from the thickening of quartz–pyrrhotite–gold mineralization in the hinge zones of centimetre- to metre-scale folds, a structural control that is well documented in the Otjikoto Mine. Mineralized shoots plunge shallowly north–northeast, suggesting a subtle inflection in the stratigraphy that hosts the Otjikoto deposit, where ore zones plunge shallowly south–southwest.

The mineralized zones are associated with a positive magnetic response that extends southward from Otjikoto as a continuous linear feature for over 4km. Testing the continuity of this magnetic feature and its possible connection with the southernmost extent of ore in the Phase 5 open pit at Otjikoto will form a significant part of the 2024 exploration program.

Epithermal Gold Discoveries in the New Khundii Metallogenic Province, Southwest Mongolia

Michael A. MacDonald[1], Gendenjamtsyn Bat-Erdene[2], Michael X. Gilles[3], Peter J. Dalton[3], Jeffrey W. Hedenquist[4]
1. Independent, Dartmouth, NS, Canada, 2. Erdene Resource Development Corporation, Ulaanbaataar, Mongolia, 3. Erdene Resource Development Corporation, Dartmouth, NS, Canada, 4. Independent, Ottawa, ON, Canada

Exploration in an underexplored region of SW Mongolia since 2005 has defined a porphyry Mo-Cu deposit (measured and indicated resources of 218 Mt at 0.057% Mo and 0.069% Cu) and discovered two epithermal gold deposits. This emerging Khundii metallogenic province, ~5000 km^2 in area in a Carboniferous-Permian island arc, is ~700 km WNW of the Oyu Tolgoi porphyry Cu-Au deposit.

After definition of the granodiorite-hosted Zuun Mod porphyry Mo-Cu deposit (297±4.8 Ma), a regional exploration program over 110,000 km^2 included compilation of geologic, geochemical, geophysical and satellite imagery information, followed by ground exploration with stream, soil, and rock chip sampling plus geologic and alteration mapping. A prospect discovered in early 2011 led to a 400×400-m soil survey that identified a Pb-, Zn-, and Au-in-soil anomaly over 1.5×5.5 km. The first drill hole within the soil anomaly discovered the Altan Nar Au-polymetallic epithermal deposit, with veins of quartz-adularia (309.7 ± 0.5 Ma), base metal sulfides, and Ca-Mg-Mn-Fe-carbonate gangue.

The Bayan Khundii gold deposit, discovered in 2015 from prospecting ~16 km SE of Altan Nar, led to adjacent deposits at Kar Mori and Ulaan SE also being discovered (https://erdene.com/). Epithermal quartz-adularia-gold veins (336.8 ± 0.5 Ma) with colloform bands and minor pyrite plus variable arsenopyrite have illite halos. Epithermal veins in the area overprint an early, unrelated alteration of residual quartz and pyrophyllite±dickite±diaspore-kaolinite. Northwest of Bayan Khundii ~3 km at Ulaan, K-feldspar and magnetite plus quartz-white mica-pyrite, related to the top of a porphyry deposit with Cu-anomalies, was eroded to the base of a lithocap and overprinted by gold-bearing epithermal veins; the porphyry prospect has yet to be drill tested at depth. The Bayan Khundii deposit has 0.5 Moz of shallow ore at 4.0 g/t Au in reserves; production is slated to start in 2025, with drill testing of adjacent deposits continuing.

Predicting Giants: The Role of Geoscientists and Mineral Systems in our World of Emerging AI

Nicole Januszczak
BHP, Toronto, ON, Canada

Mineral exploration suffers from a prediction problem: we are good at predicting mineralization, but not very good at predicting world-class mineral systems. Prospectivity models take months to build, geoscientists are under considerable cognitive load, and few techniques are used to stress test our understanding of how critical processes need to line up for world-class mineral systems to form.

AI will drastically improve our ability to predict world-class mineral systems. AI can assimilate large volumes of data, make complex decisions, and communicate how predictions are made. Exploration needs AI, but AI also needs us. Humans take empirically driven predictions from AI derived from patterns in data, and we attribute conceptual understanding of processes that govern how mineral systems form.

At BHP we are curating one of the largest cloud-hosted data lakes in the world as a key enabler for the advanced application of AI. We are investing in new spatio-temporal data sets that are revealing how world-class mineral systems form. We have built AI toolboxes, and we are pioneering the use of large language models in BHP to extract mineral systems insights stranded in unstructured legacy data.

BHP is committed to the advanced application of emerging technology, and we are preparing to use AI at scale and at pace to accelerate the delivery of high-value options to the business. New technologies take time to take hold, but early adopters will be rewarded with first-mover advantage in our competitive mineral exploration industry.

Structural Controls on Williams Brook Gold Mineralization, New Brunswick, Canada

Edawrd P. Wu[1], Stefan Kruse[1], Dominique Gagné[2], Alain Hupé[2]
1. Terrane Geoscience Inc., Fredericton, NB, Canada, 2. Puma Exploration Inc, Rimouski, QC, Canada

The Williams Brook Property, owned by Puma Exploration Inc., is located within the Appalachian Orogen in New Brunswick, Canada. The property is underlain by the Lower Devonian Wapske Formation, which comprises metasedimentary rocks locally interleaved with tuffaceous to massive rhyolite flows, mafic volcanic, and mafic intrusive rocks. The Wapske Formation is bounded to the west by the regional, dextral Rocky Brook–Millstream Fault ("RBMF"), which was active during Early Devonian Acadian orogenesis.

The Lynx Gold Zone (LGZ), a new gold discovery with diamond drill intersections including 5.55 g/t gold over 50.15 m, occurs along the contact between non-foliated rhyolite and pervasively foliated and tightly folded metasedimentary rocks. The lithological contact typically strikes southwest to south-southwest. The pervasive foliation strikes northeast or southwest and generally dips steeply to the northwest or southeast. The foliation is axial planar to gently northeast-plunging folds.

Mineralization in the LGZ is structurally controlled by the veins emplaced within the rhyolite and along the rhyolite–metasedimentary rock contact. Gold mineralization is hosted in quartz veins, typically planar within rhyolite, and folded and attenuated within the metasedimentary rocks. A total of four major vein sets were identified from trench mapping, televiewer interpretation and oriented core measurements. Auriferous veins present two main orientations: a shallow-dipping vein set and a steep vein set oriented parallel to the regional RBMF trend. The mineralized veins show mutual crosscutting relationship, suggesting that veins are syn-tectonic and were emplaced during the same deformation event. Three-dimensional structural modelling of the LGZ reveals that gold mineralization has a shallow, northeast plunge that is subparallel to the fold axis of regional folds. Based on these observations, both the structural fabrics and auriferous quartz veins of the LGZ are interpreted to have formed during a single dextral transpressive event associated with the RBMF during the Acadian orogenesis.

The Highbury Lithium Deposit, KwaZulu - Natal, South Africa

Eugene Van den Heever[2], Michael Cronwright[1]
1. ERM Global, Oxford, United Kingdom, 2. SA Lithium Pty Ltd, Cape Town, South Africa

The Highbury Lithium Deposit comprises a series of pegmatites within a granulite facies klippe of the Margate Terrane in the south of the Mzumbe Terrane of the Natal Sector of the 1200-1000Ma Namaqua – Natal Orogeny, South Africa. They are interpreted to have intruded ~1,060Ma making them broadly contemporaneous with the pegmatites of the Namaquan Orange River Pegmatite Belt (ORPB) and those of the Kibaran Belt.

Highbury is the subject of ongoing exploration by SA Lithium after being largely overlooked since its discovery in the late 1980's and described by Thomas et al (1994).

Recent exploration culminating in a total of ~21,000m of drilling and ~800kt of bulk sampling has revealed a significant lithium resource hosted by poorly zoned albite – spodumene type pegmatites of Lithium Caesium Tantalum (LCT) affinity. The Highbury pegmatites are currently the only known occurrence of LCT type pegmatites within the Natal Sector.

Lithium at Highbury is hosted primarily within a symplectic intergrowth of Spodumene and Quartz (SQI) which is interpreted to have formed after petalite because of near isobaric cooling during the waning phases of the Orogeny. Gangue consists of albite- K-feldspar and quartz with minor muscovite, garnet, graphite, and chlorite.

Highbury is unique in context of the Namaqua – Natal Orogeny in that the LCT pegmatites of the ORPB are small, complexly zoned with abundant micas and various Li bearing minerals and undergone late-stage hydrothermal alteration and lithium depletion. The Highbury pegmatites in contrast are un-zoned, with minor mica, and spodumene being the primary Li mineral. Exploration has shown the pegmatites to be continuous along strike (~1.5km) and >500m down dip with drilled pegmatite thicknesses of up to 60m. Current Resource estimates are >75Mt at <1.4% Li20 (cut-off grade dependent). Deposits of this size have yet to be discovered in the Namaqua-Natal Belt.

Insights into Enhanced Brownfield Exploration Techniques at Morenci for Improved Ore Characterization and Mineral Reserve and Resource Addition

Paul Albers[1], William Stavast[1], Logan Hill[2]
1. Freeport-McMoRan Exploration Corporation, Tucson, AZ, USA, 2. Freeport-McMoRan Morenci Operations, Morenci, AZ, USA

The mineral deposits of the Morenci district consist of copper oxide, secondary sulfide, and primary sulfide mineralization associated with a large porphyry copper system. Geologic studies indicate that a complex series of Tertiary igneous intrusive rocks were emplaced within Precambrian-age granite and overlying Paleozoic and Mesozoic sedimentary rocks. Several cycles of leaching and enrichment of the primary sulfides formed the secondary sulfide enrichment blanket and copper oxide zones currently being mined.

Morenci has had a long history of exploration, development, and production that dates back to the first records of copper mineralization identified in 1863. 43.9 billion pounds of copper have been produced since 1906, including 13.9 billion pounds of copper since 2007. World-class mining and production rates have been coupled with systematic brownfield exploration drilling programs to replenish and expand mineral reserves and resources, extend the mine life, and optimize the mine plan. Nearly 3 million feet in 2,000 drill holes of core and reverse circulation methods were drilled over the last two decades. Improved core logging workflows at a Tucson-based centralized location, coupled with technological advances, are key for enhanced orebody knowledge and characterization. The exploration drilling also serves as a valuable input to ore characterization, geometallurgical/geomechanical studies, and leaching initiatives. Detailed geological logging with the support of predictive lithology from geochemistry and geotechnical logging via high-resolution imagery has improved workflows and consistency. Drill samples include a multi-element suite analytical package coupled with mineralogy and geochemistry from XRD and XRF, which has provided significant improvements for material characterization. Enhancements in geologic resource modeling methods have been important components of this process. This presentation for enhanced brownfield exploration, including approaches taken and value gained, positions Morenci well for continual production.

Proterozoic Sediment-Hosted Mineralisation in Southeastern Australia and Links to the Belt-Purcell Basin of North America

Sheree Armistead[1], Sebastien Meffre[1], Ralph Bottrill[2], Andrew Cross[3], David Huston[3], Grace Cumming[2]
1. CODES, University of Tasmania, Hobart, TAS, Australia, 2. Mineral Resources Tasmania, Hobart, TAS, Australia, 3. Geoscience Australia, Canberra, ACT, Australia

Western Tasmania, in southeastern Australia, is well known for its large endowment of volcanogenic massive sulphide (VMS) deposits relating to the Cambrian Mount Read Volcanics and for Sn±W±magnetite mineral systems related to Devonian granitoids. The Rocky Cape region lies to the northwest of this mineral province and hosts Mesoproterozoic to Neoproterozoic metasedimentary and igneous rocks—the oldest rocks in southeastern Australia. Several historic Cu-Pb-Zn vein-hosted prospects, including the Silver Reward, Strickland, and Balfour prospects, are located within the Rocky Cape Group. Although these deposits were previously thought to be related to Devonian granites that are widespread in southeastern Australia, lead isotope data indicated apparent Mesoproterozoic model ages.

New U-Pb monazite, xenotime, and apatite data and galena Pb isotope data from these prospects indicate that sulphide mineralisation formed at c. 1350 Ma, with several resetting or remobilisation events recorded at c. 1250 Ma, c. 1100 Ma, and c. 950 Ma. Minor cobaltite was found in some of these prospects, highlighting the potential for critical metals. We have also collected new Lu-Hf garnet data (using LA-ICP-QQQ-MS) and monazite data that suggest metamorphism may have been coeval with some of the mineralisation events.

The Rocky Cape Group has been previously correlated with stratigraphy in the Belt-Purcell basin of North America. New data presented here suggest that there is also a correlation between mineralisation ages, particularly those in the Idaho Cu-Co Belt, such as the Merle Cu-Co and Spar Lake Cu-Ag deposits, presenting a new exploration target for northwestern Tasmania and, possibly, elsewhere in southeastern Australia. These new data further support a tectonic link between northwest Tasmania and Laurentia during the transition between supercontinent Nuna breakup and supercontinent Rodinia assembly.

Geological and Geochemical Evolution of the Supergiant High-Grade Resolution Cu-Mo Deposit, Arizona

Takeshy X. Coaquira, David R. Cooke, Lejun Zhang
University of Tasmania, Hobart, TAS, Australia

Copper mineralization at Resolution is related to four main paragenetic stages. Stage 1 is characterized by early chalcopyrite mineralization related to potassic alteration, along with its associated A veins, coupled with the unidirectional solidification textures recognised in early dacite porphyries, all of which provide important insights to the understanding of early hydrothermal activity in this complex system, which did endow some Cu mineralization. However, the bulk of the Cu-sulfides at Resolution are associated with overprinting white mica alteration from stage 2, which is characterized by phyllic alteration, which is extensive throughout the ore zone and overlying volcaniclastic rocks, where it produced greenish-gray to white micas and clays associated with abundant pyrite. Massive chalcopyrite veinlets, a few millimetres to several centimetres in width, are associated with phyllic alteration throughout the ore zone; they can have chalcopyrite-bearing white mica alteration halos, demonstrating the importance of this alteration type to mineralization. Stage 3 is characterized by advanced argillic alteration, with bornite and chalcocite as common Cu mineral species at Resolution, occurring dominantly as replacements of chalcopyrite. Bornite at Resolution does not follow the classic porphyry model, where it is localized in the potassic core. Instead, bornite at Resolution was added during the phyllic and later advanced argillic events. The transition with time from chalcopyrite to bornite-stable ore-forming conditions indicates that fluid chemistry was evolving to higher sulfidation states. Even higher sulfidation assemblages formed later—these are typically structurally controlled, producing chalcocite-digenite-bearing veins and magmatic-hydrothermal breccias that are closely related to APS-kaolinite-dickite ± alunite alteration. The latest paragenetic stage that enhances copper at Resolution is characterized by supergene kaolinite with steely chalcocite ore blankets. However, the bulk of high-grade ore is genetically related to hypogene enrichment processes.

Breccias in Magmatic-Hydrothermal Environments

David R. Cooke, Francisco Testa, Lejun Zhang, Victor Torres, Angela Escolme
CODES, University of Tasmania, Hobart, TAS, Australia

Large-scale, discordant breccia bodies can be significant components of magmatic-hydrothermal mineral systems. Brecciation can occur early or late during hydrothermal activity with diverse relationships to mineralization.

Early-formed breccias may form due to explosive release of high-temperature magmatic-hydrothermal fluids from a shallowly emplaced hydrous intrusive complex, with explosive brecciation creating pipes and/or dykes above the causative intrusion. The magmatic-hydrothermal fluids are initially in chemical and thermal equilibrium with the crystallising magma, producing high-temperature alteration of clasts and matrix and, in some cases, significant domains of high-grade mineralised breccia cement. These breccias define the highest-grade ores in some major porphyry Cu, Au, and Mo deposits (e.g., Rio Blanco, El Teniente, Productura, Chile; Soledad, Peru). Magmatic-hydrothermal breccias taper upwards and may contain distinctive tabular clast breccias ("shingle breccias") on their roofs and outer walls. Their distal expressions include sheeted veins with potassic or phyllic halos.

Maar-diatreme complexes are the products of catastrophic phreatomagmatic explosions that occur when magmas intrude active hydrothermal systems. Volcanic-hydrothermal breccia complexes commonly form late in the life of porphyry copper deposits (e.g., El Teniente, Chile; Tujuh Bukit, Indonesia) and early in the life of high-sulfidation (e.g., Yanacocha, Peru; Zijin, China), intermediate-sulfidation (e.g., Acupan, Philippines), and low-sulfidation epithermal Au deposits (e.g., Lihir, PNG). Some diatremes are mineralised, whereas others are barren. Concealed porphyry deposits have been discovered due to the recognition of mineralised clasts in diatreme breccias (e.g., Far Southeast, Philippines; Cuerpo 3, Chile).

Structural processes can result in the formation of mineralized and unmineralized breccias throughout the evolution of magmatic-hydrothermal systems. Tectonic-hydrothermal breccias can involve phreatic or hydraulic fracturing, abrasive wear, and/or implosion. Corrosive wear can produce solution collapse breccias in carbonate rocks and high-sulfidation epithermal environments. Overall, these diverse breccia types require systematic mapping and logging to fully unlock their exploration potential.

Prospectivity Modeling and Metallogeny of Pre-Orogenic and Orogenic Base and Precious Metals of the South Nabitah Belts, Arabian Shield

Arnaud Fontaine[1, 2], Christophe Bonnetti[1, 2], Julien Feneyrol[1, 2], Virginie Masson[1], Charles Gumiaux[3], Anne-Sylvie André-Mayer[4, 2], Rémi Bosc[1, 2]
1. Arethuse Geology, Fuveau, France, 2. LabCom Theia, Nancy, France, 3. ISTO, Université d'Orléans CNRS-BRGM, Orléans, France, 4. Géoressources UL-CNRS, Nancy, France

The Neoproterozoic Arabian Shield currently experiences an exploration revival on various commodities such as precious and base metals. Deposits hosting them cover a wide range of mineralization styles, related to the accretion history, marked by remnants of sutures amalgamated and sheared during orogeny. To optimize discovery of new deposits, prospectivity modeling has been increasingly used. In our study, we focus on VMS and orogenic gold mineralization that formed during the Nabitah-Najd orogenies. In such context, assessing this high degree of primary ore preservation related to deformation, metamorphic, and hydrothermal/supergene alteration gradients, is key for target generation in such polyphased metallogenic context. Knowledge- and data-driven approaches have been combined based on compilation, harmonization, and interpretation of GIS geoscientific as well as field data. They will allow us to (i) produce a metallogenic synthesis of the Nabitah belts; and (ii) build an integrated metallogenic model critical for predictive modeling of base and precious metal deposit prospectivity. In southern Saudi Arabia, pre-accretion arc-related VMS mineral systems are predominantly preserved along the Nabita suture zone. Orogenic gold veins mainly developed in the N- to NNE-trending Nabitah shear zones, but also along the NW-trending Najd strike-slip fault system during late orogenic extension. Field studies are used as reference sets to validate key geological factors controlling the spatial distribution of the mineralization. Favourable tectonic environments (e.g., back-arc or rifted arcs settings) and lithologies (e.g., bimodal volcanism with tholeiitic affinities), strain gradients (e.g., dilatational jogs and reactivation of early structures), occurrences and geochemical anomalies (e.g., CCPI and AI indexes), and remote sensing (e.g., structural domaining and complexity and alteration footprints) are considered to delineate the fertility and mineral potential. Such models will open new perspectives and implications for the exploration of VMS and orogenic gold in the Arabian Shield and elsewhere in Precambrian metallogenic provinces.

The Bushveld Complex and Its Mineralization, South Africa

Judith A. Kinnaird[1], Marina Yudovskaya[2]
1. CIMERA, University of the Witwatersrand, Johannesburg, South Africa, 2. IGEM, Moscow, Russia

The Complex was first shown on a map in the late 1860's although the first use of the term Bushveld Complex was in 1901. The Complex is made up of five limbs, is up to nine km thick, and covers an area of at least 90,000 km^2. The Complex is divided into a Marginal Zone, a Lower Zone, a Critical Zone—so called in the 1930's as this was where all the chromium and platinum group element (PGE) ores occurred—and an Upper Zone. Layers dip 8° to 27° towards the centre of the Complex. Recently, ultramafic intrusions below the Marginal Zone have been identified.

In 2011, when PTM drilled north of the Hout River shear zone where the Complex was thought to terminate, a thick succession of Bushveld rocks were intersected including two zones of mineralization: an upper T zone and a lower F zone. What is unique about this Waterberg project Tier 1 deposit is that the upper T reef occurs at 130-m depth at the interface between the Main Zone and the Upper Zone, hosted in a variety of lithologies, whereas the F "reef" occurs in ultramafic rocks and is shallowest at 180-m depth.

The Bushveld Complex also has the world's second largest output of vanadium from magnetitite layers of the Upper Zone. There are ~21 of these layers, and the vanadium content decreases upwards, so it is the lower magnetite layers that are the most economic. What is poorly appreciated is that the Bushveld is a world-class producer of gabbronorite and granite dimension stone, andalusite from its metamorphic aureole, and fluorite from a granite pipe. Cassiterite mineralisation from the granites has been important in the past with potential also for future extraction.

Platinum Group Metals: It Begins and Ends in the Market

Richard Stewart

Platinum group metals (PGMs) are the rarest of precious metals and yet due to their unique chemical properties are driven by market forces more commonly associated with industrial fundamentals.

Demand uses for PGMs are diverse, from medical applications to jewellery. Over the past three decade's global demand for PGMs has increased substantially from approximately 8 million ounces in 1990 to a peak at approximately 18.5 million ounces in 2018. This significant increase in demand was driven by the convergence of technological advances of the tri-metal catalyst (Pt, Pd, and Rh) that significantly enhanced the emission reduction efficiency of auto catalysts and growing global regulatory frameworks to address automobile emissions.

Over the past few years, the rapid growth of battery electric vehicles (BEVs), has driven negative market sentiment towards the outlook for PGM demand. However, the economic, technical and consumer driven constraints of sustaining recent BEV market penetration rates is increasingly being recognised, and the potential roles of other disruptive technologies to automobile power trains, including hydrogen, are garnering progressively more attention.

It is highly probable that PGM markets will undergo a long-term structural change and that both demand and supply drivers over the next three decades are going to be very different to the last. In addition to evolving regulation and technological advances, the pace and nature of that change will be impacted by evolving macrotrends, including increasingly stringent requirements for responsibly sourced metals, geopolitical tensions, the impact of artificial intelligence, and the continuous drive for global decarbonisation. Consideration of these macrotrends will be critical as the industry advances future discovery, extraction, beneficiation, and market development of PGMs to ensure their unique characteristics continue to contribute positively to society for decades to come.

Processing of the Bushveld Ores

Kirsten Youlton
SGS South Africa, Bryanston, South Africa

The two major commodities regularly recovered from the Bushveld Complex are platinum group elements (PGEs) and chromite. While these two commodities often occur together, their processing routes could not be more different, and in the instances of PGEs, the presence of Cr is even a penalty element during smelting.

Processing of commodities originating in the Bushveld has been taking place since the panning of alluvial material as early as 1924. Major advancements have been made in this time, and the current recovery process and controls are now well established. The main processing routes currently employed are PGE flotation and physical separation (gravity concentration) of chromite. As these processes are relatively well constrained, much of the ongoing process mineralogy and optimisation focuses on the identification of mineralogical variability and corresponding shortcomings to address or account for them appropriately.

Amongst other factors, grain size distribution is very relevant to PGE flotation, with both coarse- and fine-grained occurrences presenting challenges. Another important factor is the degree of association with both gangue and sulphides. Some argue that the precise species of PGE mineralisation has an impact on flotation, while others suspect that only specific types of PGEs (such as laurite) represent a significant difference in flotation response. The processing of PGEs is often tailored to the ore, employing flocculant, depressant, and other reagents to optimise conditions for ideal flotation.

There is stark contrast between the finesse of the recovery of trace PGEs and the bulk commodity of chromite. Much of the complexity of chromite processing is ensuring sufficient chromite liberation without major degrees of fines generations.

An evaluation of case studies where atypical mineralogy of both PGEs and chromite have affected the processing route taken will be presented.

Mineralogy and Petrology of the Petalite-Subtype Prof Pegmatite, Revelstoke, British Columbia, Canada

Catriona M. Breasley[1], Lee A. Groat[1], Tania Martins[2], Robert L. Linnen[3], Kyle P. Larson[4], Rhiana E. Henry[5]
1. University of British Columbia, Vancouver, BC, Canada, 2. Manitoba Geological Survey, Winnipeg, MB, Canada, 3. University of Western Ontario, London, ON, Canada, 4. University of British Columbia Okanagan, Kelowna, BC, Canada, 5. Gemological Institute of America, Carlsbad, CA, USA

The Prof pegmatite is exposed as a ~70- × 5-m dike located NW of Revelstoke, British Columbia, Canada on Boulder Mountain. Due to the abundance of petalite within it, the pegmatite is classified as a petalite subtype Li-Cs-Ta pegmatite or a group one pegmatite. The Prof pegmatite contains a suite of minerals indicative of a highly evolved pegmatitic melt including petalite, elbaite, lepidolite, and Nb-Ta oxides. It can be divided into four textural zones: (1) border, (2) intermediate, including (2.1) graphic texture dominant and (2.2) overgrowth dominant, where multiple minerals form rims around one another, (3) central, and (4) a quartz core. The border zone has a similar mineralogy to the intermediate zone and is interpreted to represent a chilled margin. The intermediate zone has a feldspar, mica, garnet, and dravite-schorl-dominant composition. The central zone hosts an evolved pegmatite core, which contains the main lithium mineralisation composed of petalite, elbaite, and lepidolite. The tourmalines, Nb-Ta oxides, and micas within the pegmatite record the geochemical evolution of the melt from more primitive Fe- and Mg-rich minerals to an Li-, Mn-, and Nb-rich assemblage indicative of an evolved geochemical system. The textures and geochemical composition of the pegmatite indicate that the melt was undercooled, highly geochemically evolved, and crystallised rapidly. Three phases of metasomatism are recognised in the Prof pegmatite: an albitization event, an Na-Li-F-rich event, and a sericitization event.

The Prof pegmatite has a similar mineralogy to known pegmatites on Mount Begbie, 15 km to the S, in particular the notable presence of the rare mineral qitianlingite, minor petalite, lepidolite, and elbaite. Together, these pegmatite bodies form part of an extensive, poorly mapped pegmatite field. Additional work is required to assess the extent and nature of mineralisation within this field.

Transtensional Tectonics Control Emplacement of Lithium-Tin Granites in the Eastern Erzgebirge (Germany/Czech Republic)

Jan Cerny[1], Sam Thiele[1], Marie Guilcher[1], Matthias Burisch[2], Uwe Lehmann[3], Henrik Kaufmann[3], Lutz Sonnabend[3], Jens Gutzmer[1]
1. Helmholtz-Zentrum Dresden-Rossendorf, Helmholtz Institute Freiberg for Resource Technology, Freiberg, Germany, 2. Mineral Systems Analysis Group, Department of Geology and Geological Engineering, Colorado School of Mines, Golden, CO, USA, 3. Landesamt für Umwelt, Landwirtschaft und Geologie (LfULG), Freiberg, Germany

The eastern Erzgebirge region contains several significant granite-related Li-mica deposits (e.g., Zinnwald-Cinovec). More than 90% of the known Li-mica occurrences of the Erzgebirge are related to late-stage intrusions (˜313-310 Ma) within the Altenberg-Teplice caldera system. In this contribution, we explore the role of transtensional tectonics and associated crustal-scale faults as the controlling factors for the emplacement and spatial distribution of small stock-like granitic intrusions within the Altenberg-Teplice Caldera. Examination of fault architectures within the wider region of the Altenberg-Teplice and Tharandt calderas and within the broadly contemporaneous western Bohemian Pennsylvanian basin (~314-312 Ma) reveals a large-scale composite pull-apart structure. This includes three pull-apart basins cleaved by a network of transform faults providing trans-crustal connectivity, which are preferentially hosting intrusive stocks and related mineralization. We propose a simplified mechanical model that invokes several second- and third-order dilatant and transfer structures linking the first-order Elbe shear zone in the northeast to the Pfahl or Danube shear zones in the southwest that have allowed the opening of this pull-apart system.

Our analysis provides insight into the intricate tectonic framework that controls the distribution and localization of Li-rich peraluminous granites and associated Li-Sn greisen systems in the Erzgebirge region, with possible applications elsewhere in the world. This approach brings a new exploration tool for identifying favorable locations for undiscovered Li-endowed granite stocks in the Eastern Erzgebirge and provides useful criteria for the exploration of Li-Sn granites in collapsed collisional orogens.

Evolution of the Magmatic System Associated with a Supergiant Porphyry Cu Deposit from Trace Element Profiles in Zircons

Massimo Chiaradia[1], Hugo Carrasco[1], Jing-Jing Zhu[2], You-Wei Chen[2], Dian-Zhong Wang[2], Angelo Aguilar[3], Sergio Pichott[3], Sergio Cubelli[3], Carolina Rodríguez[3]
1. Department of Earth Sciences - University of Geneva, Geneva, Switzerland, 2. Institute of Geochemistry - Chinese Academy of Sciences, Guiyang, China, 3. Corporación Nacional del Cobre de Chile (CODELCO), Santiago, Chile

Magmatic systems associated with supergiant porphyry Cu deposits (PCDs) undergo a multi-million-year maturation in the deep continental crust as suggested by their increasing Sr/Y, La/Yb, and Eu/Eu* values through time. This evolution is recorded also by zircon, the most used mineral to infer magma fertility in porphyry Cu systems due to its resistance to hydrothermal alteration. Nonetheless, zircon analyses are almost always limited to one or two spots and miss conveying information on changing magmatic conditions during crystallization. This information is valuable to better understand the evolution of the mineralizing magmatic system and can be obtained through detailed core-to-rim profiles.

Here we report nanoSims trace element (Eu, Gd, Ce, Ti, U) profiles in 48 zircons from 8 samples of pre- to syn-mineralization magmatic rocks associated with the supergiant Llurimagua PCD (Ecuador). The 48 zircons span a range of ages between ~23 and ~6.3 Ma. Mineralization occurs at the end of this magmatic cycle, as is typical of supergiant PCDs, between ~7.28 and ~6.22 Ma, based on molybdenite Re-Os ages.

We have quantified Eu/Gd (proxy for Eu/Eu*) zoning by calculating linear slopes in the [Eu/Gd]-[core-to-rim distance] space. Zircons between ~23 and ~17 Ma display flat Eu/Gd slopes (-0.014 to +0.017), whereas after ~17 Ma they start to have increasingly negative Eu/Gd slopes culminating at ~8.5 Ma (-0.043). After ~8.5 Ma Ma Eu/Gd slopes rapidly increase, becoming strongly positive (Eu/Gd increasing towards the rim) up to +0.093 at ~6.3. Eu/Gd values in the cores display a specular opposite trend. We interpret these coupled trends as the result of zircon crystallization from a magma that has evolved at increasingly deeper levels between ~23 and ~8.5 Ma and of subsequent zircon crystallization from a magma fractionating abundant plagioclase and degassing SO_2 at shallow level, which ultimately produced the mineralization.

National-Scale Critical Mineral Assessments to Support Accelerating Discovery in Australia

Arianne Ford, Jonathan Cloutier
Geoscience Australia, Canberra, ACT, Australia

The production of critical and strategic minerals will be key for the transition to net zero and will require increased rates of discovery to support the provision of feedstock into the earliest stages of the supply chain. Given the importance of critical and strategic minerals to the Australian national economy, a series of national-scale mineral potential assessments have been undertaken as part of the Exploring for the Future program at Geoscience Australia to support exploration, search space reduction, and discovery. These assessments focus on delineating prospective belts or districts that indicate the presence of favourable mineral system processes, derived from both new and legacy geological, geophysical, and geochemical data sets acquired as part of precompetitive geoscience programs at the national scale. Assessments have been undertaken for sediment-hosted base metal, carbonatite related rare earth element, and iron oxide-copper-gold mineral systems in Australia. Additional national-scale assessments are currently underway for high-purity silica and other rare earth element mineral systems as part of Geoscience Australia's involvement in the Australian Critical Minerals Research and Development Hub. Such assessments are underpinned by high-quality precompetitive geoscience data and mineral systems understanding, which often comes from compilation of global-scale data sets, such as the Critical Minerals in Ores database, and understanding of the formation of mineral systems in a global context, which are then integrated using appropriate methodologies, including a newly developed hybrid data- and knowledge-driven approach for mineral potential mapping, to generate robust mineral potential maps that delineate the most geologically prospective areas. This presentation will highlight Geoscience Australia's work on delivering national-scale mineral potential assessments to support an improved understanding of Australia's critical minerals inventory.

Structural Controls on the Emplacement and Mineralization of Lithium-Bearing Pegmatites

Lot Koopmans[1], Brayden St Pierre[2], Richard M. Palin[1], Nicholas J. Gardiner[3], Laurence J. Robb[1, 4]
1. University of Oxford, Oxford, United Kingdom, 2. Aure Metals Inc, Oshawa, ON, Canada, 3. University of St Andrews, St Andrews, United Kingdom, 4. DSI-NRF Centre of Excellence, University of Johannesburg, Johannesburg, South Africa

Lithium-rich pegmatites are an important subset of LCT-type pegmatites, which host economic concentrations of the Li ore minerals spodumene and petalite. Significant work has focused on understanding the source enrichment processes required to form Li-enriched melts, leading to the development of two formation models: low-degree partial melting, and high-degree fractional crystallization. However, recent work has shown that neither, in their simple sense, is able to enrich a melt sufficiently to form economic lithium concentrations. In addition, the focus on petrogenesis subverts the importance of geologic structures on the spatial distribution of pegmatites, especially as pegmatites are known to travel significant distances from their source. To work towards a mineral system model for hard-rock Li deposits, the structural context needs to be better constrained.

We present a case study on the Archean Zulu pegmatite field (ZPF) in Zimbabwe. Based on field and petrographic observations, we propose pegmatites of the ZPF are emplaced along an active Riedel-type shear zone. We find the most voluminous pegmatites (G1) are emplaced parallel to R, with subordinate pegmatites (G2) along both R' and X structures. Additionally, G2 pegmatites retain a predominantly magmatic, petalite-dominated mineralogy, suggesting a rapid and contracted cooling history. In contrast, G1 pegmatites retain structural signatures of significant reworking during sinistral shearing, coinciding with extensive albitisation and conversion of petalite to spodumene-quartz intergrowths (SQUI). We propose that G1 pegmatites experienced a protracted cooling history due to emplacement parallel to an active shear zone, which promoted the formation of SQUI.

This result shows that pegmatites with different orientations have significantly different mineralogy and volume, despite being emplaced at the same time. We propose that exploration in pegmatite fields should focus on constraining the major structures along which pegmatites are emplaced, as this will aid in defining which orientations are more favourable targets for further work.

End-Orogenic, Granite-Hosted Uranium Mineralisation: A Solution from Namibia

Paul Nex, Judith A. Kinnaird
CIMERA School of Geosciences, University of the Witwatersrand, Johannesburg, South Africa

Unique world-class granite-hosted uranium mineralisation in Namibia includes deposits in the Central Zone of the Damara Orogen such as the Rössing and Husab mines. The Central Zone is characterised by a sequence of crustally derived, voluminous granitic intrusions over a period of c. 50 Ma, granulite to upper amphibolite facies metamorphism, a prominent NE orogenic trend, and elongate domes parallel to the orogenic trend.

Since the initial exploitation of the Rossing SJ deposit in 1976, these deposits have provided a conundrum in that the most recent felsic magmatism in the orogen has been recognised as the most uraniferous. This has since been confirmed by more detailed mapping and geochemical studies at Goanikontes, Valencia, Swakop River, and the Rossing mine area.

It has been difficult to construct a viable geological model in which incompatible elements such as uranium were sequestered into late-stage, post-kinematic anatectic partial melts if earlier melts had already been extracted from the source rocks. Recent metamorphic studies of migmatites involving LA-ICP-MS trace element determinations show that the partitioning of trace elements from muscovite dehydration melting may result in uranium-enriched restite including biotite and monazite. Subsequent higher temperature metamorphism, associated with biotite dehydration melting, will release uranium into the later partial melts thereby explaining the temporal occurrence of granite-hosted uranium deposits.Alternatively a second phase of anatexis involving earlier-formed granites as the source is a possibiloity.

As the local controls on the granite-hosted uranium mineralization are similar for all primary uranium deposits over the entirety of the Central Zone, the uranium enrichment must be a regional phenomenon and linked to orogen-scale processes, which we would suggest are fundamentally caused by the conditions of anatexis within the orogen.

Diamonds from a New Witwatersrand-Like Gold Deposit Show Evidence of Archaean Subduction in the Slave Craton, Canada

Rory M. Changleng[1], Andrea Pezzera[2], D. Graham Pearson[2], Richard Stern[2], Thomas Stachel[2], Fabrizio Nestola[3], Jesse Reimink[1]
1. Pennsylvania State University, University Park, PA, USA, 2. University of Alberta, Edmonton, AB, Canada, 3. Universita di Padova, Padua, Italy

Earth's ancient tracts of continental crust – the cratons – produce 90% of Earth's gold and platinum and almost all its diamonds. One key to the long-term production and preservation of these mineral resources was the creation of thick, buoyant 'keels' of lithospheric mantle that underpin the crust. These cratonic keels represent Earth's major diamond-forming factories, however, the exact mechanisms and timing of their formation remain highly controversial. This is partly due to the scarcity of direct mantle samples from the Archaean, when cratonic keels began to form. Detrital diamonds hosted within Archaean-aged sediments represent the only undisputed direct samples of the Archaean mantle, serving as valuable geochemical windows into the processes that produced Archaean cratonic keels.

Here we report the recovery of 25 microdiamonds from the Tree River region of the Slave Craton. These are hosted within a >2.83 Ga gold-rich (grades up to 36.3 g/t) conglomerate, which bears striking similarities to the world-renowned Witwatersrand deposit, Kaapvaal Craton, South Africa. Both deposits are quartz-pebble conglomerates formed during the 2.9–2.8 Ga "golden window" and contain abundant carbon stringers and detrital pyrite, along with scarce detrital diamonds.

Carbon and nitrogen isotope values for diamonds from Tree River ($\delta^{13}C$ -19.5 to -0.3‰, $\delta^{15}N$ -1.7 to +14.5‰) exhibit far greater variation than those from the Wits ($\delta^{13}C$ -5.7 to -3.0‰, $\delta^{15}N$ -0.5 to +2.7‰). One stone represents the oldest strongly $\delta^{13}C$ depleted diamond (~-19.5‰) and contains inclusions of garnet, clinopyroxene, and magnesite, pointing to an eclogitic paragenesis. Such extreme stable isotope variation requires the input of surface-derived material into the diamond-forming fluid reservoir within the Archaean Slave cratonic keel. This requires subduction/imbrication of bio-altered oceanic crust/carbonates at 2.9 Ga, possibly to 3.5 Ga. Invoking such geodynamic regimes during the Archaean has far-reaching implications for our understanding of economic deposit formation in the Archean.

The Nosib Polymetallic Mineral Deposit in the Otavi Mountain Land of Namibia – New Deposit Type or Variation on a Theme?

Jonathon L. Dugdale[3], Allison L. Dugdale[1], Elvis Akawa[2]
1. Discover Resource Services Pty Ltd, Secret Harbour, WA, Australia, 2. Huab Energy Pty Ltd, Windhoek, Namibia, 3. Golden Deeps Ltd, West Perth, WA, Australia

Nosib polymetallic (Cu-Ag) sulphide and secondary (Cu-Pb-V-Zn-Ag) oxide deposit is located in the Otavi Mountain Land (OML) district of Namibia. The OML includes major historic mines such as Tsumeb, 40km NW of Nosib (30Mt grading 4.3% Cu, 10% Pb, 3.5% Zn produced).

The majority of the mineral deposits in the OML are primary, polymetallic (Cu-Pb-Zn) sulphide deposits hosted by Otavi Group carbonate successions of the Neoproterozoic Damara Sequence. “Tsumeb type” deposits, which include the nearby Khusib Springs deposit 15km east of Nosib, are described as pipe-like orogenic-hydrothermal deposits hosted by collapse breccias, which transgress stratigraphy.

The Nosib deposit is unusual for the OML as primary sulphide mineralisation is hosted by pebbly quartzite and phyllite/shale of the Nosib Group at the base of the Damara Sequence.

The mineralisation at Nosib is up to 50m thick and generally stratabound within the pebbly quartzite. Sulphides morphology ranges from disseminated chalcopyrite, bornite and tennantite in the matrix of the pebbly quartzite beds, to massive sulphide veins and bands conformable and cross cutting bedding. Unlike other OML carbonate hosted deposits, collapse breccias are absent and sulphides display different, generally disseminated sulphide morphology. However, the sulphide mineralisation at Nosib has similar mineralogy and geochemistry to the nearby “Tsumeb type” Khusib Springs (Cu-Ag-Zn) sulphide deposit. This indicates a similar timing and fluid source to Khusib Springs and a similar orogenic / hydrothermal mineralising event which has likely utilised porosity in interbedded units and intergranular porosity which may have also included dissolvable carbonate minerals. Veined sulphides which transgress bedding confirm a structural orogenic-hydrothermal control on mineralisation, although veining may be remobilisation of the stratabound mineralisation.

The Nosib deposit is considered to be orogenic-hydrothermal in origin but displays an unusual stratabound – pebbly quartzite host rock control due to interbed and matrix porosity rather than carbonate collapse breccia control.

A Potential Mafic-Ultramafic Magmatic Phase of the Kunene Complex in Angola and Namibia and Its Ni-Cu-PGE Fertility

Ben Hayes[1], Lorenzo Milani[2], Jérémie Lehmann[3], Trishya Owen-Smith[3], Grant Bybee[1], Hielke Jelsma[4]
1. University of the Witwatersrand, Johannesburg, South Africa, 2. University of Pretoria, Pretoria, South Africa, 3. University of Johannesburg, Johannesburg, South Africa, 4. Anglo American, Johannesburg, South Africa

New discoveries of magmatic Ni-Cu-PGE sulfide deposits are critical to meet the growing demand for these metals. Voisey's Bay, hosted in a troctolite intrusion of the Nain Plutonic Suite (NPS) of eastern Canada, is a world class example of a magmatic nickel sulfide deposit. The NPS is an anorthosite-mangerite-charnockite-granite (AMCG) suite that formed between 1350 and 1290 Ma. Voisey's Bay formed early in the evolution of the NPS at 1333 Ma. The Kunene AMCG Complex (KC) in Angola and Namibia is the largest suite of its kind and formed between 1500 and 1375 Ma at the southwestern margins of the Congo Craton. The KC is also spatially associated with a subordinate mafic-ultramafic phase that is partly exposed along its western and southwestern margins, and which has the potential to host magmatic Ni-Cu-PGE sulfide deposits. This mafic-ultramafic phase comprises a range of lithologies (dunite to gabbrodiorite) with typical coarse-grained cumulate textures hosted in relatively small-sized intrusions (~10 km^2 surface extent), typical of conduit or chonolith-like magmatic plumbing systems which are known to host some of the globally significant magmatic sulphide deposits. The mafic-ultramafic phase is locally sulfide mineralised which has led to past exploration in the region by Anglo American, African Nickel, and Gevale. Field evidence, along with strontium-neodymium and sulfur isotopic data, show that these intrusions underwent crustal contamination, highlighting the possibility that external crustal sulfur was added to the parent magmas during their emplacement in the crust. Geochronological and isotopic research to fingerprint the timing and potential genetic connection between the mafic-ultramafic phase and the KC is ongoing. However, the geological setting of the mafic-ultramafic phase, its lithological makeup, and styles of mineralisation, are broadly reminiscent of that described at Voisey's Bay, highlighting the Ni-Cu-PGE fertility of the KC and its potential as a future metallogenic province in Africa.

The Discovery and Geology of the Boundary Zone Zn-Pb-Ag Deposit, Yukon, Canada: New Perspectives on Mineralizing Processes in Sediment-Hosted Base Metals Deposits

Jack E. Milton[1], Haruna M. Grema[2, 3], Joseph M. Magnall[2], Sarah A. Gleeson[2, 3]
1. Fireweed Metals Corp., Vancouver, BC, Canada, 2. GFZ German Research Centre for Geosciences, Potsdam, Germany, 3. Institute of Geological Sciences, Freie Universität, Berlin, Germany

Fireweed Metals' new discovery, the Boundary Zone Zn-Pb-Ag deposit within the Selwyn Basin, Yukon Canada, a classic area for Economic Geology because it is the birthplace of the Selwyn SEDEX model for clastic dominated, or sediment hosted massive sulphide deposits. New insights are possible from Macmillan Pass where over 45,000 m drilling has been done between the Boundary Zone discovery and the Tom and Jason deposits.

Boundary Zone was originally discovered in the 1980s as low-grade zinc in veins and breccias. A significant new zone of stratiform laminated to massive sulphide was discovered in 2020, under cover of 5-10 metres of tills, using geological modeling, soil geochemistry, and gravity. A new resource has been defined, establishing the Macmillan Pass project as one of the world's largest undeveloped zinc deposits.

At Boundary Zone, stratiform mineralization is present at several stratigraphic intervals, spanning the late Ordovician to late Devonian. Stratiform mineralization occurred early within diagenesis as the replacement of diagenetic barite layers and biosiliceous mudstones. Stratiform mineralization is associated with rapid lateral thickness and facies changes and abundant, poorly sorted, mud-rich diamictites, interpreted as synsedimentary fault scarp slumps. Observations of sphalerite and pyrite psuedomorphs after barite support the subsurface replacement model for zinc mineralization that has challenged the SEDEX paradigm of syngenetic mineralization.

Stage two zinc mineralization at Boundary Zone occurred post-lithification, as late as the Cretaceous, Cordilleran deformation. This phase is characterized by coarse sphalerite-siderite-pyrite in veins and breccias, and also as replacement of the clasts and matrix of coarse clastic rocks. A large proportion of veins show evidence for the opening of veins within pre-existing tectonic stylolites, exploiting fracture permeability. Stage two vein mineralization accounts for a significant volume of mineralization at Boundary Zone and appears more abundant adjacent to late faults that cross-cut folding.

The Geology of the Mingomba Deposit in the Zambian Copperbelt

Shane Nielsen, Sitwala Mulonda
Kobold Metals, Lusaka, Zambia

Mingomba is a reduced facies, sedimentary-rock-hosted chalcocite-dominated deposit hosted in the Copperbelt Orebody Member (COM, aka Ore Shale) of the Lower Roan Subgroup, the lowest part of the Neoproterozoic Katangan Supergroup. On a regional scale, the deposit is part of a mineralized system continuous over ~15 kilometers from the Konkola mine in the southeast to the Musoshi mine in the northwest.

The >50,000 meters of diamond drilling at Mingomba from March 23 through April 24 has demonstrated significant lateral zonation in copper mineralogy. The drilling has identified contiguous grades averaging over 5% copper with a median true-thickness of 8.5 metres along a north-east trending zone, over two kilometers long and approximately 1.5 kilometers wide at depths of 1,150 metres to 1,650 metres. The Mingomba system remains open to the north and east.

The highest grade mineralisation is predominantly developed directly at the interface between reduced and oxidized elements of the stratigraphy. Chalcocite mineralisation occurs to the top of the Copperbelt Orebody Member, suggesting that the system is sulfur limited and the ore-fluid was extensively fertile. A second mineralized horizon is developed in the more shallow Chambishi dolomite, potentially correlative of the Mines Series in the DRC.

We hypothesize that the zone of chalcocite enrichment at Mingomba is controlled by an array of syn-sedimentary growth faults striking north-west and north-east. This potential fault array could produce a series of half-grabens or sub-basins, controlling the distribution of both the copper-bearing ore-fluid, and the distribution of the reductive rocks necessary for copper mineral precipitation.

The recent drilling campaign has demonstrated that Mingomba's size, and high-grade chalcocite-style of mineralisation, are comparable to Ivanhoe's Kakula mine in DRC. Mingomba, like Kakula, highlights the residual exploration prospectivity of the Central African Copperbelt.

The Eureka Gold Deposit: Latest Gold Discovery in an Emerging Gold District in the Northern Zone, Damara Belt, Namibia: Alteration, Mineralization and Structural Controls

Wynand Slabbert[1], Dave Underwood[2]
1. Osino Gold Exploration and Mining, Windhoek, Namibia, 2. Osino Gold Exploration and Mining, George, South Africa

Eureka is interpreted to be an orogenic gold deposit in the northern zone of the Damara belt. The host rocks belong to the Okonguarri Formation which also hosts the mineralization at the Otjikoto Gold Mine, 150km to the northeast. The formation comprises interbedded marbles and siliciclastic turbidites (calcareous sandstones, siltstones, and shales) metamorphosed to the greenschist facies within the northern zone.

During a target generation workshop in 2019, the Osino team identified this part of the Damara belt, where it bends sharply from northward strike to eastward, as a priority structural target. After systematic mapping, soil and rock chip sampling, a small outcropping gold anomaly was located in association with large garnet porphyroblasts and hematite staining in turbidites. The first drill hole was completed in late 2022. The exploration team has since drilled 36 additional diamond holes primarily into the main outcropping southern shoot. Detailed district work has identified numerous additional targets.

The coarse-grained gold at Eureka is associated with multiple fluid pulses emplacing quartz+carbonate+sulphide (pyrrhotite, pyrite and chalcopyrite) veins. Early bedding parallel quartz veins are often brecciated and displaced by later carbonate and sulphide-rich fluids, which carry higher gold grades. The early quartz veins are controlled by flexural slip between lithological units with competency contrasts that occurred during the D1 and D2 regional shortening events of the Damaran orogeny.

Wall rock alteration is dominated by an assemblage of porphyroblastic garnets and grunerite amphibole, which forms a prominent halo around the mineralization. There is no mapped intrusive body near the mineralization to explain this unusual alteration as skarn related and the source of iron to produce this iron-rich assemblage is also not evident. This provides a challenge in determining fluid source, ligand and chemical controls for the multiple fluid pulses that precipitated the gold mineralization at Eureka.

Temporal Evolution of Miocene-Pliocene Magmatic Suites Fertile for Porphyry Cu Deposits: Preliminary Results from the Tres Cerrillos Porphyry Cu Prospect, Ecuador

Hugo Carrasco[1], Massimo Chiaradia[1], Angelo Aguilar[2], Sergio Pichott[2], Sergio Cubelli[2], Ana C. Rodríguez[2], Carlos Urrutia[2]
1. University of Geneva, Geneva, Switzerland, 2. Corporación Nacional del Cobre de Chile (CODELCO), Santiago, Chile

The Tres Cerrillos prospect is located in the foothills of the Western Cordillera in northwestern Ecuador. Magmas associated with porphyry-style mineralisation in the prospect present typical calc-alkaline arc signatures. Geochemical indicators (based on whole-rock geochemistry and zircon trace elements) from porphyries and phaneritic intrusions show a fertile signature for porphyry Cu systems, where whole-rock REEs show high enrichment of LREEs, variable depletion of MREEs and HREEs, and lack of Eu negative anomaly. That suggests those magmas experienced a deep crustal evolution, with amphibole (± garnet) fractionation, prior to emplacement in the upper crust, where they exsolved mineralising fluids.

LA-ICP-MS U-Pb zircon geochronology on these intrusions combined with whole-rock geochemistry reveal a temporal chemical evolution of REEs and their fertility proxies (Sr/Y, La/Yb, Dy/Yb, V/Sc, and Zr/Y), defining four magmatic cycles during the Mid-Miocene to Pliocene, with roughly similar behaviour at ~11, ~10.5, ~8.8, and ~2.8 Ma. In each cycle, the earliest intrusions (either a pluton or a porphyry) show the flattest REE spectrum and the lowest fertility signal, followed by the emplacement of porphyries with the strongest depletion of HREEs and MREEs and the highest fertility signal. Furthermore, in the case of the magmatic cycle at ~11 Ma, where we collected more data, HREEs and MREEs get progressively less depleted as the porphyritic intrusions become younger, but still overlap the fertility range (e.g., Sr/Y values between 50-150).

We suggest this magmatic evolution from unfertile to fertile behaviour could reflect a tectonic stress regime cyclically changing from stronger compression to more neutral stress. One hypothesis for this evolution could be the subduction of the topographically irregular Carnegie Ridge beneath the Ecuadorian continental margin. Thus, fertile magmas were transferred from lower to upper crustal levels at different epochs during the Mid-Miocene to Pliocene.

Is There a Link Between Orogenic Collapse and Lithium-Bearing Pegmatites?

Tarryn K. Cawood[1], Jamie Cutts[1], Antoine Godet[2]
1. Geological Survey of Canada, Ottawa, ON, Canada, 2. Geological Survey of Canada, Quebec City, QC, Canada

Lithium-bearing pegmatites and S-type granites are widely recognized as the products of partial melting of supracrustal rocks at depth. Such melting is commonly attributed to crustal thickening during orogenesis. However, pegmatites and granites are typically emplaced during the late stages of orogeny, and many pegmatite-bearing regions also display evidence for syn- to post-orogenic extension. Such evidence includes thinned lithosphere, low-angle detachment structures, metamorphic core complexes, and basins infilled with coarse clastic sediments, along with steep metamorphic gradients and gently dipping structural fabrics, likely signifying ancient detachment faulting. Examples include the Orange River Pegmatite Belt of South Africa and Namibia, pegmatites in the Monashee Metamorphic Complex of the Canadian Cordillera, and possibly pegmatites in the Himalayas, Cote d'Ivoire, Elba Island in Italy, the Lewisian Gneiss Complex of Scotland, and the western Superior craton of Canada.

We argue that extensional collapse of thickened orogenic crust provides a key environment for the formation of lithium-bearing pegmatites. In this model, earlier subduction and/or crustal stacking during accretion and collision introduces fertile, hydrous sedimentary and volcanic source rocks to depth. Subsequent slab breakoff and/or delamination drives asthenospheric upwelling, with the associated heat pulse possibly inducing partial melting in the overlying orogenic crust and forming some early granites and pegmatites. Breakoff or delamination also gravitationally destabilizes the thickened orogenic belt. The resultant orogenic collapse involves rapid tectonic exhumation of hot, deep crust, leading to decompression melting and the formation of the main generation of granites and pegmatites, while the extensional regime and associated faults and shear zones enable melt mobility and emplacement in the mid- to upper crust. In many instances, extensional collapse is accompanied by a transition to transform tectonics, and coeval or later strike-slip faults may provide additional transport pathways and structural traps for the pegmatite-forming melts.

Petrography and Geochemistry of the Okohongo Copper-Silver Deposit, Kaoko Belt, Kunene Region, Namibia

Abner Nghoongoloka[1], Ryohei Takahashi[1], Pearlyn Manalo[1], Andrea Agangi[1], Hinako Sato[1], Manuel Nopeia[3], Nigel Blamey[4], Tavis Enno[4], Rob Bowell[5]
1. Graduate School of International Resource Sciences, Akita University, Akita, Japan, Akita, Japan, 2. Geological Survey of Namibia, Windhoek, Namibia, 3. Center for Regional Revitalization in Research and Education, Akita University,, Akita, Japan, 4. Department of Earth Sciences, Western University,, London, ON, Canada, 5. Department of Earth Science Miller Hall, Queens University, Kingstone, ON, Canada, 6. SRK Consulting, Cardiff, United Kingdom

The Okohongo copper and silver deposit has estimated resources of 10.2 million tonnes, with grades of 1.12% Cu and 17.7 g/t Ag. The deposit is located in the Opuwo district, Kunene region of Namibia, and is hosted by the ca. 760 Ma Devede and Beesvlakte Formations of the Ombombo Subgroup in the Kaoko Belt. An orebody is composed of secondary copper oxide and sulfide mineralization located along N-S disharmonic folds and NW-trending shear zones. The mineralization occurred as (1) bedding-parallel quartz-calcite veins containing chrysocolla, malachite, shattuckite, and chalcocite, which are concordantly hosted by shale; (2) semi-massive chrysocolla, malachite, shattuckite, and chalcocite; and (3) chalcocite-bearing quartz-calcite-albite stockworks, hosted in the brecciated, interbedded dolostone and shale with hematite and goethite cement. Bedding-parallel quartz-calcite veins have been observed from the surface to 50 m, semi-massive ore from 114 to 131 m, and stockwork ores occur at 200 m depth. Post-Archean Australian Shale (PAAS)-normalized REE and Y patterns of the quartz-calcite veins show a flat trend, while those of the semi-massive and stockwork ores are depleted of light REEs. Sulfur isotope ratios ($\delta^{34}S_{CDT}$) of chalcocite from the semi-massive and stockwork ores are +5.6‰ and +8.3‰, respectively. The quartz from quartz-calcite veins and the semi-massive ores hosts primary liquid and vapor inclusions with homogenization temperatures and salinity ranging from 95 to 147°C, 8.0 to 18.7 wt % NaCl eq., and 118 to 214°C, 9.1 to 17.9 wt% NaCl eq., respectively. Weighted mean of bulk gas compositions of the fluid inclusions in quartz in the quartz-calcite vein is 98.98 mol% H_2O, 0.75 mol% N_2, 0.25 mol% CO_2, 0.02 mol% CH_4, and traces of Ar, H_2S, and He. The fluid inclusion gas compositions indicate transitional shallow meteoric and magmatic signatures.

The Magmatic-Hydrothermal Nature of Granite-Hosted Uranium Mineralization on the Example of the Omahola Project in the Damara Orogen, Namibia

Alex Otto
Deep Yellow Ltd., Subiaco, WA, Australia

Within the ~540-500 Ma Damara orogen's southern Central Zone lies the world-renowned leucogranite-hosted Roessing and Husab deposits. Uranium mineralization occurs alongside sheeted leucogranite intrusions, classified into six types (A-F) by Nex et al. (2001), with types D and E being the most uranium-rich based on field relationships.

A new interpretation of existing research and new data highlight the role of magma-derived hydrothermal fluids for uranium mineralization. Field evidence indicates that leucogranite melts preferentially intrude the cores of domes (such as Ida Dome) and minor anticlinal structures, as seen at the Ongolo and MS7 deposits. Most leucogranite intrusions exhibit similar textural styles and compositions but display significant differences in relative ages across the Omahola area. The observed crosscutting relationships between different sets of leucogranites and their deformed host rocks are characteristic of highly ductile and heterogeneous host rocks with significant competency contrasts. Consequently, classifying leucogranites into different types regarding uranium mineralization may not be relevant for the study area. Another notable regional observation is the widespread occurrence (at 1-10 km scale) of skarns within the Rössing formation marbles. The distal and earlier skarn assemblage is dominated by wollastonite, scapolite, and clinopyroxene, while the proximal garnet-rich skarn is hosted in gneisses of the Khan formation and spatially associated with uranium mineralization (e.g., MS7, Ongolo, Inca, Garnet Valley). Pegmatite pods within fine-grained granite and quartz veins suggest a significant role of hydrothermal fluids during uranium mineralization. Magmatic fluids within the leucogranite melt percolated throughout the melt, concentrating uranium towards the Rössing-Khan boundary, acting as a fluid outlet. At Inca, uranium mineralization is part of an evolving hydrothermal system associated with skarn alteration, including hedenbergite, Fe actinolite, and magnetite. On a larger scale, the outermost expression of this mineral system is the Shiyela magnetite/hematite deposit approximately 25 km south of the leucogranite-hosted Ongolo deposit.

Iron Oxide-Copper-Gold Deposits: A One-Size-Fits All Genetic Model Does Not Work

Adam C. Simon
University of Michigan, Ann Arbor, MI, USA

There are enough similarities among iron oxide-copper-gold (IOCG) deposits globally that they would seem to fit a single genetic model. However, genetic models in the literature often take twists and turns to invoke processes for which there may or may not be evidence in the rocks. IOCG deposits are iron and copper rich, sulfur poor, and most contain gold, but some do not. IOCG deposits in many districts are spatially and temporally associated with igneous rocks, but not always. IOCG deposits are sometimes hosted in igneous rocks, but not always. IOCG deposits formed in orogenic and post-orogenic settings, and back-arc extension seems to have been particularly important for some districts. Drill core reveal that IOCG mineralization in some deposits transitions with depth to iron-rich, sulfur-copper-gold-poor mineralization. Stable copper, hydrogen, iron, oxygen, and sulfur isotopes fingerprint a magmatic source for the ore fluids in some deposits, but seawater-derived brines in others. Regional sodic-calcic alteration is observed in some districts, but not all. Petrographic observations of drill core indicate that some IOCG deposits, but not all, record multiple episodes of fluid infiltration with hematite followed by magnetite followed by copper-iron-sulfide in some deposits. Some deposits contain mushketovite, which is best explained by the reduction of primary specular hematite and requires at least two generations of fluid. The trace element composition magnetite, which modally dominates IOCG deposits, sometimes varies systematically with depth, and petrographic observations of magnetite textures reveal multiple episodes of hydrothermal dissolution and reprecipitation. In this presentation, I will review the characteristics of IOCG deposits from different districts and tectonic settings and try to frame a genetic model that can be adapted to explain their formation globally.

Controls on Germanium Enrichment in Zn-Pb Sediment-Hosted Mineral Deposits

Juan D. Bello Rodriguez[1], Daniel Gregory[1], Merilie A. Reynolds[2], Denis Fougerouse[3]
1. University of Toronto, Toronto, ON, Canada, 2. Northwest Territories Geological Survey, Yellowknife, NW, Canada, 3. Geoscience Atom Probe Facility, Curtin University, Perth, WA, Australia

Germanium (Ge) is commonly recovered as a by-product of zinc (Zn), which poses potential future supply risks, defining it as a critical mineral. Globally, Ge is frequently sourced from sediment-hosted Zn-Pb deposits. However deposit-scale distribution and enrichment mechanisms within these deposits remain poorly understood. Consequently, we examine the Ge-bearing Zn-Pb (Ag) Prairie Creek sediment-hosted deposit located in the Northwest Territories, Canada, using macro- to nano-scale analyses. We aim to refine the genetic model of this deposit and identify key factors influencing Ge enrichment.

This deposit exhibits two main mineralization styles: (1) stratiform and (2) quartz-carbonate vein, hosted mainly in dolostone and shales. Using whole-rock geochemical analysis, we determined that only the stratiform style contains economic values of Ge (up to 300 ppm). The results show a strong correlation between Ge and Zn, indicating that Ge content is associated with Zn-bearing mineral phases. At least two generations of sphalerite (ZnS; Sp I and Sp II) are present in the stratiform mineralization. Analysis of sphalerite chemistry using EMPA and LA-ICP-MS reveals significant Ge enrichment (up to 2,600 ppm) exclusively in Sp I. In contrast, Sp II has lower Ge concentrations (0.5 to 100 ppm). Furthermore, EMPA elemental maps show a spatial correlation between Cu and Ge within sphalerite crystals (2:1 ratio). This suggests a possible relationship between these two elements in their substitution mechanisms.

We will use electron backscatter diffraction (EBSD) and atom probe tomography (APT) to characterize the crystallographic and atomic distribution of Ge in sphalerite. In addition to isotopic analyses, this work will give insight into the sources of metals and sulfur, the nature of ore-forming fluids, and conditions for metal precipitation. This research will contribute to a better comprehension of Ge behavior in hydrothermal fluids in sediment-hosted Zn-Pb deposits, guiding mineral exploration and mining strategies targeting these deposits.

Paragenetic Evolution of Sn-dominant Lode Systems in the Pool Mining District, Cornwall

Kiara C. Brooksby[1], Robin K. Shail[1], Jens C. Andersen,[1], Jamie J. Wilkinson[3, 2]
1. Camborne School of Mines, University of Exeter, Exeter, United Kingdom, 2. Imperial College London, London, United Kingdom, 3. Natural History Museum, London, United Kingdom

SW England hosts a world-class, W-Sn-Cu-As-Zn-Pb-Li ore field associated with the Early Permian granites of the Cornubian Batholith. In the Pool Mining district, which includes South Crofty Mine, magmatic-hydrothermal mineralisation is largely associated with steeply dipping ENE-WSW striking extensional fault systems. Recent drilling by Cornish Metals Inc. has included boreholes that intercept the principal South Crofty lodes and the Great Flat lode. This study provides an updated and in depth paragenetic sequence through core logging and optical microscopy, QEMSCAN imagery, hot cathodoluminescence, and microthermometry. These new data discoveries combined with historic descriptions provide a re-evaluation and new geological model of tin lode systems within the region.

These lode systems are hosted by the Carn Brea Granite and the overlying Devonian metasedimentary and metabasic igneous rocks of the Mylor Slate Formation. The paragenetic sequence is split into six main episodes: (1) felsic sheet intrusions, (2) feldspar and quartz veins (floors), (3) quartz-tourmaline breccias and veins, (4) quartz-chlorite veins, (5) fluorite veins, and (6) cross-course quartz veins. Cassiterite was predominantly associated with the quartz-tourmaline paragenesis (reflecting deeper sampling levels) but, in some cases with the quartz-chlorite paragenesis. The quartz-tourmaline paragenesis is multiphase and includes examples of cassiterite precipitating before, during, or after tourmaline. In contrast the Great Flat lode has minimal occurrences of stage (2) with the quartz veins showing high degrees of sheering and stage (5) fluorite veins. Additionally, although the Great flat lode has numerous occurrences of tourmaline breccias they have no associated cassiterite, unlike in historic reports. Instead, cassiterite occurs within cassiterite only or cassiterite-fluorite veins which cut through early tourmaline breccias and host rock. Hot cathodoluminescence and microthermometry of cassiterite within the Great Flat lode indicates the lodes were formed by multiple generations of fluids.

Primary Rare Earth Elements Mineralization of the World-Class Mt Weld Carbonatite Deposit, Western Australia

Arthur V. de Oliveira[1], Denis Fougerouse[1], Peter D. Kinny[1], Luc-Serge Doucet[1], Anderson M. dos Santos[2], Arthur P. Deditius[2], Ganesh Bhat[3]
1. Curtin University, Perth, WA, Australia, 2. Murdoch University, Perth, WA, Australia, 3. Lynas Rare Earths, Perth, WA, Australia

Among the world's largest rare earth elements (REE) deposits is the Mt Weld deposit in Western Australia, where mining activities are currently focused on the laterite formed above a carbonatite bedrock. Because the carbonatite is buried by transported sediments (~20–40 metres thick) and regolith, a recent 1-km-long diamond drill core has provided a novel opportunity to explore the deep lithology and potential mineralisation of the Mt Weld carbonatite. Drilled in the base of the current pit, the first 63 metres of the drill core comprise of regolith and transitional carbonatite and then changes to carbonatite rock. Core logging indicates that the carbonatite body (diameter ~4 km) is zoned, with calcite carbonatite and brecciated phoscorite in the annulus and a ferroan dolomite carbonatite core. In the fresh rock, REE mineralisation is concentrated in the ferroan dolomite carbonatite, with an average of 2.12 wt % REO, higher than >80% of global REE deposits. The dominant REE ore minerals show a clear progression with depth of the core, with monazite predominantly in the fresh dolomite carbonatite from ~70 to 514 m and REE fluorocarbonates from 514 to 806 m, and Ba-Sr-Ca-REE carbonates are observed from 914.6 m to the end of the hole. Monazite occurs as small (<15 μm) fibrorradial aggregates replacing apatite and ferroan dolomite, whereas REE-fluorocarbonates are generally >30 μm, bladed aggregates interstitial to the dolomite. Replacement features of the ore minerals show that dissolution and reprecipitation were common processes during the REE mineralisation at Mt Weld, with porosity frequently associated with the mineralisation throughout the carbonatite and indicating that REE ore minerals precipitated from carbohydrothermal fluids.

A Revised Model for High Grade Ni-Co-As-Ag-U-Sb-Bi Deposits: The Product of Heating and Oxidation of Marine Evaporite Brines and Hydrocarbons from Intracratonic Basins

Jacob J. Hanley[1], Corwin Trottier[1], Jordan Burke[1], Luke Ootes[2], Zoltan Zajacz[3], Mostafa Fayek[4]
1. Department of Geology, Saint Mary's University, Halifax, NS, Canada, 2. British Columbia Geological Survey, Victoria, BC, Canada, 3. University of Geneva, Geneva, Switzerland, 4. University of Manitoba, Winnipeg, MB, Canada

High-grade, polymetallic, “five-metals” type hydrothermal vein systems represent historically important sources of silver and uranium (e.g., Cobalt, Ontario, Canada: ~460 Moz Ag; Eldorado deposit, Northwest Territories, Canada: ~13 Moz Ag; 6,000 t U_3O_8) but are now repositioned as a revelant target for exploration programs focused on a much wider variety of critical metals necessary for the transition to a post-carbon economy. In the Northwest Territories of Canada, deposits of this classification formed in the Proterozoic show widely contrasting grade and tonnage characteristics. A 5-year study revisiting these deposits and applying a variety of microanalytical methods (e.g., fluid inclusion microthermometry, stable O isotope analysis of vein minerals by SIMS, single fluid inclusion analysis by LA-ICPMS and Raman spectroscopy) leads to four key revisions to the genetic model: (i) the onset of mineralisation involved mixing and oxidation of heated, immiscible hydrocarbon fluid and marine evaporite brine introduced into structures within crystalline basement; (ii) whereas basement fluids (ancient groundwaters) and basinal brines contained only sub-ppm concentrations of ore metals, hydrocarbons were highly enriched (10s to 1000s ppm) in all relevant ore metals in this association; (iii) brines and metal-rich hydrocarbons were sourced from black shale units in former overlying intracratonic basins (e.g., Hornby Bay, Thelon, Athabasca); and (iv) fluid inclusion chemistry differentiates small, poorly-mineralized deposits that formed deeper in the basement rocks, farther from overlying sedimentary metal sources from shallower, much larger deposits that formed immediately below mature, thick basin sequences. Importantly, while basement magmatic activity may serve as a heat source driving fluid circulation, the metal endowment of this deposit style appears to be linked to only the protracted metallogenic evolution of overlying intracratonic sedimentary basins. Fluid chemistry may be used as an exploration criteria in other examples of this mineralization-style globally to differentiate between productive and unproductive "five-element" systems.

Indium in Mine Wastes: Geochemistry, Mineralogy, and Microbiology

Olivia Mejías, Laura Jackson, Anita Parbhakar-Fox
Sustainable Minerals Institute, The University of Queensland, Brisbane, QLD, Australia

Indium (In) plays an essential role in transitioning to a low-carbon economy, primarily because it serves as a key component in solar panels. Understanding the (bio)geochemistry of indium in surface environments is essential for developing effective extraction strategies. This study aims to understand indium cycling and mobility in two distinct geologic mine-waste environments, serving as an essential first step toward designing a remining process. We collected solid and aqueous mine waste samples from the Mt. Morgan (Au-Cu VHMS deposit) and Baal Gammon (Sn-Cu granite-related deposit) mines in Queensland, Australia, and subjected them to geochemical, mineralogical, and microbiological analyses. At Baal Gammon, waste rock exhibited indium concentrations above 350 ppm, with sphalerite, chalcopyrite, and stannite-kësterite identified as primary In-bearing minerals. The highest concentrations of indium are at micro- and nano-features in grain and mineral dislocation boundaries. Acidic waters contained up to 73 µm/L In, alongside elevated fluoride levels. In contrast, indium concentrations at Mt. Morgan reported indium up to 3.5 ppm in waste rock, with sphalerite and chalcopyrite as the predominant In-bearing minerals. Acidic waters here contained up to 5 µm/L In, with moderate-low fluoride levels. In both sites, indium displayed positive correlations with Cu and Ag in waste rocks, As in sediments, V in precipitates, and Pb, As, and fluoride in acidic waters. The study identified Baal Gammon as the indium-rich sulfidic waste system, along with a wide range of In-rich secondary minerals, including biogenic precipitates associated with *Alicyclobacillus fastidiosus* and *Ferroacidibacillus organovorans*. Indium cycling involves a complex interplay of anthropogenic and (bio)geochemical processes dictating its distribution and mobility. Findings suggest strategies such as selective flotation followed by leaching, prioritising As, Pb, and V removal, alongside a stepwise Fe precipitation process. Further research into the microbial communities is imperative to assess the potential for biomining and/or bioremediation of indium.

Micas as Tracers of Tungsten Mineralization Case Study: Santa Comba and Fontao Deposits NW Spain

Candela Pita Díaz[1, 2], Elena Crespo Feo[2], Sean Mc Clenaghan[3], Fernando Tornos Arroyo[1]
1. IGEO-CSIC, Madrid, Spain, 2. Complutense University of Madrid (UCM), Madrid, Spain, 3. Trinity College Dublin (TCD), Dublin, Ireland

The study focuses on the use of micas as tracers for tungsten mineralization in the Santa Comba and Fontao mines located in NW Spain. These mines are each situated within a poly-intrusive epizonal complex, conforming to the Endogranite-Stockscheider-Exogranite model. The deposits show evidence of multiple mineralizing events, each characterized by unique paragenesis and geochemical conditions. In addition to tungsten ores, muscovite and quartz are prevalent minerals across all types of mineralization.

To understand the differences in the mica composition from each mineralization and from barren intrusions, we have assessed the geochemical variations in the muscovite. Variations in trace element content, such as Li, Rb, Cs, and W, reveal differences in the sources, temperatures, and fluid composition of the fluids involved as well as the effect of fluid-rock interaction. These differences can help distinguish between mica formed in mineralized zones and those found in barren intrusions, potentially leading to a better understanding of the formation and localization of ore deposits.

Using LA-ICPMS, a systematic approach based on mica trace chemistry was used to delineate various events leading to mineralization. Three distinct mica groups were identified in each deposit, each distinguished by specific petrographic features and/or location. Analysis of their trace element composition revealed significant variations between mica groups, although homogeneous signatures were identifiable within each group.

The study highlights the relevance of trace element analysis in ubiquitous minerals such as micas as a potent tool for understanding the mineralizing processes and alteration within complex geologic settings. Moreover, it suggests the potential of this method in relating distal veins to a main magmatic body. Igneous micas were found to harbor notable concentrations of W and Sn, which could be mobilized during later hydrothermal alteration, potentially leading to ore deposit formation.

Critical Metals in Base Metal Mineralization of the Otavi Mountain Land and the Kunene Region, Namibia

Ester Shalimba[1], Rob J. Bowell[2, 3], Abner Nghoongoloka[4], Stephanie Lohmeier[5], Matthew Leybourne[2], Helke Mocke[4]
1. Department of Geosciences, University of Namibia, Keetmanshoop, Namibia, 2. Department of Geological Sciences, Kingston, ON, Canada, 3. SRK Consulting, Cardiff, United Kingdom, 4. Geological Survey Of Namibia, Windhoek, Namibia, 5. Technical University of Clausthal, Clausthal-Zellerfeld, Germany

Carbonate-hosted base metal deposits, notably those in the Otavi Mountain Land (OML) of Namibia, particularly at Khusib and Tsumeb mines, are renowned for their complex unique mineralogy rich in critical metals like Cd, Co, Ni, Ga, Ge, Mo, Se, and In. Both deposits are dominated by tennantite, galena, and associated pyrite with accessory bornite, chalcocite, enargite, chalcopyrite, and sphalerite with accessory Ag-Pb-Ge-sulfosalts. At Khusib Springs mine tennantite and chalcopyrite are the most important carriers of trace elements such as Zn, Ag, Sb, and Pb and Ag, respectively. Additionally, Khusib Springs ore reveals a diversity of Ag-bearing minerals. Conversely, Kombat mine has a similar macro-sulfide assemblage with chalcopyrite as the primary carrier of Ga, Ge, and In but negligible Ge-sulfides or complex sulfosalts and higher proportions of acanthite and molybdenite. Mineralization in the Kunene region is older than that in OML and displays significantly lower proportions of Zn-sulfides, higher proportions of Cu-sulfides, Co, Ni, and Mo. Fluid-inclusion geochemistry and mineral geothermometry indicate Kombat and Kunene mineralisation were formed at higher temperatures and from lower salinity solutions than those responsible for mineralization at Tsumeb and Khusib Springs. Geochemical data reveal a positive correlation occurs between Cd, Ga, Ge, In, and Zn concentrations and between Ag, Mo. An antipathetic relationship may occur between Co and Ga and Ge. Reconnaissance Zn isotope data (–0.19 to 0.14‰) delineate Tsumeb ore as typical Zn-rich massive sulfide ore, with values overlapping with Irish MVT deposits (–0.17 to 1.33‰). Cu isotopes are comparable to isotopic data of hypogene mineralisation for Tsumeb ore, and a potential magmatic source as well as a hydrothermal fluid for Kunene mineralisation. These variations reflect differences in fluid sources, temperatures of hydrothermal fluids, and multi-mineralisation events, with implications for exploration targeting different critical metals based on their relationships with major metals or mineralogy.

Investigating the Role of Country-Rock Assimilation on the Mineralisation of the Platreef, Turfspruit Farm, Northern Bushveld Complex, South Africa

Alessia Barbieri[1], Katie McFall[1], Richard Herrington[2], Iain McDonald[3], Bothepha Phetlhu[4], Tim Dunnett[4]
1. Department of Earth Sciences, University College London, London, United Kingdom, 2. Natural History Museum, London, United Kingdom, 3. School of Earth and Ocean Sciences, Cardiff University, Cardiff, United Kingdom, 4. Ivanhoe Mines Ltd, Mokopane, South Africa

The Northern Bushveld Complex, South Africa, hosts the world's largest deposit of platinum-group elements (PGEs), known as the Platreef. PGEs are a group of metals essential for autocatalysts, fuel cell technologies, and green hydrogen production. The down-dip extension of the Platreef in the Turfspruit Farm has an indicated mineral resource of 2 g/t of 4PE (Pt + Pd + Rh + Au), 0.16 wt% Cu grade, and 0.34 wt% Ni grade.

The thickness of the individual stratigraphic units on Turfspruit vary, and the relationship between Ni, Cu, and PGE grades is unclear and often decoupled, although correlated on a deposit scale. The thickness of mineralization is also heterogeneous, averaging 14.99m at a 2g/t 4PGE cutoff, but ranging from 2m to 93m. Traditional orthomagmatic ore-forming processes cannot fully explain the distribution of mineralization or its heterogeneous thickness.

The Turfspruit host lithologies intrude and assimilate the reactive sedimentary country rocks of the Transvaal Supergroup. These lithologies vary along strike and include dolomites, shales, banded iron formations (BIFs), and silicate sedimentary rocks. The assimilation of these lithologies formed a complex 'Footwall Assimilation Zone' (FAZ) of varying thickness that underlies the mineralization-bearing units, and consists of, hornfels, calc-silicate, and clinopyroxenites. Where the footwall assimilation was buffered, cyclical norites formed. Previous work has shown sedimentary contamination throughout the mineralized unit, but the relationship between the assimilation of these different country rock lithologies and the thickness and geometallurgy of the overlying mineralization is still poorly understood. Assimilation of reactive sedimentary country rocks, during ore formation, is likely to have released volatiles that may have remobilized PGE mineralization and influenced the grade and thickness of the mineralized units above the FAZ. We will present new petrological, geochemical, and geometallurgical data from a series of drill holes sampled above the various footwall assimilation types to test this hypothesis.

Geology, Geochemistry and Zonation of the Tsumeb Polymetallic Deposit

Rob J. Bowell[1], Ester Shalimba[2], Abner Nghoongoloka[3], Matthew Leybourne[4], Daniel Layton-Matthews[4], Stephanie G. Lohmeier[5], Hartwig E. Frimmel[6], Helke Mocke[3]
1. SRK, Cardiff, United Kingdom, 2. Geoscience Department, University of Namibia, Keetmanshoop, Namibia, 3. Geological Survey of Namibia, Windhoek, Namibia, 4. Department of Geological Sciences, Kingston, ON, Canada, 5. Technical University of Clausthal, Clausthal-Zellerfeld, Germany, 6. Dept. of Geodynamics & Geomaterials Research, Institute of Geography and Geology, University of Würzburg, Würzburg, Germany

The Tsumeb polymetallic sulfide deposit, in the Otavi Mountain Lands (OML), has a reported resource of 29 Mt of ore at an average grade of 4.3 % Cu, 10 % Pb, 3.3 % Zn, 600 g/t Cd, 450 g/t Mo, 125 g/t Ge, and 98 g/t Ag, and is the largest of similar deposits in the OML, in northeast Namibia, but relatively small by international comparison but was formally Africa's largest producer of lead. The iconic status of Tsumeb relates to the presence of over 400 known minerals in the different mineralized assemblages. The deposit is situated in a fold-thrust belt along the northern margin of the Pan-African Damara Belt hosted in carbonate rocks. Mineralization was structurally controlled, within a pipe-like structure with ore occurring as breccias. Mineralization is estimated to have formed from fluids at temperatures between 370 and 400 ^{o}C from highly saline fluids (20 – 25 wt.% $NaCl_{eq}$). Secondary inclusions have lower salinity pointing towards increasing contribution from meteoric waters. Metals are most likely from Grootfontein metamorphic complex. This unconsolidated but coarse karst fill formed prior to migration of orogenic brines. Hydraulic overpressure of these brines led to mechanical redistribution of the still loose sand and rock fragments in the karst cavities, resulting in favourable structures to host mineralization as well as intense hydrothermal alteration. A unique feature of the deposit is the presence of multiple oxidation zones related to major fractures that allowed migration of meteoric water into deeper portions of the deposit. A complex mineral paragenesis exists and shows zonation with higher Pb contents near surface, Zn and Ge concentrated in the middle depths, and Cu and Ag contents increasing with depth (over 1700 m). Supergene oxidation was not the result of a single, geological event but of multiple episodic events over time.

Application of Conventional and Unconventional Exploration Methods for Sapphire Discovery in Southeastern Madagascar: From Regional to Project Scale

Evelyn Caiza[1], Dibya J. Baral[2]
1. Central University of Ecuador, Quito, Ecuador, 2. Fura Gems inc., Antananarivo, Madagascar

The research area is located in southeastern Madagascar, within the Anosy Region, encompassing the NANTIN and SIAM licenses spanning approximately 11,910 hectares. Geologically, it spans four distinct geological domains characterized by northward structural trends, predominantly comprising foliated marbles and pyroxenites intruded by granitic bodies. Deformation and high-temperature metamorphism have led to the formation of shear zones associated with sapphire mineralization, dated to 523-510 Ma based on zircon dating. This investigation's aim is to employ both conventional and unconventional exploration methods to delineate exploration targets at a regional scale, prioritize them based on exploration significance, and advance them to a project scale to facilitate sapphire discovery.

Conventional exploration techniques, including regional geology and the assessment of sapphire mineralization occurrences, along with Remote Sensing using ASTER and SENTINEL satellite image analysis, mapping of spectral signatures associated to sapphire mineralization, and geophysical analysis employing magnetometry and radiometry, were conducted at a regional scale. These methodologies were integrated to generate mineral prediction maps using unconventional exploration techniques such as supervised (Fuzzy) and unsupervised (SOM) classifications. This comprehensive approach resulted in the identification and categorization of five new sapphire exploration zones, with Ankaramavo, Antirimena, and Satrokala designated as high-priority areas, and Ankazoavo and Andranondambo as medium-priority zones.

The prioritization of new sapphire exploration targets has emphasized the development of the Ankaramavo objective as a top-tier priority for project-scale exploration. Through geological mapping, petrographic and mineralogical characterization, and spectral signatures analysis, discernible mineralogical-spectral trends (calcite, phlogopite, enclaves, AlOH and AlOH/water indexes) indicative of sapphire presence have been identified. These findings, augmented by geochemical analyses defining the geochemical signature associated with such deposits (Al, Si, K, Rb, Y, Nb), have laid the groundwork for the establishment of a geological-geochemical model facilitating the discovery of mineralized sapphire deposits.

The Mineralogy of PGE in the Deep Critical Zone at Sandsloot, Northern Bushveld Complex, South Africa: PPGE Reef Underlain by an IPGE + Rh-Rich Base Metal Zone

Kate Canham[1], David Holwell[1], Iain McDonald[2], Hannah Hughes[3], Katie McFall[4], Andy Lloyd[5], Erin Thompson[1], Kofi Acheampong[5], Lara Du Preez[5], Matthew Loader[5]
1. Centre for Sustainable Resource Extraction, University of Leicester, Leicester, United Kingdom, 2. School of Earth and Environmental Sciences, Cardiff University, Cardiff, United Kingdom, 3. Camborne School of Mines, University of Exeter, Penryn, United Kingdom, 4. Earth Sciences Department, University College London, London, United Kingdom, 5. Anglo American South Africa, Johannesburg, South Africa

The Bushveld Complex, South Africa, is home to some of the world's largest magmatic sulfide deposits. Recent discoveries, including the down-dip extension of the Critical Zone (CZ) hosted Platreef PGE-Ni-Cu deposit, have positioned the Northern Limb of the Bushveld Complex as a frontier for PGE+Ni+Cu+Co exploration and mining in the future decades. In the Sandsloot area, the near-surface resources of the CZ are well characterised, but the recently discovered down-dip extension is not; where multiple high-grade PGE-rich zones on the scale of 10s of meters are underlain by thick base metal sulfide-rich zones. In this study we utilise PGE geochemistry, sulfide and PGM studies and S isotopes to establish the relative controls on the distinctive metal budgets of these two zones.

The PGE reef has 1-2 modal.% disseminated sulfides and is Pt+Pd+Rh(PPGE)-rich. The PGE is dominantly contained within Pt-Fe-alloys, Pt-Pd-(Pb)-bismuthotellurides and Pt-Pd-Pb-alloys, along with high Pd (<3500 ppm), Rh (<270 ppm)-bearing pentlandite. In contrast, the lower base metal (BM) zone contains higher volume, blebby to semi-massive sulfides and lower Pt+Pd grades, but significantly, high Os+Ir,+Ru(IPGE)+Rh. The PGE is predominately contained within sulfide hosted Ir-(Os)-arseno-sulfides and Ru-sulfide (62%), as well as Ru, Rh-rich (<55, 300 ppm) pentlandite and Ru-rich (<45 ppm) pyrrhotite. Within the PGE reef $\delta^{34}S$ values (+0.7-+2.4 ‰), in addition to S/Se ratios (3100-1200), reveal a magmatic control, with subsequent S-loss through fluid mobilisation. In contrast, the underlying BM zone displays an isotopically heavier, contaminated, signature (+4.2-+7.5 ‰).

The difference in PPGE to IPGE+Rh between the PGE reef and BM zone represents a disconnect between the zones. Understanding the PGE mass balance of the deep CZ at Sandsloot, along with the association of PGM and sulfide to surrounding minerals aids not only with understanding of the ore's genesis, but additionally, early understanding of geo-metallurgy has implications for processing.

Lithostratigraphic Distribution and Geochemistry of Carrollite in the World-Class Tenke-Fungurume Cu-Co Mining District, Democratic Republic of the Congo

Malena Cazorla Martínez[1], Lingli Zhou[2, 1], Murray Hitzman[1]
1. iCRAG, the Science Foundation Ireland Research Centre in Applied Geosciences, Dublin, Ireland, 2. Department of Earth Sciences, Vrije Universiteit Amsterdam, Amsterdam, Netherlands

Improving our knowledge of cobalt (Co)-rich deposits is crucial to enhance exploration success for the Co needed for the green energy transition. The Tenke-Fungurume mining district (TFM) in the Central African Copperbelt with 3.5 Mt Co (CMOC, 2022) represents one of the largest Co reserves worldwide. This study investigated the lithostratigraphic distribution and geochemistry of carrollite ($CuCo_2S_4$), the primary hypogene Co ore mineral in TFM.

Twenty-four polished sections from mineralized samples of units in the R.A.T and Mines Subgroups of the Roan Group were examined. Carrollite is most abundant in the R.S.F., followed by the R.S.C. and S.D.B. Carrollite in these samples occurs as disseminations in dolomitic siltstone in R.A.T. Grise (n=2); as disseminations in dolomite horizons in D.Strat. (n=3); as disseminations, lenses, and bed-parallel layers in dolomite-quartz-rich horizons and in quartz-dolomite veins in R.S.F. (n=11); as disseminations in massive dolomite in R.S.C. (n=3); as disseminations in dolomitic shale in S.D. (n=4); and as disseminations and lenses in dolomitic shale and quartz veins cutting bedding at high angles in S.D.B. (n=1). The variety of Co mineralization styles appears to reflect both wall rock lithology and the previous occurrence of disseminated-to-nodular or lenticular evaporite minerals.

The Cu:Co of carrollite increases systematically with stratigraphic height in the local lithostratigraphic section. The Cu:Co value is 0.36±0.05 in R.A.T. Grise and 0.35±0.01 in D.Strat., progressing upsection to 0.52±0.01 in R.S.F., 0.51±0.01 in R.S.C., 0.48±0.02 in S.D.B., and 0.51±0.01 in S.D. Iron, Se, and Mo contents in R.A.T. Grise carrollite are, respectively, 156±377ppm, 170±168ppm and 280±251ppm. Moving into the overlying stratigraphic units, Fe content in carrollite increases, whereas the Mo and Se contents decrease, by one-to-two orders of magnitude. This element systematics are attributed to the influence of redox conditions on carrollite deposition and may serve as a vectoring tool for exploration.

Mapping Under Cover: The Kalahari Copperbelt in Eastern Namibia

Branko Corner, P.R Hildebrand, M Kumar
Corner Geophysics Namibia, Windhoek, Namibia

A minerals-focused interpretation is presented of an extensive area of interest (AOI) in eastern Namibia, extending from Dordabis in the west eastwards to the Botswana border. The historically explored Kalahari Copperbelt (KCB), comprising strata of the Sinclair, Tsumis, Nosib, and Witvlei Groups, bounds the AOI in the west. Our interpretation focus extends eastwards and southwards, under the cover of Kalahari Group sediments. The following data sets were incorporated into this study: Geological Survey of Namibia (GSN) geological mapping; GSN aeromagnetic and radiometric data, flown at a 200-m line spacing; GSN gravity data; available water-borehole records, providing cover thickness and bedrock lithologies; all borrow pit lithologies; revisits of known outcrops; acquisition of helicopter-borne electromagnetic data over selected areas; soil sampling, air-core and diamond drilling.

Our integrated interpretation of these data sets has highlighted the following features:

i. In the east, a series of ENE-trending regional anticlines and synclines constituting the westward continuation of the KCB (from Botswana), comprising sequences of the following Groups: Nama, Witvlei, Nosib (Mamuno), Klein-Aub (D'Kar), and Eskadron/Doornpoort (Ngwako Pan) Formations of the Tsumis Group.

ii. A major change in KCB orientation in the western AOI sector, trending NNE, approximately west of Witvlei, associated with the late-Damaran Kudu-lineament swath.

iii. The AOI is bounded in the southwest by Sinclair Group (Kgwebe) volcano-clastics flanking the Autabib basement anticline. Our work has revealed a series of thrust-bound synclines and anticlines southeast of this, largely under cover, interpreted to have been controlled by early rifting and subsequent southeast-directed thrusting along the Autabib anticline margin. At least two of these units reveal prospective Tsumis Group sediments. Thrusted segments of Sinclair volcanics are also evident in sparse outcrop or shallow sub-outcrop.

iv. We conclude that the Tsumis Group is a likely proto-Damaran sequence, rather than part of the Meso-Neo-Proterozoic Sinclair Group

Geochemical and REE Data for Sheeted Leucogranites at the Primary Uranium Deposits of the Damara Orogen, Namibia

Guy G. Freemantle
University of the Witwatersrand, Randburg, South Africa

The inland branch of the Pan African Damara Orogeny is product of continent-continent collision and distinctly zoned. The southern Central Zone (sCZ) is characterised by highly deformed Neoproterozoic metasedimentary sequences intruded by an abundance of granitoid rocks ranging from dioritic to alkali feldspar granite in composition. sCZ metamorphism is high-T°/low-P, up to granulite facies, present structural architecture was facilitated, in part, by rheological contrasts between the Damara metasediments and unconformably underlying Mesoproterozoic and older granite-gneiss basement inliers and domes, collectively the Abbabis Metamorphic Complex (AMC). The oldest intrusives in the sCZ are calc-alkaline diorites and were emplaced from the onset of inter-craton collision; the youngest intrusives comprise suites of highly evolved, variably uraniferous, late-orogenic sheeted leucogranites (SLG). The SLG suite comprises six sub-types (types A-F) of syenogranite- through alkali-feldspar granite composition; they are mostly distinguishable through systematic field-, textural-, accessory mineral assemblage-, and selected trace-element geochemical characterisation. Two of the younger types (D, E) host economic U concentrations with strong structural and stratigraphic relationships determining the emplacement location of the deposits, reflected also in mobile element chemistry. Discrimination between uraniferous and non-uraniferous SLGs is possible with whole-rock geochemistry; however, with considerable major- and trace-element overlap. Sources of the SLGs thus appear ambiguous, with support for the basement gneisses, pelitic units of the Damara sequence, and the early calc-alkaline suites all considered as candidates. Systematic ICP-MS analyses of the SLG whole rocks have revealed contrasts of HFS- and REE abundances and ratios, discriminating uraniferous and non-uraniferous SLGs. The overall REE patterns of the uraniferous SLGs are comparable to the early calc-alkaline suites and to younger anorogenic granitoids. HFS element discrimination plots distinguish the SLGs from the older diorites and younger granitoids. Uraniferous SLGs are likely derived from different areas of basement source, not diorites, relative to the non-uraniferous SLGs.

The Antelope Deposit, Otjikoto, Namibia: An Integrated Geochemical and Mineralogical Study

Charlie B. Hamblet
Camborne School of Mines - University of Exeter, Penryn, United Kingdom

The Otjikoto Gold mine stands as the largest gold producer in Namibia. In 2022, approximately 3km south of the Otjikoto pit, the new "Antelope Deposit" was identified following drilling of targets generated from magnetic inversion data. Subsequent definition by 100m x 50m spaced drilling identified up to three separate ore shoots, with mineralisation of each shoot remaining open to the northeast and southwest. Mineralisation is predominantly vein hosted within amphibolite facies metasediments of the Northern Damara Belt, and displays a complex relationship between lithology, mineralisation, and structural deformation.

This MSc thesis employs a range of new qualitative and quantitative data derived from whole-rock geochemical analysis, ICP-MS, EPMA, and a collection of polished thin sections to develop a comprehensive overview of the genesis at the Antelope Deposit. A total of 7 drill holes have been studied across Antelope, as well as holes from Otjikoto and Wolfshag for comparison.

The integration of whole-rock geochemistry and targeted ICP-MS aims to characterise alteration composition, distribution and intensity, with a focus on its implications on Au grade and potential post-mineralisation remobilisation. Insights into alteration history and relative timing will be investigated through an EPMA study targeted at zoning growth patterns of hydrothermal garnets in mineralised zones. Growth geometry and geochemical fluctuations across garnet zones will be used as a proxy to discern fluid composition and conditions throughout mineralised periods. In addition, a detailed petrographic study aims to construct a paragenetic sequence of the mineralised zone, drawing comparisons between Antelope, Otjikoto and Wolfshag mineralisation styles.

No such dedicated study has focused on the complete characterisation of alteration and mineralisation across Antelope. This research will contribute to the mineralogical and geochemical understanding of the deposit and its genesis, as well as identifying potential vectors for further exploration on a local and regional scale.

The Role of Exploration on an Active Mine: Case of Kamoa Copper Mine

Richard Ilunga, Franck Twite
Kamoa Copper SA, Kolwezi, Democratic Republic of the Congo (Kinshasa)

The Kamoa copper deposit, 25 km west of the Kolwezi Cu-Co district in Democratic Republic of Congo, is one of the largest high-grade copper deposits in the world. Kamoa is located on the western edge of the Congolese part of the Central African Copperbelt. Two main rock units are present at Kamoa: the oxidized sandstone and siltstone of the Mwashya Subgroup (Roan Group), and the overlying reduced diamictite and interbedded siltstone-sandstone of the Grand Conglomérat unit (Nguba Group).

The copper mineralization is hosted in the lowermost lithological reduced rock overlying the Mwashya sandstone. The marker varies depending on whether the orebody is in siltstone-sandstone or diamictite, this complexity affects the production due to the orebody being hosted in multiple geological units. Exploration work including infill drilling and detailed facies studies on core were conducted in different areas of the Kamoa deposit to further understand the position of the mineralization and the geological markers for mining. Infill drilling is conducted in areas where historical holes were spaced between 200 m and 400 m. The poster will show the role of exploration on this active mine.

Geochronology and Assessment of the Age(s) of Khoemacau Cu-Ag Mineralization Event(s) in the Ghanzi-Chobe Belt Portion of the Kalahari Copper Belt

Tebogo Kelepile, Thierry Bineli Betsi, Elisha Shemang
Botswana International University of Science and Technology, Palapye, Botswana

The Khoemacau Cu-Ag deposits are located in the Ghanzi-Chobe Belt (GCB) of northwestern Botswana and are found within the metasedimentary sequence of the Ghanzi Group. The origin of this Cu-Ag mineralization is still debated because of the lack of precise and reliable ages. Therefore, there is no agreement on the genetic model for ore formation. To better understand the timing of mineralization and to elaborate on the genesis of Cu-Ag deposits in the region, we used sericite $^{40}Ar/^{39}Ar$ and in-situ biotite LA-ICP-MS Rb-Sr geochronology in samples from two mineralized zones (i.e., the Banana Zone North East Fold and Zone 5). We found that sericite intergrown with chalcocite-bornite returns ages of 669.8 ± 0.1 Ma and 588.3 ± 0.2 Ma for Zone 5 and North East Fold, respectively. Biotite intergrown with chalcocite-bornite ore from the North East Fold yields an isochron age of 565.6 ± 6.5 Ma. This data links the timing of Cu-Ag mineralization in the GCB to compressional tectonics during the Damaran Orogenesis and highlights mineralization events that were previously unknown. The ca. 670 Ma event coincides with the final break-up of the Rodinian supercontinent. The ca. 588 Ma and 566 Ma fluid pulse events correspond to syn-orogenic and syn-metamorphic Cu-Ag mineralization episodes that overlap with the window time frame of basin inversion and metamorphism in the GCB. The identification of numerous pulses of mineralization is an important factor in the formation of giant-to-supergiant sediment-hosted Cu-Ag deposits. Therefore, the possibility of such deposits occurring in the GCB cannot be dismissed. Our data confirms that mineralization took place in several stages and provides fresh insights into Cu-Ag ore genesis in the GCB. The genesis of this Cu-Ag ore can be explained by the episodic release of hydrothermal fluids from the sedimentary basin at different times.

The Growth and Expansion of Kansanshi, an Iconic Ore Deposit in the New Copper-Belt, NW Zambia

Nommia Kolala
University of Zambia, Lusaka, Zambia

North-Western Province in Zambia is labelled the New Copperbelt of Zambia as it has become the leading producer of copper in Zambia. Mining is the main economic activity of Northwestern. One of the oldest mines in the province is Kansanshi Mine, located 10 km north of Solwezi town. It is a Cu-(Au) mine that exploits a structurally controlled deposit hosted by strongly deformed Neoproterozoic metasediments in the Katangan Basin in Southern Africa and has been operational since 2005.

In efforts to revive the mine, First Quantum minerals, which owns 80% of the mine, approved the S3 Expansion at Kansanshi mine and the extension of the mining pit. The S3 Expansion is meant to boost production and extend the mine life by another two decades. The S3 Expansion, when completed, will comprise of a standalone 25 million tonnes per annum ("Mtpa") processing plant with a new larger mining fleet that will increase Kansanshi's total annual throughput to 53 Mtpa.

This $900 million project is expected to transition the current high-grade, medium-scale operation to a medium-grade, larger-scale mining operation. Upon completion, copper production from Kansanshi is expected to average approximately 250 thousand tonnes per annum for the remaining life of the mine to 2044. First production from the S3 Expansion is expected in the third quarter of 2025.

The mining pit has been extended into a new section of the orebody called the South-east dome. The open pit mine will be expanded to increase the supply of sulphide ore from the Main Pit and extend into the Southeast Dome deposit. Stripping of the South-east dome is scheduled to start this year. This should take the mine from the 135,000 tonnes of Cu produced in 2023 to 250,000 tonnes post expansion and the average ore grade will approximately be 0.58% Cu.

Petrography and Depositional Features of the Lower Argillaceous Series; Dunstan Farm, Chimanimani: Possible Charleswood Diamondiferous Grits Extension

Gerald S. Kupeta
University of Zimbabwe, Harare, Zimbabwe

This research presents a study conducted on petrographic and depositional features of the Lower Argillaceous Series [Formation] of the Umkondo Group as exposed on Dunstan Farm, Chimanimani in order to delineate the lithologies and to determine whether the study area under investigation is a possible southern extension of the Charleswood diamondiferous grit unit. Diamond mineralization is found in coarse grits, which strike north-south across Charleswood Farm. Further south on Dunstan Farm, surface geological mapping was undertaken to identify any possible extension of this grit unit. The methodologies used include a desktop study, surface geological mapping, X-Ray fluorescence analyses, petrographic analyses, heavy mineral separation analyses, data analysis and interpretation. Within the study area, six lithologies were identified namely dolerite, mudstone, sandstone, quartzite, grit and shale. The high titanium values suggest the Umkondo sediments to have a basic igneous rock parentage. The study area was subjected to Greenschist Facies metamorphism. The white quartzite of Watson, (1969) is sandstone whilst his grey quartzite is the grit unit. The the coincidence of analytical comparison of the grit unit from north to south, including petrographic and heavy mineral analyses, suggest that the study area is a possible southward extension of the Charleswood diamondiferous grit unit.

Geology of the Southwestern Part of Kakula: Structural Complexities and Evidence of Tectonic Inversion

Serge M. Makenda, Franck Twite
Kamoa copper SA, Kolwezi, Democratic Republic of the Congo (Kinshasa)

The Kakula Copper deposit part of the Kamoa-Kakula Complex is located approximately 25km west of Kolwezi. The deposit contains 540Mt of indicated resource at 2.67% Cu and 90Mt of measured resource at 3.1% Cu (1% Cu cut-off).

The southwestern part of the Kakula deposit shows a complex structural architecture between the Kakula dome and the current mining zone. The deposition of the Grand Conglomérat formations occurred in an actively forming rift basin, which controlled the geometry, grade distribution and thickness variation. The main growth fault was formed during the extension of the Kakula sub-basin. Near the Kakula Dome, there is evidence of steep beds with 40° average dips, small-scale folds, fractures, thrust faults and intense foliation. The zone of high foliation is often associated with quartz and carbonate veins bearing copper sulfides. These structures are evidence of tectonic inversion related to the Kolwezian phase of the Panafrican Orogeny. Detailed underground mapping coupled with cover drilling constitute very effective tools to investigate the complexity of the deposit. The poster shows the position and extent of the geological variations in the southwestern area of the Kakula mine.

Fingerprinting Mineralization in the Sediment-Hosted Cu-Ag System: Constraints from the Khoemacau Zone 5 Cu-Ag Deposit in NE Botswana's Portion of the Kalahari Copper Belt

Tefo Masena, Thierry Bineli Betsi, Tebogo Kelepile
Botswana International University of Science and Technology, Palapye, Botswana

The Ghanzi Group's Neoproterozoic sedimentary rocks are known to house the Khoemacau Zone 5 Cu-Ag deposit, located in the Ghanzi-Chobe Belt. Despite lacking visible signs of alteration, these host rocks may still display hydrothermal overprints that are difficult to detect, potentially impeding undercover exploration success. As mineral deposit exploration becomes more challenging, it becomes necessary to employ novel approaches that are reliable and cost-effective. One such approach is to use various geoscientific proxies as exploration tools to direct mineralization. We utilized a combination of alteration types, changes in mineralogy, and rock chemistry from barren to mineralized rocks to develop geochemical and mineralogical vectors for the Cu-Ag mineralized system. We employed inductively coupled plasma-optical emission spectrometry, X-ray fluorescence spectrometry, X-ray diffraction spectrometry, and multivariate statistical analyses (including PCA and AHC).

Copper is highly correlated with Mn and Mo and, in some cases, with Co and As. This suggests saturation in sulfide minerals such as molybdenite, cobaltite, and arsenopyrite. Copper, Mn, Sr, Mo, As, Li, and Y concentrations increase proximal to the mineralized zone. Within the mineralized zone, increased Cu content is associated with elevated As, Co, Li, Mn, Mo, and Sr concentrations. Beyond the mineralized zone, As, Cu, Mn, Mo, and Y concentrations decrease proportionately to their proximity to the mineralized zone, but Li and Sr content increases locally. The mineralized zone is characterized by high concentrations of calcite, chlorite, and mica, suggesting a possible correlation between high Cu concentrations and the processes of sericitization, carbonatization, and chloritization. Significant albitization occurs beyond the mineralized zone. As a result, higher Mn, Mo, Li, Sr, and local As and Co can be utilized as indicators of mineralized zones. Chloritization, sericitization, and carbonatization are examples of changes that can serve as mineralization vectors.

Reassessing the Bulk Geochemistry and Origin of Hematitic Iron Ores from the Northern Cape Province of South Africa

Neo G. Moabi[1, 2], Bjorn von der Heyden[2], Janne Blichert-Toft[3], Fanny Thibon[3], Harilaos Tsikos[2, 4]
1. Council for Geoscience, Pretoria, South Africa, 2. Stellenbosch University, Stellenbosch, South Africa, 3. Ecole Normale Supérieure de Lyon, Lyon, France, 4. University of Patras, Patras, Greece

Massive hematite iron ores from the Northern Cape Province of South Africa display a geographical distribution controlled by a regional unconformity contact separating chemical sediments (BIF, carbonates) of the Paleoproterozoic Transvaal Supergroup from shales and associated siliciclastic sedimentary rocks of the younger Olifantshoek Supergroup. This stratigraphic framework has been widely used in support of an ancient lateritic origin for these ores. We performed bulk geochemical (major-minor and trace elements, including REE) and iron isotope ratio analyses of 30 representative iron ore samples from across the Northern Cape region, to constrain the composition of likely protoliths to the ores and test the paleolateritic model of ore formation. Results were assessed against average bulk-rock compositions of Transvaal BIF and Olifantshoek shale as end-member precursor candidates to iron ore geochemistry.

Ore textures vary from massive to laminated and breccia-like. The ores are dominated by hematite (>90% modal Fe_2O_3 in most samples), with minor Si, Al, K and/or Mn making up the remainder. PAAS-normalized trace element spidergrams occupy the entire compositional space defined by the two chosen end-member protoliths (BIF and shale). Similarly, REE spidergrams generally mimic the seawater-like average BIF pattern at low REE abundances, while samples with higher REE concentrations show flatter, shale-like patterns with slight relative depletion in heavy REE. Trace elements that appear to show atypical relative enrichments include Ba, Sr, Pb and less so Cu and Zn. Selected detrital proxy ratios (K/Rb, Ti/Nb) compare closest to those of shale rather than of BIF. Iron isotope data exhibit narrow variability ($d^{57}Fe$: -0,2±0.3‰) with only six samples registering values below -1‰. Collectively, our results do not support a common origin of the iron ores from geochemically conservative lateritic weathering of BIF, but instead point to a possible hydrothermal replacement origin involving variable contributions of shale as key additional precursor material.

Stratigraphic Position Change of the Mineralization at the Kamoa Deposit

Innocent Mushobekwa[1], Franck Twite[2]
1. KAMOA COPPER SA, Kolwezi, Democratic Republic of the Congo (Kinshasa), 2. KAMOA COPPER SA, Kolwezi, Democratic Republic of the Congo (Kinshasa)

The Kamoa-Kakula complex hosts two known copper target areas, namely Kamoa and Kakula. Kamoa is one of the sediment-hosted stratiform copper deposits that is located 25 Km west of Kolwezi town in the Congolese Copperbelt. The deposit contains 759 million tonnes of ore at 2.59% Cu in Indicated resource (1% Cu cut-off grade). Two main rock units are present at Kamoa: sandstone and siltstone of the Mwashya Subgroup (Roan Group), and overlying diamictite with interbedded siltstone-sandstone of the Grand Conglomérat unit (Nguba Group). The Mwashya Subgroup forms the oxidized lower strata, and the pyritic rocks of the basal diamictite form the reduced mineralization host.

The orebody is generally hosted in the lowermost reduced lithological unit that overlies the Mwashya sandstone, these include the Kamoa Pyritic Siltstone (Ng1.1.2) and the three subunits of the Basal Diamictite (Ng1.1.1). These three subunits are composed of a clast-rich diamictite (Ng1.1.1.1) at the base overlain by a siltstone-sandstone (Ng1.1.1.2) and a clast-poor diamictite (Ng1.1.1.3). The mineralization above the cut-off grade is rarely constrained in one stratigraphic unit; it generally overlaps between the siltstone-sandstone and the clast-poor diamictite. The clast-rich diamictite is mineralized towards the southwest of the Mupaka Fault where the Basal Diamictite thickens considerably with multiple siltstones. There's no unique stratigraphic unit that can be tracked as a host rock for grade control purposes at Kamoa. Thus, the underground geological mapping and the surface drill core review contribute significantly to defining the footwall markers and delineating the boundaries for the different facies at the mine. The poster will highlight the stratigraphic position of the mineralization and the footwall geological markers that are critical to the optimization of the mine production.

The Effect of the Paarl Fault on the Main Sulphide Zone, Selukwe Subchamber, Great Dyke

Takudzwa Mwagura[1]
1. University of Zimbabwe, Harare, Zimbabwe, 2. Geological Society of Zimbabwe, Harare, Zimbabwe

Platinum group elements (PGEs) are amongst the valuable precious resources in the world. The mining of low-grade PGEs is damaging to any economic system. The drive for optimum and profitable grades can however be affected by geological structures.

This research assessed the impact of the Paarl Fault on the main sulphide zone (MSZ) in the Selukwe Subchamber of the Great Dyke. The main objectives of the study were to assess the effect of the Paarl Fault on the MSZ, the PGE concentration and distribution in the Paarl block, and the alteration mineral assemblages associated with the fault. The lithological units that were identified include gabbronorite, websterite, orthopyroxenite, pegmatoidal pyroxenite and websterite, xenolith outcrops, and serpentinite. Assay results from sampled drillcores were analysed and grouped into two datasets, dataset Paarl A (254m away from the fault) and dataset Paarl B (448m away from the of fault). Dataset A had an average grade of 4.38g/t and a standard deviation of 0.556g/t, with the highest grade of 5.28g/t observed in drillcore PAR49. Dataset B had an average grade of 3.83g/t and a standard deviation of 0.196g/t, with the lowest grade being recorded of 3.64g/t in drillcore PAR42. Within the MSZ, clinopyroxene, orthopyroxene, and olivine showed alteration patterns, with the formation of serpentine also observed mainly in PAR40 drillcore thin sections.

With the assay values obtained in this project, the evaluation of the results suggested that the Paarl Fault could have acted as a conduit for mineralizing fluids, which resulted in high PGE values (g/t) in proximity to the fault. Alteration assemblages show consistency with the effects of faulting on host rocks, whereby the hydrothermal fluids from the fault could have promoted alteration processes resulting in the formation of secondary minerals and different textures which could result in PGE barren zones impacted when mining.

A Supergene-Hydrothermal Origin of the BIF-Hosted High-Grade Iron Ores in the Mbarga Prospect, Mbalam Iron Ore District, Southern Cameroon, Congo Craton

George L. Ngiamte[1, 2], Eleanor Green[2], Olugbenga A. Okunlola[1], Roland Maas[2], Alan Greig[2], Cheo E. Suh[3, 4]
1. Mineral Exploration Geoscience Program, Pan African University Life and Earth Sciences Institute (including Health and Agriculture), University of Ibadan, P.M.B 20, Ibadan, Nigeria, Ibadan, Nigeria, 2. School of Earth Sciences, The University of Melbourne, Parkville, Victoria 3010, Australia, Melbourne, VIC, Australia, 3. Department of Geology, Mining and Environmental Science, The University of Bamenda, P.O. Box 39, Bambili, North West Region, Cameroon, Bambili-Bamenda, Cameroon, 4. Economic Geology Unit, Department of Geology, University of Buea, P.O. Box 63 Buea, South West Region, Cameroon, Buea, Cameroon

The Mbarga prospect lies within the Mbalam iron ore district of the Ntem Complex on the Congo Craton (CC). It is characterized by Banded Iron Formations (BIFs), high-grade supergene, and specularite ores. This study presents mineralogical, geochemical, and isotopic datasets on these ores to determine their origin. Ore microscopic studies indicate that the BIFs are of the oxide facies type, with magnetite showing alteration to hematite-martite. The supergene ores consist of hematite + martite + goethite ± gibbsite ± magnetite ± quartz, while the specularite ores are composed of hematite + martite ± quartz. Magnetite microchemistry suggests formation under low-*T* hydrothermal conditions (~200–300 °C) with high fO_2. Geochemical analyses suggest that the supergene and specularite ores have higher Fe_2O_3 (88.27 to ~100 wt%) and lower SiO_2 (<0.01 to 0.18 wt%) contents than the BIFs (31.95 wt% Fe_2O_3, 67.16 wt% SiO_2). The enrichment of Fe in the supergene ores is attributed to the depletion of major oxides and trace elements due to weathering and supergene enrichment, while the high Fe content in the specularite ores stems from the precipitation of iron-rich, but trace- and REE-deficient hydrothermal fluids. The slightly higher Al_2O_3 content and positive Ce anomalies in the supergene ores suggest retention of Al-bearing minerals (gibbsite) and reveal highly oxidative conditions during martitization. Stable isotope analyses reveal that the supergene and specularite ores have $\delta^{18}O$ values of -2.5 to -0.3‰ and -2.0 to -3.4‰, and δ^2H values of -75 to -123‰ and -70 to -119‰, respectively, suggesting the involvement of isotopically light-evolved meteoric water in their formation. In contrast, the BIFs exhibit heavier $\delta^{18}O$ (8.5 to 10.2‰) and δ^2H (-85 to -91‰) values, suggesting formation from mixed magmatic-metamorphic fluid sources. A “polygenic-supergene-hydrothermal” model is suggested for the formation of the Mbarga BIF-hosted iron ores.

From Deposition to Diagenesis and Beyond – Tracing the Origins of Manganese-Lutites of the Kalahari Manganese Field

Valentin Oge[1], Louis L. Coetzee[2], Bradley M. Guy[1], Albertus J. Smith[3], Jens Gutzmer[1]
1. Helmholtz Institute Freiberg for Resource Technology, Freiberg, Germany, 2. Teck Metals Ltd., Trail, BC, Canada, 3. PPM & CIMERA, University of Johannesburg, Johannesburg, South Africa

The Kalahari Manganese Field (KMF) of the Northern Cape Province of South Africa is undoubtedly one of Africa's iconic ore deposits. Hosting about 74% of all known minable manganese ores globally, it is by a far margin the largest land-based Mn deposit. More than 90% of the resource can be described as mangano-lutite, e.g., a microcrystalline, finely laminated sedimentary rock typically marked by an abundance of mm-sized ovoids and containing between 30 and 40 wt. % Mn. Across the entire KMF, this mangano-lutite occurs in three well-defined symmetrical units that are intercalated and in gradational contact with Superior-type banded iron formations of the 2.42 Ga Hotazel Formation. The latter forms part of the virtually undeformed and unmetamorphosed volcano-sedimentary succession of the Postmasburg Group, Transvaal Supergroup. Despite the economic relevance of the mangano-lutites as the world's single-most important source of manganese, there is a striking lack of mineralogical and mineral paragenetic data available in the published literature. This contribution provides a summary of recent mineralogical and mineral paragenetic studies on mangano-lutites from the southern part of the main Kalahari deposit as well as the Avontuur deposits further to the North. The results do not only provide an unprecedented insight into the complex diagenetic evolution of the mangano-lutites, but also reveal surprising differences in mineral assemblages in different parts of the KMF. These differences are thought to be due to systematic lateral variations in physicochemical conditions of manganese and iron precipitation during the deposition of metalliferous muds in a shallow, complexly structured marginal marine basin just prior or at the onset to the Great Oxidation Event.

Sequence Stratigraphic Correlation and Depositional Environment Mapping in the Katangan Basin of Central Africa: Implications for Mineral Exploration

Martin Purkiss[1], Michael C. Daly[2]
1. Ivanhoe Mines, Sandton, South Africa, 2. University of Oxford, Oxford, United Kingdom

The Neoproterozoic Katangan basin of Central Africa contains the Katangan Supergroup, recording ~300 Myr of depositional history spanning the Tonian, Cryogenian and Ediacaran. The basin hosts the Central African Copperbelt: the world's largest sediment-hosted copper province.

Lithostratigraphic schemes are abundant across the Katangan basin. The value of lithostratigraphy is limited, however, on account of diachroneity being imparted by correlating on the basis of lithology, rather than time. A sequence-driven stratigraphic framework can alleviate this, providing predictive capability for the distribution of mineral system components in underexplored parts of a basin.

Proposed here is a revision of the Katangan Supergroup based on sequence stratigraphy, and its division into four 2nd-3rd-order sequences. The sequence stratigraphic correlation is based on detailed sedimentological logging of Katangan stratigraphy across the basin, supplemented by data from literature. It encompasses predominantly 2nd-3rd-order cycles and follows a modified version of depositional sequence model III. The sequence stratigraphic correlation is incorporated into a new structural interpretation of the basin to generate depositional environment maps for each major stage of the basin's tectonostratigraphic evolution.

Mapping of gross depositional environments constrains the distribution of metal sources, reductants, fluid pathways, and evaporites in space and time, relating them to key periods in the Katangan basin's history.

The revised stratigraphic scheme allows for more reliable correlations to be made across the basin where lithologies are repeated or too ambiguous to be correlated lithostratigraphically (e.g. Grand and Petit Conglomerat). Mineralised horizons such as the 'ore shale' are shown to likely be diachronous. Deposition of metal sources is predominantly restricted to periods of rifting (Lower Roan, Mwashia Groups), with reductant and evaporite deposition tending to coincide with post-rift thermal subsidence (Upper Roan, Nguba Groups).

The proposed stratigraphic scheme provides an improved framework to better understand the Katangan basin's evolution and exceptional metal endowment.

Cu and Fe Isotope Contrasts Between the Paleoproterozoic Haib Porphyry Copper Deposit and Representative Phanerozoic Porphyry Systems

Bibi Ayesha Jogee[1], Grant Bybee[1], Ryan Mathur[2], Laurence Robb[1, 3]
1. University of the Witwaterstand, Johannesburg, South Africa, 2. Juniata College, Pennsylvania, PA, USA, 3. University of Oxford, Oxford, United Kingdom

Abstract:
Porphyry copper deposits are the world's primary source of copper — an essential critical metal in the green energy sector. These deposits were typically formed in the Phanerozoic Eon and occur principally at convergent plate boundaries. The Haib copper-molybdenum deposit (dated at 1920 to 1880 Ma) is one of the few porphyry copper deposits to occur in the Precambrian of Africa and is exposed at mid- to upper-crustal levels. The deposit, host to disseminated-, massive- and vein- type Cu-Mo mineralization, is situated in the Richtersveld Subprovince of the Namaqua-Natal Province. In this work we use new geochemical data, including Fe and Cu isotopes, to better understand the metallogenesis and unique temporal position of the Haib deposit. Results show that the mineralization is characterized by two phases: disseminated-massive (heavier δ^{65}Cu of 0.06–0.21 ‰ and δ^{56}Fe of 0.02–0.06 ‰) and vein-type (lighter δ^{65}Cu of -0.08–0.11 ‰ and δ^{56}Fe of 0.11– -0.01 ‰). The δ^{65}Cu signatures of the disseminated chalcopyrite grains overlap with magmatic-hydrothermal or hypogene sulphides (-1 to 1 ‰). The results indicate that the ore-forming fluid evolved and progressively precipitated various generations of sulphide minerals with fractionated δ^{65}Cu and δ^{56}Fe isotope signatures, due to sequential mineral crystallization and temperature changes related to variable equilibration of minerals and fluids. The overlap of the δ^{56}Fe signature of the disseminated sulphides with the δ^{56}Fe signature of the mantle suggests a mantle source for the copper. The limited range in the δ^{65}Cu and δ^{56}Fe signatures of the Paleoproterozoic Haib deposit, relative to Phanerozoic deposits, is consistent with the deeper levels of this porphyry system, suggesting that the shallower levels, where isotopic fractionation is likely to have occurred, have been eroded away. Cu and Fe isotopes can therefore provide important exploration proxies for estimating the crustal level of porphyry systems.

The World-Class Shamva Gold Deposit, NE Zimbabwe

Ahmet K. Sener
Ariana Resources plc, London, United Kingdom

The >5Moz Shamva gold deposit is located within the Harare Greenstone Terrane in northeastern Zimbabwe. As the third greatest historical gold producer in Zimbabwe, the mine has operated for >100 years producing c.2.6Moz of gold from 19.4Mt @ 4.2g/t Au. Approximately 40.8Mt @ 1.96g/t Au (2.6Moz) remains, suggesting that Shamva may represent the largest gold deposit in Zimbabwe.

Shamva is hosted by a unique sequence of Lower Shamvaian Group (c.2.67 Ga) sedimentary rocks metamorphosed to upper greenschist facies. Ubiquitously pyritiferous and silicified clastic and volcaniclastic debrites and epiclastic conglomerates occur within a tightly folded <2 km wide stratigraphic unit which extends ENE-WSW over a strike length of 5.5 km. The intrusion of several syndepositional andesitic-dacitic feldspar porphyries (2,672±12Ma) in to an unconsolidated volcanosedimentary sequence and coincident generation of a sub-aqueous exhalative system may have caused an anomalous trace-element geochemistry.

The ore zones at Shamva are restricted to an ENE-WSW trending brittle-ductile transpressional duplex system. This system is c.250-300 m wide by c.2.5 km long, tapering at both eastern and western ends, forming an attenuated en échelon array of dominantly ENE to E trending shears that are, in places, dissected/wrenched apart in a sinistral sense by essentially synchronous but sequentially later NE-trending shears. Large orebodies occur at the intersections of the ENE- and major NE-trending shear zones, and primarily dip (N)NW at about 75 degrees, but also dip steeply to the (S)SE in places. Zetamoidal ore zones, developed along attenuated link shears within the thrust-duplex system are characteristic and form sigmoidal structures in cross-section.

Widespread carbonitisation, potassium metasomatism and sulphidisation occur within all ore zones. Sulphides constitute 1-12% (average 4.5%) by volume, with a late pyrite, arsenopyrite and gold0.6-silver0.4 assemblage corresponding with the majority of ore zones, the deposition of which occurred at 250-450°C at 1-3 kbar.

The Implication of Metamorphisim and Deformation of the Vioolsdrif Domain on the Haib Porphyry Copper-Molybdenum Deposit, Namibia

Shafiishuna Silvanus
University of Namibia, Keetmanshoop, Namibia

The Haib Copper deposit, located in the south of Namibia in the ǁKaras Region,is hosted mainly by the high-K calc-alkaline intrusions of the Vioolsdrif Suite and sub- aerial volcanic rocks of the Orange River Group along the western part of the Paleoproterozoic Richtersveld Magmatic Arc (RMA) of Vioolsdrift Domain, Namaqualand Metamorphic Province. Porphyry-style copper mineralization is developed in and around a quartz-feldspar porphyry (hereafter QFP) intrusion, which is a minor phase of a large composite batholith formed by plutons of the Vioolsdrif Intrusive Suite (VIS). The QFP is a minor constituent of the VIS, the only other occurrences being the Tatasberg prospect in the northeast. Much of the Haib copper deposit area is inaccessible and not well studied thus leading to many outstanding questions with regard to whether the deposit was affected highly by subsequent deformation and metamorphism episodes experienced in the domain and the extend of the deposit erosion. In this study, we characterize the mineralization of ore types, grade of metamorphism and structural controls of the Haib deposit to understand the evolution of the deposit. We employed EPMA (electron microprobe analyzer), XRF, petrography and core and field observation. The study shows that the deposit experienced low grade metamorphism but is highly remobilised into preferred structurally controlled sites.

Magmatic Stratigraphy of the Deep Platreef at Sandsloot, Northern Bushveld Complex: Carbonate Contamination and Controls on Ni-Cu-PGE Mineralisation

Erin S. Thompson[1], David A. Holwell[1], Marc Reichow[1], Iain McDonald[2], Thomas Blenkinsop[2], Hannah Hughes[3], Katie McFall[4], Kate R. Canham[1], Matthew A. Loader[5], Lara Du Preez[6], Kofi Acheamponga[6], Andy Lloyd[6]
1. Centre for Sustainable Resource Extraction, University of Leicester, Leicester, United Kingdom, 2. School of Earth and Environmental Sciences, Cardiff University, Cardiff, United Kingdom, 3. Camborne School of Mines, University of Exeter, Penryn, United Kingdom, 4. Department of Earth Sciences, University College London, London, United Kingdom, 5. Anglo American UK, 17 Charterhouse Street, London, United Kingdom, 6. Anglo American South Africa, 144 Oxford Road, Rosebank, Johannesburg, South Africa

The Platreef, northern limb of the Bushveld Complex, South Africa, is widely regarded as one of the world's largest resources of platinum-group elements (PGE). These PGE, as well as significant amounts of base metals (Ni, Cu, Co) also found in the northern limb, are essential for the growth of many sustainable and environmentally friendly technologies. The Platreef is part of a complex package of Critical Zone (CZ) rocks which differ from the wider Bushveld in that high grade PGE mineralization is spread over a much greater thicknesses compared to the Merensky Reef. Additionally, it lies north of the Thabazimbi-Murchison lineament, and rests directly on the metasedimentary Transvaal Supergroup. The resulting footwall interactions are known to have had a significant impact on the fluctuating geochemical and mineralogical signatures of the northern limb magmas, as well as their Ni-Cu-PGE grades.

In this study, the petrology, bulk geochemistry, including Nd-Sr-Hf-Pb isotopic compositions, and silicate mineral chemistry of samples from the Platreef at Sandsloot are examined, with the aim of establishing the magmatic stratigraphy of the deeper Platreef, down-dip from the Mogalakwena open pit. Here, the Platreef is composed of up to five pyroxenitic packages, with significant PGE and base metal mineralisation observed in two. Varying CaO/Al_2O_3, and Eu/Eu* ratios, Pb and Sr anomalies are associated with different units and allude to varying degrees of contamination. Variations in the parental magma compositions for each of the identified units are more difficult to distinguish. Hybrid skarn textures are observed within the Base Metal Zone (BMZ), indicating the importance of interactions with the underlying Malmani Dolomite in forming economic Ni-Cu-PGE mineralisation at this location. Significant PGE mineralisation is also commonly accompanied by extensive hydrothermal alteration, which is not observed in barren units, and suggests that rock-fluid interactions may be key to concentrating PGE grades at Sandsloot.

Geometallurgical Evaluation of the Lowermost, Low-Grade Manganese Bed in the Kalahari Manganese Field, South Africa

Louis van Eeden
Stellenbosch University, Stellenbosch, South Africa

The Kalahari Manganese Field (KMF), situated in the Northern Cape of South Africa, boasts a significant portion, approximately 77%, of the world's known land-based manganese reserves. The focus of the study is on the lower most Mn seam, in the eastern portion of the KMF, at one of the mines containing ore of lower grade, approximately <36 wt% Mn.

Macroscopic visual logging identified five distinct textures within the ore, each containing a unique composition. This study unravels the intricate relationship between these textural features, the geochemical characteristics, and the comminution behaviour of the ore within the low-grade mine of the KMF.

Utilizing a random forest regression algorithm, the study elucidates the relationship between the geochemical and textural variability within the ore. Findings from this analysis underscore a direct relationship between manganese grade and the textural attributes, facilitating the creation of a predictive Mn model, using the five textures as the input variables to prevent off-seam underground mining. Furthermore, the comminution behaviour of each texture is evaluated through relative crushability measurements using a laboratory-scale jaw crusher. Integrated analyses reveal the mineralogy as the driving factor for differences in crushability among the textures.

Mineralogical analyses, conducted through scanning electron microscopy and reflective microscopy techniques, identified two primary ore types. Unaltered braunite lutite, containing carbonate ovoids and laminae of diagenetic origin, and altered ore, characterized by an influx of bedding-parallel hausmannite due to metasomatic influx during late-stage, low-grade metamorphism.

The strong correlation between the textural variation and geochemical properties of the ore, together with the varying comminution behaviour and mineralogy, will aid in developing detailed models of the geometallurgy and genesis of the Mn ore at the mine, to not only improve the understanding of the ore but also improve the processing effectiveness and increase the life of mine.

Timing Diagenetic Processes in the Zambian Copperbelt: Insights from U-Pb Dating of Detrital Ilmenite and Authigenic Hematite

Victor I. Vincent[1, 2], Koen Torremans[1, 2], Darwinaji Subarkah[3], Sarah Gilbert[4], Juraj Farkas[3], Aileen L. Doran[1, 2], Alan S. Collins[3], Simon Jones[1, 2], Jon Stacey[5], Murray Hitzman[1, 2]
1. Faculty of Earth Sciences, University College Dublin, Dublin, Ireland, 2. Science Foundation Ireland Research Centre in Applied Geosciences (iCRAG), University College Dublin, Dublin, Ireland, 3. Department of Earth Sciences, University of Adelaide, SA 5005, Adelaide, SA, Australia, 4. Adelaide Microscopy, University of Adelaide, SA 5005, Adelaide, SA, Australia, 5. Tangila Exploration, Lusaka, Zambia

The Lubambe-Mingomba (formerly Lubambe Extension) Cu-Co deposit is located in the northern Zambian Copperbelt (ZCB). The basal Mindola Clastics and the overlying mineralized “Ore Shale” of the Neoproterozoic Roan Supergroup were deposited during the mid to late Tonian (~890 to 765 Ma). Detailed petrographic analysis show remnant detrital ilmenite with exsolved hematite and rutile in the basal clastics consistent with early diagenetic breakdown in sedimentary basins. Our study attempts to constrain the timing of ilmenite-hematite breakdown and hence early diagenetic processes preceding complex multi-phased sulfide mineralization in the ZCB.

LA-ICP-MS U-Pb ages from exsolved hematite grains with no remnant ilmenite return post-diagenetic ages between 700-680 Ma suggesting resetting of the U-Pb geochronometer. Rim-core ages from hematite grains with remnant ilmenite record an earlier (~650 Ma) post-depositional age overprinted by a later (~534 Ma) Cambrian event concomitant with the peak of the regional Lufilian Orogeny. In-situ U-Pb dating of ilmenite yielded systematically younger ages (497 to 483 Ma), likely linked to post-Lufilian cooling. Our results produce ilmenite U-Pb ages within uncertainty of other post-orogenic U-Pb ages from apatite and rutile in the same samples.

Our data shows no preserved detrital and diagenetic ages suggesting episodic post-diagenetic alteration of hematite and ilmenite in the ZCB that matches pulsed sulfide (re-)growth models suggested by other workers for the Central African Copperbelt [2]. The episodic breakdown of detrital ilmenite was key to the release of Ti crucial for the formation and stability of abundant rutile and may have increased the Fe budget in metal-laden brines linked to Cu-Co-Fe sulfide mineralization in the ZCB. Furthermore, our study highlights the utility of in-situ ilmenite and hematite U-Pb dating in constraining diagenesis and post-compaction alteration in sedimentary basins.

EPMA, Raman spectroscopy and EBSD analysis track physiochemical changes during hypogene to supergene mineralization in the Lubambe-Mingomba Cu-Co deposit

Victor I. Vincent[1, 2], Koen Torremans[1, 2], Richard Unitt[2, 3], Christian Burlet[4], Aileen L. Doran[1, 2], Simon Jones[1, 2], Jon Stacey[5], Murray Hitzman[1, 2]
1. Faculty of Earth Sciences, University College Dublin, Dublin, Ireland, 2. Science Foundation Ireland Research Centre in Applied Geosciences (iCRAG), University College Dublin, Dublin, Ireland, 3. Department of Biological, Earth and Environmental Sciences, University College Cork, North Mall Campus, Cork, Ireland, 4. Royal Belgian Institute of Natural Sciences, Brussels, Belgium, 5. Tangila Exploration, Lusaka, Zambia

Lubambe-Mingomba (formerly Lubambe extension) is a supergiant sediment-hosted Cu-Co deposit in the Zambian Copperbelt (ZCB). Hypogene and supergene ores are hosted in Lower Roan sediments of the Neoproterozoic Katangan Supergroup. Petrographic analysis of ore-bearing intervals reveals several generations of Cu minerals hosted in disseminated stratiform sulfides, bedding-parallel and irregular sulfide-bearing veins. Supergene fluids caused in-situ alteration and replacement of earlier-formed hypogene chalcopyrite, bornite (bn), and chalcocite (cc) by supergene cc phases and covellite (cov). Electron probe micro-analyzer (EPMA), raman spectroscopy and electron back-scattered diffraction (EBSD) measurements were carried out on these sulfide phases to provide insights into the transition from hypogene to supergene mineralization conditions in the ZCB.

$Cu_{2-x}S$ sulfide generations show clear variations in Cu:S ratios ranging between 1.92 ± 0.05 and 1.10 ± 0.10. Raman spectroscopy of hypogene phases reveal systematic phonon shifts (measured in cm^{-1}) for increasing x in $Cu_{2-x}S$, decreasing from a phonon shift of 302 cm^{-1} for milkish-white supergene cc (1.98 ± 0.12), 305 cm^{-1} for whitish-blue hypogene cc (1.88 ± 0.10) to 310 cm^{1} for light-blue altered hypogene cc (1.70 ± 0.10). Similar trends are seen for dark blue supergene phases, with $Cu_{2-x}S$ values of 1.60 ± 0.09 (shift of 271 cm^{-1}) and 1.00 ± 0.05 (shift of 266 cm^{-1}). Supergene altered cc and bn phases are identified by sharp stretch bands in the Raman spectra at 472-475cm^{-1}, interpreted to be due to the appearance of S-S bonds. EBSD mapping of these phases confirms that differences in crystal symmetry coincide with chemical variation, allowing for further distinction of sulfides formed under hypogene-supergene conditions.

Our results also show that spectroscopic techniques have the capacity to accurately distinguish between various hypogene and supergene sulphide phases. In summary, supergiant Cu-Co deposits such as Lubambe-Mingomba are formed and upgraded by multiple generations of hypogene and supergene Cu(+Fe)-Co sulfides.

The Hidden Talents of Quartz and Mica: Indicating Metallogenic Potential in A-Type Granitic Systems

Leo Vonopartis[1], Paul Nex[1], Judith Kinnaird[1], Laurence Robb[1, 2]
1. University of the Witwatersrand, Johannesburg, South Africa, 2. University of Oxford, Oxford, United Kingdom

Many minerals have been used as mineralisation vectors; however, there are few vectors for Sn and other critical metals hosted in granites. Quartz and mica are ubiquitous in granitic assemblages and are promising candidates as tools in mineral exploration. The Zaaiplaats Tin Field (ZTF) of the Bushveld Complex is a well-known expression of endogranitic Sn-mineralisation and is used in a comparison of mineralised and unmineralised A-Type granites. The ZTF comprises the Nebo Granite at the base that grades into the Bobbejaankop Granite then the Lease Granite, which forms the cupola. These granites exhibit an increasing degree of hydrothermal alteration respectively, with mineralisation restricted to the Bobbejaankop and Lease phases.

Trace elements in quartz can record the evolution of a granite and the point of fluid-saturation. The quartz from the ZTF exhibits an increase in Al/Ti and Ge/Ti ratios from the base to the cupola, illustrating sequential fractionation and increasing fluid-rock interaction. The quartz exhibits a shift from a magmatic fractionation-controlled to a hydrothermally-controlled system, identifying the point of fluid-saturation and the most fractionated, and potentially mineralised, lithologies. Comparing the mica compositions from the ZTF with other mineralised and unmineralised A-Type granites highlights differences in their evolution and metallogenic potential. Mineralised and unmineralised granites display distinct mica characteristics. The ZTF and other mineralised granites exhibit a Fe-Li trend of trioctahedral micas and a common occurrence of alteration-associated dioctahedral micas. In contrast, unmineralised granites have Mg-rich trioctahedral micas and a lack of dioctahedral micas. Differences in mica chemistry, evolution, and alteration affect the presence of volatiles and sequestration of metals, such as Sn. Therefore, mica compositions indicate the potential of an A-Type granitic system to have developed late-stage mineralisation. Using these minerals in concert can inform mineral exploration by indicating the degree of fractionation and metallogenic potential of a granitic system.

Hypabyssal and Volcanic Rocks of the Western Foreland Shelf, Democratic Republic of Congo

James Yates, Amanda M. Stoltze
Ivanhoe Mines Exploration, Kolwezi, Democratic Republic of the Congo (Kinshasa)

The Western Foreland Shelf (WFS) of the Central African Copper Belt (CACB) in the Democratic Republic of Congo (DRC) is a sedimentary rift basin that can be traced from the DRC, through NW Zambia into Angola. The WFS is host to the Kamoa-Kakula Cu complex in the DRC and as such its evolution is of interest. Regionally extensive mafic intrusions and volcanic rocks are found along the length of the WFS and have been attributed to the onset of rifting.

The relative age for the volcanic mafic rocks has been established as prior to the onset of Nguba sedimentation at the western edge of the shelf and extending into early Nguba sedimentation on the eastern side of the shelf shown by the occurrence of peprites. Intrusive rocks have been harder to attribute to a relative age as they all occur prior to Nguba sedimentation but with an intrusive relationship to the Mesoproterozoic Kibaran basement.

Whole-rock geochemistry was used to characterise a range of hypabyssal and volcanic rocks found in the Western Foreland of the DRC also with a selection of hypabyssal and plutonic rocks from the adjacent Kibanan belt. The results of the geochemistry and thin section investigation identify the hypabyssal and volcanic rocks from the WFS as formed as part of an evolving magmatic sequence. The basalts have evidence of plagioclase fractionation identified by a negative Eu anomaly. The peperites from the eastern side of the WFS proximal to Kamoa also have a geochemical composition consistent with the basalts from the western edge of the shelf. Plutonic and hypabyssal rocks from the Kibaran basement have a more varied composition. Some consistent with the hypabyssal rocks of the WFS and some with significantly varied composition suggesting they were formed during different magmatic events.

The Significance of In Situ Sr-Nd in Apatite for Constraining Magma Sources and the Origin of Platinum Mineralisation in the Bushveld, South Africa

Peace Zowa, Ben Hayes, Grant M. Bybee
University of the Witwatersrand, Johannesburg, South Africa

The magma source(s), mode of magma emplacement, and metallogenesis of the Rustenburg Layered Suite (RLS) of the Bushveld Complex remain controversial despite decades of research. There is an agreement, however, that the RLS was formed through (1) the mixing of multiple pulses of magma in a chemically evolving magma chamber and (2) there was significant pre-emplacement crustal and/or lithospheric mantle contamination of the parent magmas. This is evident in bulk rock variations in Sr-Nd isotopes. On the other hand, Lu-Hf isotopes of accessory zircon show a limited range in εHf (at 2.06 Ga) of -8.6 ± 1.2 throughout the entire RLS. These homogeneous εHf isotopic values indicate decoupling between the major rock-forming silicates and the accessory minerals. The bulk rock Sr-Nd isotopic data therefore provide only a partial story about the solidification history of the RLS. Competing models have been suggested to explain the homogeneous εHf isotopic signature, including melting/contamination of asthenosphere-derived Bushveld magmas by the sub-continental lithospheric mantle and high-level crustal contamination of the RLS by fluids derived from the Transvaal Supergroup. Apatite is also an accessory mineral in the RLS that crystallised in close association with zircon. We analysed Sr-Nd isotopes in apatite using LA-MC-ICP-MS and show that apatite records homogeneous isotopic compositions (average initial $^{87}Sr/^{86}Sr$ of 0.7088 and εNd of ~-7 ± 1.4 at 2.06 Ga). Our Sr-Nd isotopic data in apatite corroborate the existence of isotopic decoupling in the RLS cumulates. We speculate on the origin of the isotopic decoupling with hypotheses including (i) melt percolation in the crystal mush that led to isotopic homogenisation of the trapped melt from which zircon and apatite crystallised and (ii) the emplacement of isotopically distinct cargoes of silicate crystals in an isotopically homogeneous carrier liquid during the emplacement of magmas into the RLS chamber.

Gold Remobilization in Gossans of the Usakos Dome in the Southern Central Zone (sCZ) of the Damara Orogen, Namibia

Abia Amukongo
Geosciences Deprtment, University of Namibia, Keetmanshoop, Namibia

The Usakos Dome is located within the South Central Zone (sCZ) of the Damara belt, and it's characterized by a metamorphosed shelf-to-continental slope-type rock sequence known as the Swakop group, characterized by a metamorphosed host-rock sequence at amphibolite-facies conditions. The area exhibits a geological composition of a thin layer dominated by marble of the Okawayo Formation overlying the Spes Bona Formation, followed by quartz-biotite schists of the Oberwasser Formation. Within the Okawayo are the oxidized sulfide facies of interest to this study, gossans. Gossans are iron-rich regolith that are produced by the oxidative weathering of primary metal sulfides and they represent an evolving geochemical system, that may influence gold solubility and precipitation. This active gold geochemical cycle occurs during the transition from sulfide- to oxide facies, often coinciding with gossan formation. Understanding the dynamics of gossans is essential, particularly in areas like the Usakos Dome within the Damara belt, where gold mobility within these formations may have significant implications for mineral exploration. The study involves geological mapping in conjunction with petrography and various geochemical analysis for lithogeochemistry, petrography and geochemistry. The geological mapping reveals that the gossans in the Usakos dome are found at the contact of the Okawayo marble and Obberwasser, with a stronger presence in the Okawayo marble. These gossan outcrops consist mainly of weathered quartz-carbonate veins containing a significant amount of sulfide minerals. Thin sections reveal the mineral composition of gossans within the area typically including goethite, hematite, malachite, siderite, and quartz, the assay results show an average gold concentration of 0.2 grams per ton.

Invisible Gold Mineralization within Korivipalle Dolostone and Dhone Dolerites: Implications from EPMA and LA-ICP-MS Analysis of Pyrite

Ramesh C. Behera[1], Sahendra Singh[1], K S V. Kumar[2], Vinod K. Tirkey[3], Ramakrishna Reddy[5], S Shalivahan[6]
1. INDIAN INSTITUTE OF TECHNOLOGY (INDIAN SCHOOL OF MINES) DHANBAD, DHANBAD, JH, India, 2. URANIUM CORPORATION OF INDIA LIMITED, TUMMALAPALLE, AP, India, 3. St. Xavier's College, RANCHI, JH, India, 4. SENIOR RESEARCH FELLOW, DHANBAD, JH, India, 5. Loyola College, Pulivendla, AP, India, 6. Indian Institute of Petroleum & Energy, Visakhapatnam, AP, India

The study focuses on auriferous sulfides hosted in the Vempalle dolostone to elucidate the genesis of invisible gold mineralization in the Korivipalle area of the Cuddapah Basin, southern India. Petrographic examination, SEM, and EPMA data reveal various sulfide phases including pyrite, chalcopyrite, and galena. Textural characteristics and Co/Ni values differentiate two pyrite generations: sedimentary pyrite (Py1) and hydrothermal pyrite (Py2). Py1 grains are finer and anhedral, while Py2 grains are coarser and euhedral, associated with quartz calcite veins. The Au vs As diagram obtained from Laser Ablation Inductively Coupled Mass Spectrometry analysis suggests invisible gold occurs both as solid solution and nano inclusions within sulfides. Nano-gold inclusions in pyrite form due to increased solubility during crystallization and exsolution of Au nanoparticles during later mobilization. Py2 exhibits higher gold concentration than Py1 due to dissolution and remobilization by later hydrothermal fluid. Organic activities provided the redox environment for sulfide precipitation, evident by their presence along oolitic boundaries.

In Dhone, a dolerite dyke intrudes the dolostones, rich in pyrite, chalcopyrite, and magnetite, with accessory phases of covellite, galena, and martite. Texture-based classification identifies two pyrite types: coarser, porous Py1 and finer, euhedral Py2. Py1 and chalcopyrite are mainly located along vertical to near-vertical fracture zones, likely formed during the dyke's cooling phase. This is evidenced by crosscutting relationships of Py1 with host silicates minerals. Py2 originates from a higher temperature fluid, inferred from enrichment of high-temperature elements like Bi, Te, Co. Py2 has lower Au content due to gold instability at higher temperatures. All points on the Au vs As diagram lie below the gold solubility curve, indicating solid solution mode of occurrence. In conclusion, gold was initially present within the Py1 phase and later remobilized by a high-temperature fluid, resulting in a lesser concentration in Py2 compared to Py1.

Liquid Bismuth Collector as a Gold Concentration Mechanism in a Late-Variscan Gneiss Dome: Martinamor Case of Study

Daniel Bermejo-López[1], Lorena Ortega[1], Santos Barrios[2], Pedro Castiñeiras[1], Elena Crespo[1], Juan Gómez-Barreiro[2]
1. Complutense University of Madrid, Madrid, Spain, 2. University of Salamanca, Salamanca, Spain

The Morille-Martinamor Sn-W mining district (Salamanca, Spain) is situated within the Martinamor late-Variscan gneiss dome. This dome evolved during the late Variscan gravitational collapse, leading to the formation of various styles of Sn and W mineralization during its exhumation. These Sn and W deposits were mined, particularly during the Second World War and in the 1970s and 1980s. Historical records indicate the occurrence of gold in the tin and tungsten concentrates. Additionally, alluvial gold can be found nowadays in the district streams. Limited descriptions of gold in stibnite-bearing quartz veins have been documented in the area.

This study focuses on arsenopyrite-bearing veins from Sn and W mines, in which gold has been identified. This occurs within a Bi-Te-Au association occurring as inclusions within both arsenopyrite and quartz. Gold in these inclusions occur in form of electrum (25-50 % Ag) and is mainly related to native bismuth and Bi-sulfotellurides (such as joseite-A, ikunolite, nevskite, and an unidentified Bi-sulfotelluride), and occasionally with Bi-Ag-Pb sulfosalts and sulfides, such as sphalerite and stannite. In most cases, these inclusions exhibit droplet-like morphologies and are displayed following inclusion trails. A thermometric study was also conducted on these inclusions using the Zn/Fe partition thermometer between sphalerite and stannite and it revealed a calculated temperature range from 263 to 277°C. The melting temperature of bismuth (271°C) that falls within the estimated range of temperatures, along with the textural study suggest that the transportation of native bismuth was in the form of immiscible melt droplets in hydrothermal fluids, at least during the arsenopyrite mineralization stage. Consequently, the liquid bismuth collector model likely acted as the mechanism for gold concentration, scavenging gold from the fluids and depositing it as electrum.

Structural Evolution of the Magino Gold Mine, Wawa Subprovince, Ontario: An Overprinted Archean Intrusion-Related Deposit

Ian Campos[1, 3], Bruno Lafrance[1], Ross Sherlock[1], Stefan Kruse[2]
1. Mineral Exploration Research Centre (MERC), Harquail School of Earth Sciences, Laurentian University, Sudbury, ON, Canada, 2. Terrane Geoscience Inc., Halifax, NS, Canada, 3. Agnico Eagle Mines Limited, Montreal, QC, Canada

The Magino gold deposit, located within the Michipicoten greenstone belt of the Wawa subprovince (Ontario, Canada), is a past-producing underground mine that has recently begun production as an open-pit gold mine with total resources of 5.4 Moz of Au. Gold is primarily hosted within the ca. 2724 Ma Webb Lake stock (WLS), a steeply dipping, tabular multiphase tonalitic intrusion oriented parallel to the strong regional S_2 foliation along the Goudreau Lake deformation zone. The Magino deposit underwent two gold mineralization events (Au_1, Au_2) and three episodes of ductile deformation (D_2, D_3, D_4). The main gold event (Au_1) is expressed by pervasive biotite-phengite-quartz-pyrite alteration of the stock and the emplacement of auriferous sugary quartz veins. The veins have a molybdenite Re-Os age of 2731 ± 6.9 Ma, which overlaps within error with the age of the WLS. Quartz-feldspar porphyry dikes, cogenetic with the WLS, cut the veins, both of which are transposed, stretched, and folded parallel to the S_2 foliation within high strain corridors in the WLS, supporting a pre-D_2 timing for the Au_1 event. These high-strain corridors formed by strain localization along Au_1 hydrothermally altered zones during D_2 N–S shortening. The Au_2 event is associated with structurally controlled syn-D_2, N-trending quartz-tourmaline-carbonate veins with albite-paragonite-ankerite-pyrite alteration selvages. The D_3 event resulted in dextral reactivation of the high-strain corridors. D_4 vertical loading produced a subhorizontal crenulation cleavage, which is overprinted by chloritoid porphyroblasts that grew during late upper greenschist-facies peak metamorphism. The Magino deposit represents a paragenetically complex Archean intrusion-related gold system, which was structurally modified and overprinted by a second gold mineralization event.

Geochemistry, Petrography and Mineralogy of Intrusive Metamorphic Dykes and Their Relation with Gold Occurances in the South of Minas Gerais, Brazil

Gabriela O. Carvalho, Gustavo L. Pires, Silvia R. Medeiros
Universidade Federal do Rio de Janeiro, Rio de Janeiro, RJ, Brazil

The Tiradentes Formation, located in the south of the São Francisco Craton, Minas Gerais, Brazil, consists of a succession of quartzites and metapelites deposited in a Mesoproterozoic basin of an aulacogenic rift type, the São João Del Rei Basin. This is overlain by a succession of metacarbonates and metapelites which form the base of the Carandaí Basin. The group was deformed and metamorphized under greenschist facies conditions during the Brazilian orogeny in the Neoproterozoic. The Fm. Tiradentes is cut by mafic dykes associated with the rifting that originated the Carandaí Basin. The importance of the study is due to the history of gold extraction in the region in the 17th century, with a lack of in-depth studies about the mineralization. The aim of the work is to describe key outcrops, as well as to collect 29 samples for petrography, X-ray diffraction and lithogeochemistry, in order to understand the genesis of the magmatism and the correlation with the mineralization. The results showed that the dykes are tabular bodies of metric to decametric thickness with an ENE-WSW orientation and cut the rocks of the Fm. Tiradentes, but do not intrude the rocks of the Carandaí Basin. In the outcrops, the rocks are locally cut by quartz veins parallel to the foliation. From the petrography, three lithologies were separated: chlorite, porphyritic aphanitic metabasite and porphyritic metabasite. DRX analysis identified chamosite, quartz, epidote, albite, magnetite, pyrite and rutile. The quartz are interpreted as metamorphic and the occurrence of gold is associated with portions of the dykes with quartz veins parallel to the cleavage. During the description of mineral grains, gold was identified in the quartz veins near the dykes, but no gold grains were observed in the mafic rocks. In this context, the dykes may have acted as a physical-chemical barrier to the mineralization.

The Buritica Gold Deposit, Middle Cauca Metallogenic Belt, Colombia

Daniel A. Castano Madrigal
University of Texas at El Paso, El Paso, TX, USA

The geology of Colombia is dominated by the collision between oceanic island arc systems and the South American craton. During the Miocene, several porphyry and epithermal deposits were formed in the Middle Cauca metallogenic belt associated with subduction-zone magmatism between 6 and 8 Ma, including Buritica, Marmato, Quinchia, and La Colosa.

The Buritica deposit is located west of the Cauca-Romeral fault system, and it may be classified as an intermediate-sulfidation epithermal Au deposit. The regional geology is characterized by volcano-sedimentary units of the Early Cretaceous Barroso Formation intruded by Late Cretaceous Santa Fe tonalite and Buritica Tonalite, and Miocene andesite, diorite, and intrusive/hydrothermal breccias of the Buritica Intrusive Complex, the latter dated at 7.4 Ma. The age of gold mineralization based on hydrothermal white mica is 6 to 8 Ma. Two main vein systems are present at Buritica, Yaragua, and Veta Sur. Overall, twenty-four individual veins cut all lithologies and present alteration halos of varying intensity. Three stages of mineralization are identified at Buritica, Stage 0 is an early porphyry style mineralization characterized by potassic alteration and low-grade gold, recognized in hydrothermal breccia clasts overprinted by white mica alteration. Stage 1 presents the highest gold concentration as is characterized by veins and breccias with sphalerite, galena, chalcopyrite, and pyrite, plus quartz and carbonate gangue. Stage 2 is characterized by carbonate, tetrahedrite, and gold in the form of veins and breccias that cut Stage 1. Low-grade ore shoots with 1-3 g/ton gold are also present. Drill hole geochemistry shows a high relationship between gold and zinc, copper, and silver. The average Au:Ag is 1.60, and boiling is proposed as the main control on gold mineralization, it was probably triggered by the pressure drop from lithostatic to hydrostatic between stages 1 and 2.

Geneses of Gold Mineralization in the Mugomo and Chifumbazi Prospects of the Mualádzi Gold Field, Tete Province, Northern Mozambique

Euclesia P. Cossa[1], Andrea Agangi[2], Manuel Nopeia[3], Vicente Manjate[4]
1. Tete Polytechnic Institute, Tete, Mozambique, 2. Akita University, Akita, Japan, 3. Center for Regional Revitalization in Research and Education, Akita University, Japan, Akita, Japan, 4. National Institute of Mines, Maputo, Mozambique

The Mualádzi gold field is one of several gold occurrences in the northwestern part of Tete province, northern Mozambique. The Mualádzi gold field is hosted in a terrane composed of metavolcanic rocks of the Proterozoic Mualádzi Group, part of the Southern Irumide Belt. This study aims to describe the gold mineralization and the associated hydrothermal alteration in the Mugomo and Chifumbazi prospects, based on petrography, bulk chemistry, fluid inclusion, and sulfur isotope analyses.

The gold mineralization in this gold field occurs mainly as dissemination in the metavolcanic rocks of the Mugomo prospect. Part of the gold mineralization is also associated with epidote, quartz and chlorite veins and veinlets hosted by metavolcanic rocks and quartzite, as well as milky quartz, quartz+carbonate, and chlorite veins hosted by sericite-quartz-schists of the Chifumbazi prospect.

Of the three mineralization stages identified in the area, the most important ones are stages I in the Mugomo prospect and stage II on the Chifumbazi prospect. The stage I is characterized by visible gold in pyrite disseminated in the metavolcanic rocks, whose ores contain up to 2 ppm Au. Quartz veins in the stage II of the Chifumbazi prospect consist of visible gold associated with pyrite and chalcopyrite, and contain up to 5 ppm Au. Homogenization temperatures and salinity of fluid inclusions in quartz veins of the Mugomo prospect range from 200 to 430 °C and from 0.0 to 15.0 wt.% NaCl eq. While those in Stage II of the Chifumbazi prospect range from 216 to 369 °C and from 0.7 to 15.3 wt.% NaCl eq., respectively.

The Mugomo prospect was formed by H_2O-NaCl-CH_4 fluids, while the Chifumbazi prospect was formed by H_2O-NaCl-CO_2 fluids. Sulfur isotopic compositions of sulfides are higher in the Mugomo prospect (+2.5 to +3.8 ‰) than those in the Chifumbazi prospect (+1.0 to +2.5 ‰).

Chlorite Composition as a Tracer for Mineral Exploration of the Disseminated Gold Deposit

Tao Cui[1], Kunfeng Qiu[1], Lejun Zhang[2]
1. China University of Geosciences, Beijing, Beijing, China, 2. University of Tasmania, Hobart, TAS, Australia

Orogenic gold deposits are one of the most important styles of gold mineralization. Despite extensive research that has been published, some crucial questions regarding metallogenic process of disseminated ore remain unresolved. Much of this uncertainty is due to the absence of fluid inclusions, which significantly complicates understanding the mechanisms of mineral precipitation. Chlorite is one of the most widely developed altered minerals in orogenic gold deposits. In this contribution, we conducted petrographic and geochemical analyses on chlorite in barren and ore-bearing marble, granite, and leptynite to better understand the hydrothermal alteration processes, Au precipitation mechanisms, and to provide insights into ore-forming tracer. Ripidolite was identified as the primary chlorite phase in the wall rocks, replacing biotite, hornblende, and plagioclase through dissolution-crystallization processes, which suggests relatively low water-to-rock ratio conditions. In contrast, the chlorite in the ores consists primarily brunsvigite and pyrochlore, often coexisting with pyrite, with no residual precursor minerals, which suggests formation from dissolution-migration-crystallization processes under high water-to-rock ratio environment. Mineral chemistry analyses show that chlorite in ores exhibits relatively higher Mg / (Fe + Mg) ratio, which possibly indicating a relatively higher oxygen fugacity in a reducing environment. A comprehensive investigation revealed that the relative increase in fO_2 leads to a drop in the concentration of HS^- and consequently, metal precipitation. Chlorite formed during this process also exhibits increased Mn concentration and decreased Ti, Ni, Co, and Rb concentrations, with a lower Rb/Sr ratio, indicating the limited mobility of weakly reactive elements within the ore-forming fluid. Thus, we conclude that Mg / (Fe +Mg) ratio, Rb/Sr ratio, and concentrations of Ti, Ni, Co, and Rb in chlorite could potentially serve as an ore-forming indicator in the Guocheng gold deposit.

Morphochemical and Isotopic Analysis (δ65Cu) of Gold from the Serra Dourada and Bananal Placer Deposits in the Aguapeí Belt, Southwest of the Amazonian Craton, Brazil

Kelvin dos Santos Alves[1], Juan G. Barreiro[1], Santos B. Sánchez[1], Ryan Mathur[2]
1. Universidad de Salamanca, Salamanca, Spain, 2. Department of geology, Juanita College, Huntingdon, PA, USA

And understanding of the composition of gold and its morphochemical evolution in supergene environment is important for the development of prospecting techniques for new deposits. In in this work, we study the morphochemical and isotopic footprint ($\delta^{65}Cu$) of gold from the Serra Dourada (SD-primary) and Bananal placer (BP-alluvial) deposits. The latter formed after the erosion and transport of gold from SD. In both deposits, the core of the gold (Au1) is formed by a ternary alloy of Au-Ag-Cu, with variable composition and isotopic footprint ($\delta^{65}Cu$), divided into 3 geochemical groups: SDG, composed of $Au_{99-100}Ag_{0-0.1}Cu_{0-0.1}$-($\delta^{65}Cu$ = + 0.30%), formed by SD crystals, and groups BG1 and BG2, formed by BP nuggets, with compositions $Au_{98-99}Ag_{1-2}Cu_{0-0.03}$-($\delta^{65}Cu$ = + 0.10%) and $Au_{87-99}Ag_{1-13}Cu_{0-0.17}$-($\delta^{65}Cu$ = - 1.45%), respectively. These values are typical of hypogene gold and indicate a compositional and isotopic zoning of the primary deposit. In addition to Au1, the BP nuggets contain golds Au2 and Au3, which exhibit high purity (aprox. 100% Au wt) and signs of leaching (Au2), dissolution (Au3) and reprecipitation (Au3), result of the chemical refinement of gold in supergene environments. Morphologically, the size of gold in SD and BP varies between 1 and 11 mm and comprises 3 morphological populations. SDP, formed by SD crystals, exhibits primary morphologies. In BP the nuggets comprise two morphological groups differentiated by the contrast the degree of morphological maturity of the nuggets, caused by different transport conditions. BP1 is formed by nuggets with primary morphological features. BP2 is formed by rounded nuggets. The mean values of the flatness index related to the transport distance Cailleux Flatness Index y Shilo are 2.11-1.11 and 2.67-1.67 for BP1 and BP2. These values suggest a transport distance between 2 and 4 km from their source, consistent with the distance between SD and BP.

Geochemical Analysis of Native Gold from the Roman Mining Area of Pino del Oro (Zamora, Spain)

Vanessa L. Escobar Duche, Santos Barrios Sánchez, Kelvin dos Santos Alves, Juan Gómez Barreiro
Universidad de Salamanca, Salamanca, Spain

In the Pino del Oro area there are two types of gold deposits: 1) hydrothermal veins with quartz and sulphides related to the Villalcampo shear zone that present abundant remains of Roman mining and 2) Quaternary placer deposits. Previous research works focused on the evaluation of its content, but did not characterize its geochemical composition. In this study, we have determined its "geochemical signature", which provides information about the primary deposit and could be used in the traceability of Roman archaeological pieces.

Six sediment samples along the Fuentelarraya creek (secondary) and 3 soil samples from the old mining works (primary) were washed with a sluice box and a pan. A total of 454 gold particles and 1 lode quartz fragment with visible gold were obtained. 660 analyses were carried out in gold using EMPA, while the mineral inclusions were determined by SEM. Two types of Au:Ag gold alloys have been identified: 1) Au1: Au75.32-83.44:Ag24.64-16.53 (Hg0.01-0.04, Te0.01) reflecting the geochemical signature of the primary source, and 2) Au2: Au99.49 resulting from a composition modified by supergene processes, mainly for placer particles. Mineral inclusions include arsenopyrite, galena, sphalerite, pyrrhotite, pyrite, chalcopyrite, cobaltite, gersdorffite, ullmanite and berthierite. Placer particles show higher Ag contents than particles from soils, which indicates that there is a vertical compositional zoning in the primary deposit. Previous studies of fluid inclusions indicate that gold precipitation took place at <200ºC. The low fineness of Au1 could be indicative of a low formation temperature, coinciding with those results. Mineral inclusions are not associated with a specific composition of gold, but they would have previously precipitated between 250°C and 390°C.

Thanks to a Florián de Ocampo 2022 research grant.

Geological Mapping and Reinterpretation of the Gold Prospect "Descoberto" Located Near the World-Class Cuiabá Orogenic Gold Deposit, Minas Gerais, Brazil

Beatriz A. F. Andrade[1], Carolina P. N. Moreto[1], Ricardo O. A. Mabub[2]
1. University of Campinas (UNICAMP), Institute of Geosciences, Campinas, SP, Brazil, 2. AngloGold Ashanti Córrego do Sítio Mineração S.A, Sabará, MG, Brazil

The Iron Quadrangle is a world-renowned mineral district known for its iron, manganese, gold, gems, and industry minerals deposits. It is located in the southern part of the São Francisco Craton in Minas Gerais, Brazil, and responsible mining and studies of gold prospects are encouraged in the region due to its historical gold extraction since the 18th century. However, several questions remain open, with emphasis on a better understanding of the structural architecture that hosts the mineralization at the deposit scale. The study area is within the geological context of the metavolcanosedimentary sequence of the Archean Rio das Velhas greenstone belt, with mineralization predominantly hosted in quartz veins. The Descoberto prospect area presents a scarcity of outcrops and is inserted in an area of dense vegetation cover, with a high degree of weathering, steep reliefs, and well-developed soils due to the tropical climate. In this context, the integration of geochemical, geophysical, and SRTM data with field mapping greatly assists in geological interpretation at local and regional scales. Detailed cartographic studies have been conducted in the region in the past; however, the most recent interpretation of structural geology and drillhole data suggests a new cartographic interpretation of the area related to a geometry change of the main structure where the prospect is hosted. Although the latter interpretation is debatable, the integration of existing historical data into this research is valuable, along with newly acquired data. For this research, pre-field cartographic products were carried out, and geophysical surveys of Induced Polarization (IP), regional geophysical surveys, multi-element soil analyses, as well as a set of geological interpretations from field mapping and drilling campaigns, were considered.

Tellurium and Selenium in and Epithermal Gold Deposit in the Chatkal-Kurama Region (Uzbekistan)

Danis I. Filimon[1], Rustam A. Khalmatov[2], Anna Gioncada[1]
1. Earth Sciences Department, University of Pisa, Pisa, Italy, 2. Institute of Geology and Geophysics, University of Geological Sciences, Tashkent, Uzbekistan

The average concentration of tellurium (Te) and selenium (Se) in the Earth's crust is low, being 100 parts per billion (ppb) and only 2 ppb, respectively. Similarly to other Group VI elements, they can occur in the form of mineral groups such as tellurides and selenides. A characteristic feature of these elements is their potential association with gold (Au), silver (Ag), bismuth (Bi), lead (Pb), and copper (Cu) as manifested in gold-(silver)-telluride-(selenide) deposits.

The Chatkal-Kurama region presents outcrops of the Valerianov-Beltau-Kurama Magmatic Arc, an insular arc formed by the subduction of the Turkestan Ocean. Connected with Paleozoic magmatic activity, several porphyry, epithermal, and skarn ore deposits formed. Kochbulak and Kyzylalmasay represent two epithermal deposits, respectively, with Au-Te and Au-Se linked to the last intense phase of magmatism (late Carboniferous-early Permian). The geochemical anomaly of Te and Se is evident in the complex mineralogy of the deposits, which binds tellurium (hessite, petzite, calaverite) and selenium (naumannite, aguilarite) directly to Au and Ag, which also occur in native form, electrum, in sulfides and sulfosalts (freibergite, polybasite). The mineralizing phases are characterized by different events, as often happens in these cases, but the events that bind Au-Te and Au-Se are well-defined in the paragenetic sequences.

The different geochemical behavior of Te and Se controls their association with Au, Ag, and the resulting complex geochemistry of the ore deposits. Understanding the genesis of the Au-Te and Au-Se deposits in Chatkal-Kurama may contribute to clarifying whether they could behave as useful elements for mineral exploration of epithermal gold-(silver)-telluride-(selenide) deposits.

Characterization of the Gold Hill Low Sulfidation Epithermal Au-Ag Vein Deposit, Northern Nye County, Nevada

Silas L. Goetz, Simon Jowitt
University of Nevada, Reno, Reno, NV, USA

The Gold Hill low-sulfidation epithermal (LSE) vein system is an exceptionally well-preserved and extensive Oligocene-age deposit located within Round Mountain mining district of northern Nye County, Nevada, USA Gold Hill seven kilometers north of the world-class Round Mountain LSE deposit, which has produced >16 million ounces of gold to date. Although there is a close spatial and temporal relationship between Gold Hill and Round Mountain, the genetic relationship (if any) between these two systems remains unclear. This study presents new data that characterizes the ore and gangue mineralogy of at least three stages of mineralization at Gold Hill as follows: Stage I), Pyrite-breccias and quartz- adularia- K-feldspar alteration of wallrock angular clasts polymictic-monomictic breccias with pyrite +/- As-rich rims. Stage II), white-tan-pink medium-coarse grained, massive-ghost-bladed- saccrodial- quartz veins with pyrite +/- marcasite +/- electrum +/- Cu-Zn metal +/- chalcopyrite, and Stage III), fine-grained crustiform-colloform quartz-chalcedony veins with pyrite, electrum, proustite-pyrargyrite, argentiferous tetrahedrite-tennantite, and minor arsenopyrite. Economic Au-Ag mineralization is restricted to the base of the boiling zone ~450 m below the base of the sinter. Wallrock alteration envelops high grade portions of the vein-hosted orebodies and typically consists of a quartz-adularia-illite-pyrite assemblage that replaces primary feldspars. This alteration transitions to a K-feldspar-interlayered illite/smectite-quartz +/- sericite assemblage adjacent to inferred high-temperature upwelling zones. Pyrite, chalcopyrite and acanthite samples from the Gold Hill deposit have $\delta^{34}S$ values that range from -14.4‰ to +16.6‰, including a pyrite-chalcopyrite mineral pair that yielded a temperature estimate of 226°C. Gold Hill appears to be a unique vein system, outboard of the Round Mountain deposit, that shows distinct base-metal anomalies associated with high-grade veins, a subvertical nature of the vein system, and a close spatial association to intrusive dike units, indicating potential for other additional deposits within the Round Mountain mining district.

Understanding Microstructure of Gold Particles in Placers. A New Vector for Transport Intensity?

Juan Gomez-Barreiro[1], Santos Barrios Sánchez[1], Inéz Puente-Orench[2], Juan Morales Sánchez Migallón[1]
1. University of Salamanca, Salamanca, Spain, 2. Institute Laue-Langevin, Grenoble, France

Gold particles in alluvial environment are subjected to several processed that modify their primary signature. Weathering combined with several cycles of erosion and transport result into a heterogeneous modification of both composition and morphology. Besides, biomineralization could contribute to particle growth in supergene environments under specific circumstances. While de-alloying processes in supergene context and morphology-to-distance relationships have been investigated, little is known about the microstructural imprint in gold particles in alluvial deposits. The identification of microstructural vectors and its correlation with other factors could be critical to refine distance-to bedrock primary sources. In this work we explore gold microstructure of alluvial gold particles with distinct Corey Shape factor from Central Spain by a combination of nondestructive synchrotron and neutron diffraction analysis. Our results discuss the potential of those techniques on the identification of particle population mixing, quantitative description of individual gold particles and transport intensity.

Chemical Composition of Garnets and the Relation to Mineralizing Fluids in the Tucano Gold Deposit, Amapá State, Brazil

Deniro F. Gonçalves Costa[1], Gabriel A. Rodrigues Soares[1], Rosaline C. Figueiredo e Silva[1], Mahyra F. Tedeschi[1], Steffen G. Hagemann[2]
1. Graduate Program in Geology, Federal University of Minas Gerais (UFMG), Belo Horizonte, MG, Brazil,
2. School of Earth Sciences/Centre for Exploration Targeting, The University of Western Australia, Crawley, WA, Australia

The Tucano gold deposit, located in the northeastern segment of the Amazon Craton, within the Guiana Shield (Amapá State, Brazil) is a structurally controlled deposit hosted in amphibolite facies metasedimentary rocks of the Paleoproterozoic Serra do Navio greenstone belt, mainly in marble and banded iron formation (BIF). Mineralization is controlled by the N-S trending brittle-ductile Urucum shear zone and associated subsidiary structures. Hydrothermal alteration is characterized by skarn-type calc-silicate assemblages replacing metamorphic minerals and overprinting precursors' fabric. Since garnets are important minerals to consider further relation with mineralizing fluids, this work aims to investigate their chemical composition indicating similarities with gold mineralization. Petrographic investigation identified gold (< 100 µm) as interstitial or fracture-filling in silicates (amphibole, clinopyroxene, garnet). A petrological study was conducted on select diamond drill core samples analyzed by electron microprobe (EPMA), aiming to investigate the compositional variation between minerals from the metamorphic, igneous, and hydrothermal rocks on the area, with focus on garnet. X-ray semi-quantitative maps were reduced with XMapTools. Hydrothermal garnet shows a large compositional range, but in general it contains higher grossular (4 – 34%) and some minor andradite (up to 8%), when compared to metamorphic and magmatic ones. A Garnet-biotite geothermometry yielded the estimate for peak metamorphic temperature: 592 ± 25 °C (1 σ, n = 5). The same garnet crystals resulted in pressure estimates of 4.1 ± 0.6 kbars (1 σ, n = 5) using the geobarometer. Orogenic hypozonal gold mineralization at Tucano is characterized by high temperature hydrothermal minerals such as clinopyroxene, garnet, amphiboles, phlogopite and loellingite. Wide distal alteration zones suggest a high fluid/rock ratio, which is further manifested in the alteration mineralogy – clinopyroxene-garnet-amphiboles-phlogopite-magnetite-pyrrhotite – indicating an intense fluid-buffered composition.

Genesis and Mineralogy of the Golddigger Property in British Columbia, Canada

Randall Karcher[1], Simon Kocher[1], Holly Stein[2], Roger Rosmus[3], Simone Pujatti[3], Manuele Lazzarotto[3], Thomas Monecke[1]
1. Center to Advance the Science of Exploration to Reclamation in Mining, Department of Geology and Geological Engineering, Colorado School of Mines, Golden, CO, USA, 2. AIRIE Innosphere Ventures, Fort Collins, CO, USA, 3. Goliath Resources, Toronto, ON, Canada

The Golddigger property is a recent discovery in the Golden Triangle of British Columbia which has drawn interest for its high-grade gold. At Golddigger multiple major structurally controlled quartz veins with semi-massive to massive sulfide accumulations are hosted by Upper-Hazelton sedimentary and volcanic rocks. Ongoing research on the deposit is focused on developing a deposit model to aid with future exploration on the property and in the surrounding region. Methods include petrography of surface and drill core samples by transmitted and reflected light microscopy, cathodoluminescence microscopy, scanning electron microscopy, fluid inclusion petrography, U-Pb geochronology, and Re-Os geochronology. Petrography and fluid inclusion analysis of the quartz veins indicate formation close to the brittle-ductile transition, with hydrothermal fluids having high CO_2 content, indicating entrapment pressures of >500-600 bar (>5-6 km paleodepth). Ore minerals observed include major pyrrhotite, chalcopyrite, pyrite, arsenopyrite, sphalerite, and galena with lesser molybdenite, Bi-Te minerals, Sb minerals, native silver, and native gold. The paragenic sequence at Golddigger is consistent with that of a magmatic-hydrothermal system. Petrographic analysis suggests the presence of spatial mineralogical gradients, which is supported by statistical analysis of geochemical data demonstrating a southerly depletion of Sb and enrichment of Bi. U-Pb geochronology from early-stage garnet alteration and Re-Os geochronology of molybdenite inclusions withing pyrrhotite both suggest mineralization to have occurred in the Eocene (50-55 Ma), potentially linked to the Alice Arm Intrusive suite which is known to be associated with molybdenum and polymetallic deposits in the region. Depth of mineralization, ore mineral assemblage, apparent zoning, and a coincident date with intrusive activity known for polymetallic mineralization all suggest Golddigger to be a reduced intrusion-related gold deposit. This discovery is significant due to the deposit type's distinct characteristics relative to orogenic or epithermal deposits which have implications for exploration.

An Assessment of the Grade Control Spacing at QKR Mine and the Effect on Au Grade Estimation

Ndinelao N. Kashidinge
None, Swakopmund, Namibia

The Navachab gold mine has three plants, with two plants feeding the third plant. Lower grade materials with average gold grades of Au 0.6 - 1.30g/t are fed to the pre-concentration plant. This plant produces materials with high Au concentration, which are fed to the ARGO plant. The ARGO plant mills the material to finer particle sizes and pumps it to the main CIP plant for leaching. Medium to high grade materials (1.30g/t to ceiling) are fed to the main primary crusher. The crushed materials are stored in a silo going to the SAG mill to be milled to the required size fraction. These materials join the ARGO material and are all leached together. The primary crusher material known as CIP ore, has shown less variance when comparing feed grades to plant reported grades. From January to June 2023, the plant reported grades (Head grades) have been higher than the feed grades. However, from July to October 2023, the head grades have been reporting less than the feed grades. This has prompted the grade control geologist to assess the current drilling pattern, suggesting a reduction in the distance between holes in the X (north) direction from 12.5m to 6.25m. This assessment aimed to confirm the continuity of mineralization in the x direction to delineate ore and waste materials. The grade control drill pattern for the assessment will change to X(north) 6.25m by Y(east) 6.25m, while sampling procedures remain the same. The geologist also agreed to change the pattern to investigate if the current grade control pattern (12.5m by 6.25m) determined by geostatistics analysis is efficient for sparsely distributed ore surrounded by waste, specifically the pre-concentrated ore (PCP), which has recorded for the year 2023 head grades and feed grades discrepancies averaging at 20%.

Studying the Internal Structure of Native Gold Using the EBSD Method

Aleksandr Kozin
Institute of Mineralogy, Urals Branch, Russian Academy of Sciences, Miass, Russia

The internal structure of placers gold was studied using the EBSD method to determine changes in its structure under exogenous conditions.

Samples were collected from placers in the Middle and Southern Urals and Eastern Kazakhstan, associated with gold-quartz type deposits.

Aggregates of placer gold are characterized by a granular structure; single crystals are less common. Grains with a size of 30-300 μm have an isometric shape and smoothly curved boundaries. Particles from short-distance diluvial placers are practically not deformed and have not experienced significant exogenous changes. The degree of disruption of the primary structure of gold increases as placers move away from the primary sources. Plastic deformations are expressed by large zones of misorientation of the crystal structure and twining. An important feature of gold grains that have experienced deformation during transportation is the development of high-purity gold rims in the peripheral zones of the particles. The rims have a fine-crystalline mosaic structure and are composed of polyhedral grains of pure gold. The thickness of the rims usually does not exceed 10-30 μm. The rims are homogeneous in structure and lack signs of deformation, while the core parts are significantly deformed.

The regular increase in the number of plastic deformations of gold with increasing distance from the lode source confirms the logical assumption that the main reason for the deformation of placer gold is transportation in a hydrodynamic environment. The fine-grained mosaic structure of gold rich rims, the absence of deformations of the grains composing them indicate the formation of rims as a result of recrystallization of deformed zones of native gold aggregates, accompanied by the removal of impurities.

Mg Isotopes of Carbonate from the Shuiyindong Carlin-Type Gold Deposits of the Youjiang Basin (Southwest China) Provide Insights into Gold Genesis

Jialong Liu[1], Xiaoye Jin[1], Xiaodong Deng[2], Jianwei Li[1, 2]
1. School of Earth Resources, China University of Geosciences, Wuhan, China, 2. State Key Laboratory of Geological Processes and Mineral Resources& School of Earth Resources, China University of Geosciences, Wuhan, China

The Youjiang basin in SW China hosts the world's second largest Carlin-type gold province after Nevada in USA. Gold deposits in the Youjiang basin are mostly stratabound and hosted in the Permian marine limestone and intercalations of volcaniclastic sedimentary sequences, with minor ores in Triassic carbonates and clastic rocks. The localization and distribution of the orebodies are structurally controlled by gentle folds and associated brittle faults. Gold mineralization is closely associated with hydrothermal alteration, including decarbonatization, silicification, sulfidation, dolomitization, and argillization. Despite extensive research for decades, genesis of most gold deposits in the Youjiang basin remains elusive.

Shuiyindong, the largest Carlin-type gold deposits with over 300 tonnes of proven gold reserves in Youjiang basin, is characterized by abundant hydrothermal dolomite associated with gold mineralization. Here we present Mg isotopes of hydrothermal dolomite and its precursor in the ore-hosting rocks, combined with mineralogical and geochemical data, to provide better understanding in the genesis and process of gold mineralization. The ore-related dolomite tends to decrease from the main ore zone to the ore-hosting rocks. There is no systematic difference in CaO and MgO between the ores and wall rocks, but dolomite in the ores has much higher $\delta^{26}Mg$ values (-1.52‰ to -0.77‰) compared to the wall rocks (-2.37‰ to -2.01‰). Furthermore, a positive correlation between $\delta^{26}Mg$ values and gold contents, with $\delta^{26}Mg$ values increasing progressively from wall rocks ($\delta^{26}Mg$ = -2.21±0.05‰, 2SD) through the weakly altered rocks periphery ($\delta^{26}Mg$ = -1.52±0.08‰, 2SD) to the gold ores center ($\delta^{26}Mg$ = -0.77±0.08‰, 2SD), suggests a fractionation of Mg isotopes between dolomite and ore fluid. This correlation indicates that Mg isotopes are not only an indicaor for fluid-rock interactions but also serve as a novel tracer for gold mineralization process. Our finding has implications for ore-forming mechanisms of Carlin-type gold deposits.

Relationships Between the Tectonic Evolution and the Formation of Polymetallic Vein Deposits Along the Caledonian-Appalachian Belt

Taija Torvela, Rob Chapman, James Shaw, Georgian Manuc, Shane Webb, Lucia Savastano
University of Leeds, Ores and Mineralization Group, Leeds, United Kingdom

The Caledonian-Appalachian Belt (CAB) hosts several economic and sub-economic polymetallic (Au-Ag-Pb-Zn-Cu) vein systems and is, therefore, an excellent natural laboratory to study the variations in the characteristics of ore deposit and the timing of their formation, within an orogenic belt. We present some of the preliminary outcomes of a large-scale study on the variability of the vein-hosted mineralisation within the CAB, covering c. 20 different polymetallic vein systems within the Grampian Terrane and its equivalents of CAB in Scotland, Northern Ireland, Ireland and Newfoundland.

The multi-disciplinary study combines structural data, age determinations, vein textural mapping and paragenetic interpretation, gold geochemical characterisation, and stable isotope data. We show that although the occurrences show many superficial structural and mineralogical similarities, there are differences between the structural styles, mineralogy and timing of the mineralisation in different areas. At least three, possibly four, distinct mineralisation stages are identified between Ordovician and Early Devonian; however, the main mineralisation pulse is identified at around 420-410 Ma. 'Magmatic' (possibly porphyry/epithermal) and 'orogenic' mineralisation styles are found associated with specific areas which in turn are characterized by distinct structural styles that can be linked to the overall tectonic evolution of the Grampian Terrane around this time.

The study shows what are the critical windows in time and space for mineralisation, whilst demonstrating that the local structural style and ore deposit type can vary significantly even with coeval deposits, depending on the wider tectonic controls on the mineral system. The coeval nature and mineralogical similarities of the 420-410 Ma vein systems across the area suggest that the entire terrane was fertile and that the sources for the main metals were likely similar for all vein systems.

The Structural Setting of Navachab Gold Mine and Satellite Orebodies: Insights for Exploration

Richard R. Manyanga
QKR Namibia Navachab Gold Mine, Karibib, Namibia

The Navachab Gold Deposit, central Namibia, is situated in amphibolite-facies, marble-schist-dominated metasediments of the Neoproterozoic Damara Sequence in the Pan-African Damara Belt. The deposit is hosted by continental-shelf-type metasediments, primarily biotite-schists, marbles, and calcsilicates. Gold mineralization is hosted by two main sets of auriferous quartz veins that are developed on the subvertical northwestern limb of the regional-scale anticline, the Karibib dome (KD). The orientation, relative timing and progressive deformation of quartz veins indicate that veining occurred during folding-and-fold amplification of the KD. This deformation forms part of the main phase of northwest-directed collisional tectonics in the Damara belt at ~530-510 Ma. The two main vein sets include bedding-parallel, shallowly plunging ore lenses/shoots at the base of a prominent marble unit, and a laterally extensive package of shallowly dipping quartz veins that truncate host rocks at high angles. The bedding-parallel lenses represent dilational jogs that opened during flexural flow along bedding-parallel slip planes during the amplification of the KD. The KD is an asymmetric anticline with a steep limb dipping 75°NW, with siliciclastic and carbonate rock formations of contrasting rheology. Flexural flow and associated bedding-parallel fluid infiltration was concentrated proximal to the contact between marbles and the underlying siliciclastic formations. This contact represented a pronounced rheological contrast, which increased slip rates during fold amplification. The shallowly dipping, sheeted-quartz veins were emplaced during the fold lock-up of the KD. Quartz veining occurred when the KD's northwestern limb rotated subvertically, with the halting of bedding-parallel flexural slip. Extensional fracturing for veining was facilitated by the short-lived supralithostatic fluid pressures. The high-amplitude fluid pressure associated with extensional fracturing likely triggered gold precipitation in the shallow-dipping veins. Crosscutting relationships between the two main vein systems indicate that the two vein sets represent separate entities and succeeded each other rather than forming an interconnected fracture mesh.

Characteristics of the Ore-Forming Fluids of the Giant Karagba-Chauffeur-Durba (KCD) Gold Deposit, Northeastern Democratic Republic of the Congo (DRC)

Rowin P. Marais, Bjorn P. von der Heyden
Stellenbosch University, Stellenbosch, South Africa

The giant KCD gold deposit with more than ten Moz Au, is situated in the Neoarchean Moto Greenstone Belt within the Kibali gold district, DRC. Situated on the Congo Craton, this district is one of the largest gold-producing regions in Africa. Characterized as an orogenic-type gold deposit, the mineralization in this district is a result of complex deformation events, which included thrust faulting, shearing, and folding. Gold mineralisation preferentially localises on reactive banded iron formation (BIF) units, and although the trapping mechanism is thus predictable, the characteristics of the ore fluid have not been fully constrained.

The study objective is to provide detailed characterisation of the ore fluid physicochemsitry inferred from petrographic and chemical evaluations of the wall-rock alteration, and from direct observations of fluid inclusions (microthermometry and Raman spectroscopy). The results from twelve samples collected from four mineralized lodes represented by highly-altered quartz-carbonate-pyrite damage zones, were investigated using 22 fluid inclusion sections, eighteen thin sections and 52 polish mounts. Fluid inclusion results indicate a H_20-NaCl-CO_2-CH_4-N_2 fluid with emplacement temperatures ranging from 156°C - 375°C (avg. 265°C). Petrographic analyses show that the siderite- and pyrite-rich alteration signatures are dominated by carbonic and sulphuric reactions with the iron-rich minerals present in the host lithologies. These reactions suggest the presence of an aqueo-carbonic, sulphur-rich fluid responsible for the transport and ultimate precipitation of gold. These detailed insights into the fluid physicochemistry, coupled with an evolving knowledge of the local structural geology, provide a firmer basis for predicting the localised conditions under which ore shoots are likely to form, thus contributing towards improved exploration paradigms for the area.

Mapping Cryptic Telescoping Events in Porphyry-Epithermal Systems Through Integrated Microtextural-Chemical Pyrite Analysis at the Bonanza-Type Brucejack Au-Ag Deposit

Duncan F. McLeish[1, 2], Anthony E. Williams-Jones[1], James R. Clark[1]
1. McGill University, Montreal, QC, Canada, 2. Newmont Corporation, Vancouver, BC, Canada

Telescoping, the mechanism by which shallow, low-temperature hydrothermal alteration and mineralisation are superimposed onto deep, high-temperature hydrothermal assemblages, is thought to have played a key role in forming some of the world's largest porphyry-epithermal gold deposits. Although the causes of telescoping, namely rapid, erosion-driven syn-hydrothermal degradation of paleosurfaces and sector collapse of volcanic edifices, are now well recognised, relatively little is known about (1) the prevalence of telescoping during the different stages of porphyry-epithermal deposit formation, and (2) the evolution of telescoped architectural characteristics, especially in multi-pulsed hydrothermal systems. These knowledge gaps exist because mineral assemblages related to early alteration events are commonly obliterated during later overprinting. Recent research has, however, shown that the intra-grain microtextural and trace element record from hydrothermal pyrite can provide a window onto the early, pre-telescoped alteration history of a deposit, in addition to capturing information on the nature of the telescoping process. We have investigated hydrothermal pyrite from Brucejack, a strongly telescoped, bonanza-type epithermal Au-Ag deposit in northwest British Columbia, Canada, to gain insight into the early history of alteration. Backscatter SEM imaging and LA-ICP-MS analyses of pyrite from two vertical sample transects reveal an evolution from magmatic- (elevated Co-Ni; cubic growth zonation) to epithermal- (elevated As-Au; colloform, oscillatory growth zonation) type fluids. We show that this evolution, which is otherwise cryptic due to overprinting by subsequent alteration and emplacement of bonanza-style epithermal veins, was caused by telescoping. Three-dimensional modelling of the microtextural and trace element signatures in pyrite demonstrates that this telescoping led to >1-km vertical and lateral compression of the Brucejack hydrothermal system. Furthermore, we document a previously unrecognised telescoping-reversal event in the deposit, as evidenced by localised mantling of Co-Ni on As-Au growth zones.

Minor and Trace Elements in Native Gold: Controls and Metallogenic Implications

Julian D. Melo Gómez[1], Evan Hastie[2], Harold Gibson[1]
1. Laurentian University, Sudbury, ON, Canada, 2. Ontario Geological Survey, Sudbury, ON, Canada

Gold is one of the most critical metals in human history, and still, few investigations have directly studied the composition of native gold as a guide to explain gold systems. Ontario's (Canada) gold deposits serve as a natural example due to their abundance and distribution. The results presented constrain the geochemical signature of native gold using EPMA and LA-ICP-MS in-situ techniques for 299 samples from 70 gold deposits across Ontario. Generally, Ag, Cu, and Hg occur over 100 ppm, with Sb, Pd, Cd, and Bi typically as trace elements between 0.001 and 100 ppm. Gold geochemical signature varies at the craton scale but is consistent at the camp scale: Red Lake, Geraldton-Beardmore, and Confederation Lake samples have similar elemental values and ratios but differ from those in the Abitibi, Temagami, or Wawa greenstone belts. In the Abitibi greenstone belt, differences were identified between samples from the Larder Lake-Cadillac Fault zone versus those from the Porcupine-Destor Fault zone; the latter have higher concentrations of Sb, Cd, and Cu relative to the former. These spatially distinct trace element signatures are independent of local host rocks and deposit types, which have significant implications regarding the controls on the metal content of gold deposits and ore-forming mechanisms, that may reflect district-scale mineral system differences such as source and transport, metamorphism, and remobilization, among others. These data also have important practical applications for industry, for instance, in using elemental vectors that reflect local gold composition and/or in metallurgical processing.

Textural and Mineralogical Characterization of Gold Deposits of the Vumba Greenstone Belt of Northeastern Botswana

Lebogang Mokane[1], Andrea Agangi[1], Ryohei Takahashi[1], Pearlyn Manalo[1], Kelebogile Phili[1], Manuel Nopeia[1], Thierry Bineli Betsi[2], Tebogo Kelepile[2], Zibisani Bagai[3]
1. Akita University, Akita, Japan, 2. Botswana International University of Science and Technology, Palapye, Botswana, 3. University of Botswana, Gaborone, Botswana

The *ca.* 2.7 Ga Vumba Greenstone Belt (VGB), located in the southwestern part of the Zimbabwe Craton in northeastern Botswana, is endowed with various commodities, such as gold and base metals found in the Somerset, White Pidgeon, Eldorado, and Arab deposits. The VGB has been affected by several episodes of metamorphism, with metamorphic grades spanning from greenschist to granulite facies. Previous geological studies have not revealed the mineralization characteristics, timing of the gold mineralization, and relationships among the gold deposits. This study focusses on textures and mode of occurences of ore minerals among the gold deposits mentioned above using optical microscopy, SEM-EDS and EPMA. In the White Pidgeon deposit, sulfides occur as disseminated along the foliation in the biotite schist and amphibolite, and gold occurs as inclusions of electrum in arsenopyrite and native gold in quartz-siderite veins. The Somerset and Arab deposits show some similarities in terms of the mineralization styles, and they are characterized by arsenopyrite that is rich in inclusions of chalcopyrite, pyrrhotite, galena, ilmenite, rutile, and electrum. The sulfides occur in different textural positions: (1) disseminated, (2) vein-hosted, (3) distributed along the foliation in the biotite-garnet schist, garnet-mica schist, amphibolite, and garnet amphibolite, and (4) occasionally as inclusions in garnet, and as fracture-filling in host rocks. The Eldorado deposit contains gold-bearing sulfides in quartz veins and disseminated in the amphibolite along the foliation. Native gold and bismuth telluride occur as inclusions in arsenopyrite. Hydrothermal ilmenite and rutile are coeval with pyrrhotite, arsenopyrite, electrum and chalcopyrite in all deposits. Ilmenite is partially replaced by titanite at the edges and is variably enriched in MnO from 0.97wt% to 8.28wt% and lesser amount of MgO (up to 0.68wt%), Al2O3 (up to 0.12wt%), and CaO (up to 1.10wt%).

Links Between Mafic Volcanism and Porphyry Gold Mineralization: Insights from the Maricunga Belt, Chile

Allyson N. Murray, Hannah C. Baker, J. Tomás Ovalle, Adam C. Simon
University of Michigan, Ann Arbor, MI, USA

The Oligocene to early Miocene Maricunga Belt, located in northern Chile (28°S to 26°S) is renowned for its abundance of porphyry gold and epithermal gold deposits emplaced during short pulses of extension within a long-lived compressional regime. The deposits are typically found within intermediate subvolcanic intrusive centers, and some deposits are copper poor, but others contain both gold and copper. Recent work suggests that the source magmas responsible for mineralization contain an inherent gold-enriched geochemical signature. We will explore the most primitive magmas in the area to gain insights into the magma source beneath the Maricunga Belt and assess its role in the generation of magmas with high fertility to form gold porphyry deposits.

Here, we present a comprehensive study of mafic lavas (basalts to basaltic andesites), including the Segerstrom lavas, in the Maricunga belt, which erupted contemporaneously with gold mineralization. We utilize petrographic analysis, scanning electron microscopy with energy dispersive spectroscopy, and electron microprobe analysis, of both the whole rock and phenocrysts, including olivine, amphibole, pyroxene, and plagioclase. Additionally, we document olivine and amphibole-hosted melt inclusions. We assess the source of the mafic magmas, as either generated via partial melting of either an asthenospheric mantle magmas or subcontinental lithospheric mantle during extension and evaluate recorded redox conditions and degrees of partial melting. By providing a holistic overview of the mafic lavas we gain deeper understanding of the deep mantle-crust processes in the generation of magmas forming porphyry gold deposits both in the Maricunga Belt and gold porphyry deposits more broadly.

Epithermal Au-Ag Mineralization at Galim-Legalgorou, Cameroon Volcanic Line (CVL): Constraints from Ore Mineralogy, Mineral Chemistry of Electrum and Fluid Inclusion Data

Terence Cho Ngang[1, 2], Cheo Emmanuel Suh[1, 3], Thomas Wagner[2], Tasin Godlove Bafon[1], Tobias Fusswinkel[2], Akumbom Vishiti[4], Eric-Claude Essono Owona[5]
1. Department of Geology, Mining and Environmental Science, Faculty of Science, The University of Bamenda, Bamenda, Cameroon, 2. Institute of Mineralogy and Economic Geology, RWTH Aachen University, Aachen, Germany, 3. Economic Geology Unit, Department of Geology, Faculty of Science, University of Buea Cameroon, Buea, Cameroon, 4. Department of Civil Engineering, The University Institute of Technology (IUT), University of Douala, Cameroon, Douala, Cameroon, 5. Department of Earth Sciences, University of Yaoundé I, Cameroon, Yaounde, Cameroon

The Au-Ag mineralization at Galim-Legalgorou (North Cameroon) is spatially associated with tertiary volcanic rocks of the Cameroon volcanic line (CVL), and likely to represent an epithermal system in a region where all previously studied gold occurrences have been classified as orogenic. Despite many years of artisanal mining operations and limited exploration, the origin of this mineralization remains unknown. In this contribution we present ore mineral and alteration associations, mineral microchemistry of electrum and fluid inclusion data to constrain the origin and evolution of the mineralization. Electrum contains high Ag (53.1 wt%) with fineness values that vary from 379 and 721 with an average of 455. A cumulative percentile plot of Ag concentrations in the cores of all electrum particles, and the observed step-change from Au-rich cores to more Ag-rich mantles in zoned electrum particles, suggest that the mineralization was deposited from two distinct hydrothermal fluid influxes of variable Au/Ag ratios. Fluid inclusions associated with sealed fractures in amethysts yielded homogenization temperatures between 215.6 ^{0}C and 308.5 ^{0}C (average 284.9 ^{0}C), with low salinities (0 to 5 wt% NaCl equivalence) and coexisting vapor-rich and liquid-rich inclusions suggesting that the mineralization experienced transient periods of hydrostatic fluid pressure, and that boiling may have accompanied ore deposition. These data, in combination with observed ore mineral association (electrum, Se/Tl-acanthite, pyrite, Fe-rich sphalerite, ± galena) and the presence of colloform banded chalcedony and hydrothermal breccia, are consistent with features of low sulfidation epithermal deposits. This study, therefore, demonstrates for the first time, the presence of significant low sulfidation epithermal Ag-Au mineralization on the CVL, an intracontinental rift region only previously known for orogenic gold mineralization.

Provenance of the Placer Gold-Bearing Banket Conglomerate, Iduapriem, Ghana: Constraints from Morphology and Oxygen Isotope Characteristics of Quartz Pebbles

Edwin A. Obeng, Kofi A. Ansah
University of Mines and Technology, Tarkwa, Ghana

Economic gold mineralisation of the Tarkwaian Group in Ghana is restricted to the Banket quartz-pebble conglomerate. No oxygen isotopic studies have been conducted to unravel the provenance for the gold-bearing Banket conglomerate of Ghana. Quartz is very resistive to alteration and can retain its original isotopic composition, even after low temperature fluvial transport and overprinting low-grade metamorphism. Hence, the current study at the Iduapriem Mine in Tarkwa area analysed oxygen isotopic values in the quartz pebbles that constitute the placer gold-bearing Banket conglomeratic reefs A, B, and C. The oxygen isotope data, together with petrography and geochemical assay values, were used to elucidate the provenance of the Banket conglomerate and also the environment of deposition of quartz veins oxygen isotopic studies that could supply the quartz pebbles in the gold-bearing Banket conglomerates based on temperatures of 350-450°C for hydrothermal gold mineralisation and oxygen isotope fractionation. The results show the oxygen isotope data of the Banket conglomerate quartz pebbles range from +11.2 to +13.2‰ to be lower than the oxygen isotope data for quartz veins related to gold deposits along the left flank of the Ashanti metavolcanic belt. Since the oxygen isotopes values are considered representative of the original formation characteristics, this difference implies that the Banket conglomerates were not erosional derivatives of the existing Birimian lode-gold deposits. It is likely that older, potentially large early Paleoproterozoic gold-rich terranes existed in Ghana prior to the formation of the gold-bearing Banket conglomerate and the existing Birimian lode gold-deposits on the west flank of the Ashanti metavolcanic belt. Based on the structural and geophysical works of previous workers, the inferred sources from the east of the Ashanti Volcanic belt could explain the reason for the very few existing gold deposits to the east of the Ashanti metavolcanic Belt, in contrast to the west.

A Reappraisal of the Structures, Alteration and Tectonic Setting of the Rambi Gold Target, Western Kenya

Maryce N. Obindah
The University of Nairobi, Nairobi, Kenya

Rambi gold prospect, located in southwestern Busia-Kakamega Greenstone Belt, has been worked for gold since the 1930s. As promulgated by the colonialists, the mineralization model conforms to a quartz-vein-hosted gold system within a shear zone. This study is set to reappraise the gold prospectivity of the Rambi target through an integrated mordernized exploration approach. It is geared towards confirming the existence of gold ore in the target as per the colonial records. Airborne magnetic and geochemical datasets are the main methods used to assess the geology and geological structural patterns that lead to probable gold mineralization zones in the prospect. In this study, the shear zone in the Rambi target has been delineated from an interpretation of airborne geophysical data. The magnetic data component of the airborne data was processed to accentuate the structural fabric of the area. The major lineaments in the prospect trend NE with less equal offsets in the orthogonal direction, which have formed new target zones in the prospect.These shear lineaments reveal two main deformation episodes in the area as supported by analysis from the fabric diagrams . The lineaments are grouped into D2 and D3 tectonic episodes that trend in the ENE-WSW and WNW-ESE direction, respectively. From the data retrieved from the recent 2018 exploration excercise, the colonial mined reef structurally lines up with the D2 structures. These structures form the two prospective internal shear zones, which are under-explored by both colonialists and other companies that have previously undertaken exploration activities over the target. This discloses more potential sites for increased exploration. The major fracture systems that play a significant role in the location of gold ore in the study area are the internal shear zones.

Assessing Structural Impact on Gold Mineralization Using Aeromagnetic, Radiometric Data Sets and Field Observations: AngloGold Ashanti Iduapriem Mine, Ghana

Newman T. Odonkor, Prince O. Amponsah, Samuel Nunoo, Marian S. Sapah
University of Ghana, Accra, Ghana

The impact of structures on gold mineralization at the AngloGold Ashanti Iduapriem Mine in southwestern Ghana was investigated by combining aeromagnetic and radiometric data sets with field observations. The geophysical data processing focused on delineating lithological boundaries and structures that may potentially influence gold mineralization within the mine. The application of enhancement filtering techniques to magnetic data, along with ternary radiometric imaging, facilitated the mapping of various lithologies belonging to the Tarkwaian Group and the Birimian Supergroup. The former group constitute the Kawere conglomerate, Huni sandstone, Tarkwa phyllites, gold-bearing Banket conglomerate, and greenstones ascribed to the Birimian Sefwi Group.

Three main deductions were made from this research: (i) the field observations indicate gold mineralization, primarily associated with the Banket conglomerates marked by distinctive sedimentological features such as preserved bedding, trough cross-bedding, and particulate gold along inter-grain boundaries, suggest a sedimentological control of gold distribution; (ii) the lineament analysis from the analytical signal and tilt derivative revealed two principal tectonic events, predominantly NW-SE and minor N-S trending regional structures, corresponding to the D1 and D2 deformational phases of the Eburnean Orogenic Event; and (iii) these structural transpositions which affected the primary sedimentary features, resulted in late-stage structures that appeared to have remobilized the earlier gold occurrences.

Geochemistry, Sulfur Isotopes, and Fluid Inclusion Characteristics of the Neoarchean Gold Deposits of the Tati and Vumba Greenstone Belts of NE Botswana

Kelebogile Phili[1], Ryohei Takahashi[1], Andrea Agangi[1], Pearlyn Manalo[1], Lebogang Mokane[1], Manuel Nopeia[1], Hinako Sato[1], Thierry Bineli-Betsi[2], Tebogo Kelepile[2], Nigel Blamey[3], Tavis Enno[3], Zibisani Bagai[4], Kemmonye Baliki[5]
1. Akita University, Akita City, Japan, 2. Botswana International University of Science and Technology, Palapye, Botswana, 3. Western University, London, ON, Canada, 4. University of Botswana, Gaborone, Botswana, 5. Mupane Gold Mine, Francistown, Botswana

The Tati Greenstone Belt (TGB) and the Vumba Greenstone Belt (VGB) are Neoarchean terranes located in northeastern Botswana, along the southwestern margin of the Zimbabwe Craton. These belts are composed mostly of mafic and ultramafic meta-volcanic rocks intruded by granitoids; they host several gold deposits and minor Ni-Cu-PGE deposits. Despite early discoveries in the 1800s, research on the fluid evolution and genesis of gold mineralization within the TGB and VGB remains limited. This study aims to constrain the genesis of gold mineralization in the Mupane and Golden Eagle gold deposits in the TGB, as well as the White Pidgeon and Somerset gold deposits in the VGB, based on sulfide geochemistry, sulfur isotopes, and fluid inclusion characteristics. Gold-bearing sulfide ores in both belts are mainly observed as disseminations in strongly altered schist and amphibolite host rocks and in quartz-calcite veins that crosscut these rocks. Native gold and electrum in the TGB deposits mainly occur as inclusions or intergrowths in arsenopyrite and sphalerite. In contrast, in the VGB, both native gold and electrum occur as inclusions within arsenopyrite and are occasionally intergrown with chalcopyrite and pyrrhotite. The $\delta^{34}S$ values of pyrrhotite, arsenopyrite, and pyrite from the studied deposits range from -0.7 to +4.0 ‰, suggesting a potential magmatic origin of sulfur. Quartz-hosted primary fluid inclusions are classified into single-phase carbonic, two-phase aqueous, and three-phase aqueous-carbonic inclusions; the aqueous inclusions homogenize at 120 to 346 °C and their salinity ranges from 1.3 to 10.5 wt.% NaCl equivalent. The coexistence of the three fluid inclusion types in the same quartz crystals suggests coeval entrapment of likely immiscible fluids. Bulk gas analysis of fluid inclusions in quartz and calcite revealed that H_2O (59.6 to 95.2 mol%) is the dominant component, with elevated contents of CO_2 (0.92 to 39.9 mol%) and CH_4 (0.33 to 4.33 mol%).

Precious and Base Metal Occurrence in Amphibolites within the Buenos Aires Complex, Tandilia System, Argentina

Belén Marone[1, 2], Federico J. Saponara[1, 2], María F. Lajoinie[1, 2], Mabel E. Lanfranchini[1, 3]
1. Instituto de Recursos Minerales (INREMI), La Plata, Argentina, 2. Consejo Nacional de Investigaciones Científicas y Técnicas (CONICET), La Plata, Argentina, 3. Comisión de Investigaciones Científicas de la Provincia de Buenos Aires, La Plata, Argentina

Amphibolites from the Paleoproterozoic basement of the Tandilia System (Buenos Aires province) occur as 'sill-type' intercalations, dike-shaped bodies, or as xenoliths associated with gneisses and migmatites. Recent geochemical studies, conducted on whole rock, have determined the first occurrence of Au, in association with Ag, As, Se, Hg and Sb, in these amphibolites. Gold concentrations vary from 1 to 2 ppb in xenolithic and sill-type amphibolites. Across all metabasite bodies anomalous contents of silver (from 0.01 to 0.26 ppm), arsenic (0.2 to 1.6 ppm) and selenium (0.2 to 0.5 ppm) have been found. Additionally, mercury (0.019 ppm) and antimony (0.08 ppm) were detected in a single xenolith and a sill-type amphibolite, respectively. Mercury was also found through Scanning Electron Microscopy and X-Ray microanalysis, associated with Fe sulphides, on stratiform amphibolites. Significant values have been established for additional base metals, including lead (0.5 to 24.5 ppm), zinc (99 to 166 ppm), and copper (11.8 to 900 ppm) within all metabasites. Some of these amphibolites exhibit an affinity to Mid-Ocean Ridge Basalt (MORB), while others display characteristics indicative of a volcanic arc origin, possibly associated with the Transamazonian Orogeny (Paleoproterozoic). It is plausible that the MORB-affiliated ones formed initially. Although no ore deposits have been defined yet in the basement of Buenos Aires, the tectonic setting, the identification of these metals, and the comparison of this occurrence with other areas, along with further detailed studies, will allow for the postulation of these metabasites as a potential source in a model of orogenic gold deposit.

Evidence of Sulfide Melting in Pegmatite Veins Containing Precious and Base Metals in the Paleoproterozoic Basement of the Tandilia System, Buenos Aires, Argentina

Federico J. Saponara[1, 2], Belén Marone[1, 2], María F. Lajoinie[1, 2], Mabel E. Lanfranchini[1, 3]
1. Instituto de Recursos Minerales (INREMI), La Plata, Argentina, 2. Consejo Nacional de Investigaciones Científicas y Técnicas (CONICET), La Plata, Argentina, 3. Comisión de Investigaciones Científicas de la Provincia de Buenos Aires (CIC-PBA), La Plata, Argentina

Igneous and metamorphic rocks within the Buenos Aires Complex (Tandilia System basement) are intruded by a set of pegmatite veins composed of quartz, plagioclase, tourmaline, garnet, biotite, muscovite and black and white microcline containing sulfide inclusions. These bodies are derived from the fractionated melt evolution of leucosomes and leucogranites formed during an amphibolite-granulite facies metamorphic event. The aim of this study is to describe the first evidence of sulfide melting rich in precious and base metals in the Tandilia System basement and its comparison with other world examples. Examination under chalcopetrographic and electronic microscopes (SEM-EDS) revealed that microclines contain multiphase and monomineralic sulfide inclusions along with isolated concentrations of sulfosalts and other Low-Melting-Point Chalcophile Elements (LMCEs), while quartz contains sulfide inclusions trails. Based on the theoretical formula reconstruction, multiphase inclusions contain daughter mineral phases of chalcopyrite-galena-molybdenite-pyrite-bismuthinite, chalcopyrite-galena-molybdenite-pyrite, argentite-pyrargyrite and stibnite-pyrite-sphalerite. Furthermore, monomineralic sulfide inclusions of galena, sphalerite, argentite, pyrite and molybdenite were found in microclines. The LMCEs are mostly distributed in concentrated pockets of sulfides and sulfosalts, that have irregular shapes, comprising arsenopyrite, stibnite, miargyrite, pyrargyrite and gold. Also, LMCEs such as antimony, arsenic, silver, gold and bismuth are disseminated in the microcline crystals. These observations are similar to those found in metamorphized ore bodies such as Broken Hill, Australia; Rajpura-Dariba, India and Aggeneys, South Africa. In these cases, the presence of multiphase and monomineralic inclusions of sulfides, as well as isolated concentration of sulfosalts and other LMCEs were interpreted as the result of partial melting of sulfide bodies rich in base and precious metals during the metamorphic peak. In this sense, the information provided in this research is relevant for the definition of future mineral deposits in Proterozoic igneous-metamorphic terranes and for the improvement of genetic modeling.

Chlorite Thermometry and Its Implication to the Genesis of Gold Mineralization at Paramanahalli, Dharwar Craton, India

Manju Sati[1], Rajagopal Krishnamurthi[2], Sakthi S. Chinnasamy[3], Ajit K. Sahoo[4]
1. Research Scholar, Department of Earth Sciences, IIT Roorkee, Roorkee, UK, India, 2. Professor, Department of Earth Sciences, IIT Roorkee, Roorkee, UK, India, 3. Associate Professor, Department of Earth Sciences, IIT Powai, Mumbai, Mumbai, MH, India, 4. Assistant Professor, Department of Geology, BHU, Varanasi, Varanasi, UP, India

Gold mineralization at Paramanahalli, Chitradurga Greenstone Belt (CGB) in Dharwar Craton (DC) of Peninsular India is shear zone controlled and confined to altered metabasalt and Banded Iron Formation (BIF). Chlorite (Fe-rich) + quartz + ankerite + rutile + monazite ± pyrite ± gold is the mineral association found in the mineralized zone. Chloritization, carbonitization, silicification, and sulfidation are the major alteration processes inferred in the area and can be attributed to wall rock-fluid interaction. Detailed petrographic studies, textural association and mineral chemistry of the altered wall rocks indicate three different types of chlorite phases. *Chlorite-1* is an Mg-rich (clinochlore) found in schistose metabasalt, whereas *Chlorite-2* is an Fe-rich (chamosite) and closely associated with gold in carbonated metabasalt. *Chlorite-3* has higher FeO content (i.e., brunsvigite), found to be with pyrite and gold, and occurs only in mineralized BIF. The concentration of Mg, Fe, and Al (IV) in chlorite may vary according to the geological settings. The close association of hydrothermal chlorite (2 and 3), pyrite and gold grains revealed their cogenetic nature. Thus, chlorite thermometry was chosen to estimate the temperature of formation of minerals developed during hydrothermal activity. *Empirical methods* based on tetrahedral Al content and amount of octahedral vacancy (Mg, Fe^{3+} and Fe^{2+}) and *thermodynamic methods* based on intra-crystalline exchange reactions among the chlorite components were used. The results indicate a temperature range of ~310-360˚C and this can be proposed as the temperature of hydrothermal fluids responsible for gold mineralization at Paramanahalli. The temperature of formation of chlorites have also been found to be similar to fluid inclusion micro-thermometry (~280-382˚C) of gold bearing quartz vein from the study area. Hence, the above methods of study are useful to understand the gold metallogeny with the discovery and development of new gold deposits/prospects in the similar geological setting.

Vein Textures at the Moss Low-Sulfidation Epithermal Gold Deposit, Arizona: Constraints on the Processes of Mineral Deposition

Greg Seitter[1], Thomas Monecke[2], Patrick Wood[3]
1. Colorado School of Mines, Golden, CO, USA, 2. Colorado School of Mines, Golden, CO, USA, 3. Elevation Gold, Bullhead City, AZ, USA

The Miocene low-sulfidation epithermal deposits of the Oatman district in northwestern Arizona are located in the south-central portion of the Basin and Range province. Discovered in the late 1800s, mining in the district has yielded over two million ounces of gold. Historically, high-grade ores have reached concentrations exceeding 30 g/t Au. Open-pit mining operations are currently active at the Moss deposit on the west side of the Oatman district. Hosted in the Peach Springs Tuff and the Moss monzonite porphyry, the mining targets at the Moss mine comprise the Moss, Ruth, and West veins. The Peach Springs Tuff is a product of an eruption event from the Silver Creek caldera ~18.8 Ma. Free gold is the dominant presentation of gold in the colloform quartz of this system. Small dendritic aggregates of the free gold and other ore minerals hosted by fine-grained quartz have formed in distinct ginguro bands. The textures are similar to those encountered in low-sulfidation epithermal deposits formed by the flashing of the hydrothermal liquids. Bladed calcite -- a frequent gangue texture commonly replaced by quartz -- suggests the gentle boiling of the hydrothermal fluids in addition to flashing. Late open-space filling of the veins, including the formation of comb quartz, appears to have occurred during periods of steady-state flow. Further petrographic analysis will enhance our understanding of the processes resulting in diverse quartz textures in the veins and the textural relationships between the ore and gangue minerals.

The Discovery Under Cover of the 1.3-Moz Dokwe Gold Project, SW Zimbabwe

Ahmet K. Sener
Ariana Resources plc, London, United Kingdom

The c.1.3-Moz Dokwe gold deposit is located within a previously undefined Archaean terrane, extending from the Maitengwe Greenstone Belt in Botswana and linking up with the Bulawayo-Bubi Greenstone Belt in Zimbabwe. As the Archaean units are overlain by Karoo and Kalahari sedimentary units of up to 25-40m in thickness, an innovative geochemical method was utilised to detect the deposit under cover. The east-northeast striking greenstone belt is accompanied by complex folding and thrust imbrication and is dissected by a series of major subparallel sinistral shear zones.

At the Dokwe Gold Project area, the barren sedimentary cover is dominated by calcrete, with a few metres of sand at the surface, and mudstone and sandstone located towards the base. The basement Archaean volcanic sequence comprises a series of quartz-rich volcaniclastic units, tuffs, and agglomerates, that grade into felsic irregular rhyolitic flows; intermediate vesicular dacite; agglomerates and andesites, which were subject to greenschist facies metamorphism. The sequence appears intruded by near syn-depositional quartz porphyries and later by dolerite. Brittle deformation, characterised by fracturing, is common in felsic tuff and porphyry units whilst more ductile deformation characterises the dacitic and andesitic units.

Dokwe North is characterised as a large low-grade deposit containing relatively few quartz veins, containing several very high-grade zones including visible gold. Dokwe Central is a smaller higher-grade pipe-like deposit containing abundant quartz veins and defined by several long high-grade zones. The two deposits appear to be strongly structurally controlled, occupying two distinct structural domains within a broad ENE trending shear zone. Gold mineralisation at Dokwe North is associated with silicified zones containing thin quartz-carbonate pyrite veins in addition to disseminated and fine-grained pyrite. Much of the economic gold mineralisation occurs in the dacitic unit and in the overlying felsic tuff, with lesser mineralisation in the quartz porphyry and andesitic units.

The Karibib Region as an Emerging Gold District in the Damara Belt, Namibia: Linking Local Fluid Sinks to Regional Fluid Flow

Josia Shilunga, Alexander Kisters
Stellenbosch University, Stellenbosch, South Africa

The area surrounding the Karibib town in central Namibia within the high-grade metamorphic terrain of the Pan-African Damara orogenic Belt is rapidly emerging as a significant gold district. Established deposits such as the Navachab gold mine (>4 Moz), the recently discovered Twin Hills deposit (>3 Moz), and numerous historically mined, small-scale prospects and recently delineated prospects in the region underline the prospectivity of the region. This has sparked a dramatic increase in the exploration efforts in the Karibib region. However, the range of mineralization styles, different lithological controls and varied structural positions, have posed challenges to exploration efforts. Consequently, there remains a lack of a comprehensive conceptual model that accounts for the different styles of mineralization and that may guide exploration. This presentation provides an overview of both recently discovered and existing gold deposits and prospects in the Karibib region. The study aims to identify similarities or variations among individual deposits and prospects that highlight the underlying structural and lithological controls of the regional fluid flow and gold mineralization. Gold deposits and prospects are hosted by the amphibolite-facies, mixed carbonate-siliciclastic metasedimentary sequence of the middle and upper Swakop Group. Gold-sulfide (pyrrhotite > arsenopyrite > pyrite > chalcopyrite) mineralization is associated with quartz vein systems and in parts disseminated. The vein systems vary systematically between prospects and deposits but, show close geometrical relationships with the main generation of east-northeasterly trending folds and thrusts that formed during the main phase of northwest-southeast directed, subhorizontal shortening ($D_{2/3}$) between ca. 530-510 Ma. Economic-grade gold mineralization corresponds to a combination of lithological and structural controls. This includes (1) contacts between rock types and formations of distinctly rheological contrasts, (2) high strain zones, where strain localization promoted fracturing and veining, and (3) regional-scale deviations of wall-rock structures from the main east-northeasterly Damara Belt trend.

Geology and Gold Mineralization at the Leeville Underground Complex, Carlin, Nevada

Koray Tasbicen[1], Elizabeth A. Holley[1], Kendle Fraley[2], Elizabeth Stock[3], Carl Pollard[4], Joseph Becker[2]
1. Colorado School of Mines, Golden, CO, USA, 2. Nevada Gold Mines LLC, Elko, NV, USA, 3. Barrick Gold Corporation, Toronto, ON, Canada, 4. Barrick Gold Corporation, Elko, NV, USA

The Leeville Underground Complex, within the Carlin district, Eureka County, Nevada, hosts the Four Corners, West Leeville, Turf and North Leeville high-grade Carlin-Type Gold Deposits.

The complex is confined to the hangingwall of the NNW-trending Leeville Fault, bounded by the WNW-trending Castle Reef Fault to the south and the NNE-trending Basin Bounding normal fault to the west. Mineralization is commonly stratabound hosted in calcareous siltstones of the Devonian Rodeo Creek Formation and micritic limestones of the Popovich Formation. Favorable stratigraphy gently plunges to the north and is covered by >200 meters of Ordovician Vinini Formation comprised of siliciclastics of the Roberts Mountain allochthon.

Since the late Devonian, multiple pre-mineral compressive events have locally folded and thrusted stratigraphy, generating a NNE and NNW-trending structural framework and creating fluid pathways for Eocene mineralization. Carlin-type fluids are acidic, gold-bearing, sulfur-rich and low temperature. Mineralization is associated with sulfidation, decalcification and silicification, often occurring as tabular silica-sulfide breccia, which are spatially related to the high-angle feeder and low-angle lateral fluid flow structures.

Throughout the complex, high-grade centers of gold mineralization are also spatially associated with pre-mineral Mesozoic dikes. A Mesozoic Stock, to the west of the complex, is currently interpreted to be the northern margin of a larger buried intrusive complex interpreted source of diking and metasomatism in the district. Host stratigraphy on the west side of the complex exhibits pre-mineral metasomatism manifested as low-temperature hornfels.

The study will initially document the geology at the Leeville Complex through detailed observations and a series of crosssections. Detailed petrography, mineralogy and geochemistry will constrain the genetic relationship between magmatism, structures and mineralization. This study will identify the cumulative factors controlling the Eocene gold mineralization. The expected outcomes will improve the ore body knowledge in the Leeville Complex and exploration targeting in northern Nevada.

Kinetics of Silica Deposition as a Control on Silica-to-Ore Ratio in Mineralized Epithermal Veins

Lauren R. Terry[1], Thomas Monecke[1], Cliff Buck[2]
1. Colorado School of Mines, Golden, CO, USA, 2. Coso Operating Company, Inyokern, CA, USA

Banded veins in high-grade, low-sulfidation epithermal deposits are characterized by a variety of microtextures. Bands containing ore minerals display distinctive textural characteristics indicative of gentle boiling or flashing. This includes bands that originally were entirely composed of silica microspheres and contain dendritic ore minerals. The noncrystalline silica is typically recrystallized to quartz. These bands are thought to have formed during flash vaporization, as this process results in the generation of large amounts of vapor, causing deposition of silica and carbonates, as well as ore minerals from the resulting supersaturated liquid. In geothermal systems, which are modern analogs to epithermal systems, flashing occurs at the back-pressure plate precipitating scales. In geothermal power plants such as the Coso geothermal system in California, the flashed liquid is cooled and subsequently injected underground in reinjection wells. Scaling at Coso is pronounced in these wells and not restricted to the back-pressure plate where a drop in pressure occurs. In this study, petrographic investigations were performed on scales forming in the reinjection wells. These consist of banded accumulations of noncrystalline silica forming microspheres. The silica microspheres range up to several micrometers in size and are closely packed. Although silica supersaturation in the liquid was probably already reached during flash vaporization, the deposition of silica occurs along the entire subsequent flow-path of the flashed liquid and in the reinjection wells. As the silica is not deposited at the flash point only, the kinetics of silica deposition may be an important control on the silica to ore ratio that develops in mineralized bands in epithermal veins.

What Drove the Leaching of Gold from the Mesoarchen Land Surface: A Global Climate Modelling Approach

Lisa Wasitschek, Hartwig E. Frimmel
University of Würzburg, Würzburg, Germany

Currently, the most plausible genetic model for the world's largest gold province, the Witwatersrand Basin in South Africa, is based on the assumption that most of the gold was leached from the Archean hinterland, then fixed by microbial mats, most of which, in turn, became mechanically reworked only to release minute gold particles to concentrate into rich paleoplacers. While this model explains neatly the spatial distribution of the conglomerate-hosted gold in the Witwatersrand, it fails to explain the stark contrast in gold endowment between the 2.96 – 2.91 Ga West Rand Group and the 2.90 – 2.78 Ga Central Rand Group. Conglomerates in the former are almost barren; those in the latter contain close to 30% of all known gold in the world. It has been shown before that prior to 2.90 Ga, chemical weathering rates were low, whereas after 2.90 Ga, they were very high, thus providing a basic precondition for the formation of the rich Witwatersrand goldfields. In this contribution, we discuss the possible causes of this drastic change in chemical weathering rates and try to assess the most important driving forces using global climate models with intermediate complexity (Planet Simulator). General agreement exists on a high concentration of free greenhouse gases in the Archean atmosphere to offset the effect of a lower solar intensity, but estimates on the surface temperature vary widely. The change in chemical weathering rates at around 2.9 Ga could be the result of global climate change, maybe triggered by rapid growth of continental crust and thus a transition from predominantly submarine to more subaerial volcanism. Alternatively, it can be also explained by the paleogeographic position of the Kaapvaal Craton, that is, its drifting from high to low latitudes.

Direct Dating of the Mineralization by the Single Grain (U,Th)-He Pyrite Geochronology

Olga Yakubovich
Saint Petersburg University, St. Petersburg, Russia

Genetic models which are based on the geodynamical settings are essential for exploration strategy. The age of mineralization is one of the most effective ways to link the hydrothermal process to the geological settings. Routine dating of hydrothermal events is complicated by the lack of proper minerals for geochronological studies.

Pyrite is abundant in most of the Au, Ag, Cu, Pb, Zn, Sn, W, and Mo deposits and is known to be crystallized at different stages of ore formation. Therefore, pyrite geochronology has a range of potential applications, which include dating of ore replacement and remobilization events. However, the wide use of Re-Os, Rb-Sr, Sm-Nd, and Ar-Ar isotope systems for the dating of pyrite is limited by several isotope-geochemical reasons related to low concentrations of parental isotopes.

Recent advances in understanding the He thermal retentivity in pyrite allow us to propose this mineral as a (U,Th)-He geochronometer [1]. Single grain (U,Th)-He pyrite dating was successfully tested on several deposits [2,3].

Herein we would like to present and discuss the capabilities of this technique to study ore processes on the example of several gold deposits located in Russia. Based on >70 measurements of (U,Th)-He pyrite ages from these localities we can conclude that this approach is a powerful tool to study the chronology of hydrothermal impulses within the ore-bearing areas.

Funding: RSF 22-77-10088

1. Yakubovich et al. *Petrology* **2019**, *27* (1), 59–78.

2. Yakubovich et al. U-Th-He Geochronology of Pyrite from the Uzelga VMS Deposit (South Urals)—New Perspectives for Direct Dating of the Ore-Forming Processes. *Minerals* **2020**, *10* (7), 1–20.

3. Yakubovich et al. U-Th-He Geochronology of Pyrite from Alteration of the Au-Fe-Skarn Novogodnee-Monto Deposit (Polar Urals, Russia)—The Next Step in the Development of a New Approach for Direct Dating of Ore-Forming Processes. *Geosciences* **2021**, *11* (10), 408.

Hydrothermal Alteration and Mineralization Footprint of the Čoka Rakita High-Grade Gold Skarn Deposit, Timok, Serbia

Jelena Zivanovic[1], Dragana Davidovic[1], Mladen Zdravkovic[1], Nutu Groza[1], Ivan Strmbanovic[1], Dragan Milosevic[1], Snezana Janjic[1], Olivera Dukic-Mijucic[1], Aleksandar Sestak[1], Vuk Milic[1], Ivana Suzic[1], Vaso Kitanovic[1], Dragana Drobnjak[1], Bojana Kotlajic[1], Tijana Mitic[1], Janja Mihajlovic[1], Nikola Novicic[1], Stefan Metodiev[1], Kuncho Kunchev[1], Kalin Kouzmanov[2], Aleksandar Pacevski[3], Paul Ivascanu[1], Istvan Marton[1]
1. Dundee Precious Metals Inc., Bor, Serbia, 2. Department of Earth Science, University of Geneva, Geneva, Switzerland, 3. Faculty of Mining and Geology, University of Belgrade, Belgrade, Serbia

The Čoka Rakita deposit is located in the northwestern part of the Late Cretaceous Timok Magmatic Complex, a metallogenic district in Serbia hosting world-class Cu-Au porphyry (Majdanpek, Veliki Krivelj) and high-sulfidation deposits (Bor, Čukaru Peki). Čoka Rakita is a recent high-grade gold skarn discovery, with unique mineralogical and grade characteristics not only for the Timok district, but for the entire Western Tethyan Belt, with maiden inferred mineral resources of 1.78 Moz of gold.

The main economic mineralization at Čoka Rakita is related to exoskarns, developed as massive stratabound bodies formed by the replacement of a calcareous clastic sedimentary sequence in the hanging wall of a sill-like porphyritic monzodiorite and east of a major monzonitic intrusion. Mineralization also extends into the endoskarn zone of the monzodiorite, up to several meters from the intrusive contact.

Other mineralization types at Čoka Rakita include: (1) porphyry Au-Cu-Mo mineralization in the potassically altered monzodiorite and the epiclastic sequence; and (2) stratabound Cu-Au mineralization at a deeper stratigraphic level within a marble-skarn altered limestone and the overlying conglomerates.

Gold mineralization is dominantly associated to andradite-grossular garnet skarn with minor pyroxene, and K-feldspar, overprinted by retrograde assemblage consisting of epidote, chlorite, calcite, albite and minor actinolite, late-stage andradite and sulfides. Gold occurs in several distinct mineral associations, the main ones being: (1) native gold and pyrite-dominant mineralization, with minor chalcopyrite-bornite; and (2) gold and pyrite-pyrrhotite-magnetite mineralization with minor chalcopyrite-sphalerite-galena-marcasite-tetrahedrite-tennantite-Pb(-Sb) sulfosalts. Gold appears as disseminations and often as visible aggregates, reaching few millimeters in size. Gold is hosted by various gangue and sulfide minerals, mainly by garnet, occupying its pores and microfractures, or less commonly in pyrite, epidote, K-feldspar and albite. Gold almost always contains silver in various proportions (up to 10 wt.% Ag), as revealed by preliminary SEM-EDS analyses.

Toward Automatic Airborne Magnetic Mapping – VAMPair Colibri

Laurent Ameglio[1], Francis Galluser[2]
1. EXIGE Australia Pty Ltd, Sydney, NSW, Australia, 2. Coleoptere s.a.r.l., Strasbourg, France

Airborne magnetics is an effective and routinely used technology for natural resources exploration. It is however commercially provided by specialised contractors implying poor agility, responsiveness and cost effectiveness for exploration programs. The use of innovative technologies such as drones and non-absolute scalar magnetometers, along with data processing automation and artificial intelligence, can change the face of the modern green- and brown-field mineral sectors.

We integrated a true backpack transportable, ultra-light (~1 kg), small (size of a hand), quick to deploy (5 mn), easy to use and safe (no towed sensors), single person operated, affordable, UAV-born (DJI Mavic series), fluxgate magnetics mapping platform—VAMP^{air} Colibri.

Fluxgate magnetic sensors are more versatile and a better value-for-cost than traditional scalar magnetic sensors. They still provide similar data quality with the unique added values of: light weight, lower power requirement, and correction of remanent and induced fields from other equipment carried with the sensor. Their drift issue is resolved with a robust calibration procedure used onboard satellites. Results of surveys in South Africa are presented.

The flying hardware comes with a tablet. All operations aspects are managed by a newly developed software—SUAVE. While survey design still requires minimal inputs from the user for the time being, all other aspects of magnetic mapping such as flight management (e.g., generation of reflight navigation files), magnetic data quality control (including data compilation between flights) and infield magnetic mapping, including targets generation, are all automatized.

The Colibri is destined to become a companion of the geologist's hammer, compass, and magnifying glass for geological investigations. Also, paired with third-party 3D holographic services (such Microsoft Azure with HoloLens) it will provide a field-to-boardroom direct link for exploration companies. With the growing offer of drone nest technologies, it will also become a complete autonomous solution for the mining industry.

Using InSAR to Detect Soil Expansion: A New Way to Explore for Kimberlites

Kamogelo Baloyi, Stephanie Enslin-Scheiber
University of the Witwatersrand, Johannesburg, South Africa

Kimberlites are the most important sources of diamonds in the world; however, they may easily be buried, weathered or simply inaccessible, and accordingly hard to explore for. This creates the need for innovative exploration methods that can at least be used to narrow down the probable hidden occurrences of kimberlites. Hamilton and Webb noticed a seasonal variation in the surface elevation over a known kimberlite pipe. They related the elevation increase to an increase in the moisture being stored in weathered kimberlite clays during the rainy seasons; and the elevation decrease to the clays drying out during the dry seasons.

InSAR techniques can be used to detect these variations during green-field exploration. The technique involves the transmission of electromagnetic waves onto the earth's surface by an orbiting satellite. These waves are reflected to the satellite and the distance they travel is calculated as wavelength phases that are a function of the two-way travel time of the wave back to the satellite. If applied correctly in this project, InSAR should highlight areas of seasonal surface deformation resembling that in Hamilton and Webb's 2003 paper. These areas can then be studied further to determine their quality and feasibility status.This will lead to significant and frequent future discoveries of diamonds, often in inaccessible areas and at limited cost, using available data and resources.

MOSMIN: Multiscale Observation Services for Mining Related Deposits

René Booysen, Sandra Lorenz, Moritz Kirsch, Richard Gloaguen
Helmholtz Institute Freiberg for Resource Technology, Freiberg, Germany

The shift toward a sustainable economy has led to a heightened demand for raw materials, typically obtained through mining operations. Mining operations produce byproducts like rock waste, tailings, and stockpiles, which pose environmental and safety hazards. It's crucial to effectively manage these materials throughout their lifespan, prioritizing stability and the prevention of water and soil contamination. Earth-observation (EO)-based techniques are rarely used for monitoring these deposits, and multi-sensor field data are commonly not integrated despite recent technological advancements. We aim to establish holistic, full-site services for geotechnical and environmental monitoring, along with the valorization of mining-derived deposits. This will be achieved through an integrated approach utilizing both EO data and in situ geophysical data. The work will be accomplished under the MOSMIN project: "Multiscale Observation Services for Mining Related Deposits," and funded by the European Union Agency for the Space Programme (EUSPA) with project number 101131740. MOSMIN services will use Copernicus EO data for time-resolved, spatially extensive, remote monitoring of ground deformation and surface composition. Cutting-edge algorithms for change detection will pinpoint displacements and identify environmental hazards. Satellite data will be integrated with real-time, high-resolution data obtained from drones and sensors installed on site, leveraging the power of machine learning for fusion and resolution enhancement of multi-scale, multi-source data. Novel geophysical techniques such as distributed fibre-optic sensing will provide subsurface information to identify potential risks such as internal deformation and seepage. MOSMIN will collaborate with international partners and mining companies to leverage pilot sites located in the EU, Chile, and Zambia. These sites will serve as testing grounds for the development of comprehensive monitoring services. The MOSMIN integrative services and tools will improve the efficiency and reliability of monitoring, maximise resource utilisation, and help mitigate environmental risks and the impact of mining operations. Submitted on behalf of the MOSMIN Consortium.

Hydrothermal Alteration Zones Estimated by ASTER Satellite Image and ASD TerraSpec on the Eastern Zone of Cerro de Pasco, Perú

Evelyn Caiza[1, 2]
1. Central University of Ecuador, Quito, Ecuador, 2. MINEXFIELD, Quito, Ecuador

The research area is located in Pasco Region of Peru, covering an area of 2570 km² and geologically on the eastern zone of the polymetallic deposits associated with Miocene volcanic diatremes belt. This investigation aims to target new hydrothermal alteration zones by interpretation of an ASTER satellite image.

The geological, tectonic, and structural setting of the studied area creates a suitable environment for the formation of hydrothermal systems. With the presence of a mountain system with low vegetation cover, it is suitable to use an ASTER VNIR-SWIR multispectral image. The analysis and procedures on the satellite image were compared with the SWIR spectral signatures obtained from the ASD Terraspec in the mine sites of the investigated area, allowing us to define areas of enhanced lithological and alteration signatures defining new exploration zones.

Four spectral techniques were used to identify hydrothermal alteration zones. The first technique involved mapping clay minerals and iron oxides using the combination of RGB bands. The second technique used band ratios to determine mineral alteration associations. The third technique used SWIR spectral indexes to corroborate the spectral results. Finally, the last technique was the spectral mapping using the Spectral Angle Mapper (SAM) method, which needs a special procedure, because it uses both field data obtained from hand samples using the ASD Terraspec with spectral signatures from ASTER image.

The application of this last technique permitted to corroborate the mineral alteration zones obtained with the previous methods and also to refine the hydrothermal alteration halos based on mineralogical associations characteristic of mineral deposits. And taking into consideration all the metallogenetic, geological, and structural environment, 10 new exploration zones were defined. Seven for high sulfidation polymetallic systems with advanced argillic and intermediate argillic alteration, and three for porphyry systems with phyllic, intermediate argillic, advanced argillic, and propylitic alteration.

Deep Learning-Based Predictive Modeling of Mineral Prospectivity with a Class-Balanced Focal Loss Function for Tackling Training Data Inequality

Zhiqiang Zhang[1], Gongwen Wang[2], Emmanuel John Carranza[3]
1. Hebei Key Laboratory of Strategic Critical Mineral Resources, Hebei GEO University, Shijiazhuang, Hebei 050031, China, 2. School of Earth Sciences and Resources, China University of Geosciences, Beijing 100083, China, 3. University of the Free State, Bloemfontein, South Africa

Because mineralization rarely happens, mineralized localities are extensively less than barren ones. Therefore, a set of geoscience data used for training of a model for predictive mineral prospectivity mapping (PMPM) for exploration targeting is normally imbalanced and poses a problem for PMPM by machine learning (ML) or deep learning (DL). Besides the huge inequality in the amount of training class sample data for mineralized and barren localities, the training data also carry redundant information (i.e., similar attributes), which also impinges on the performance of ML- or DL-based PMPM. This presentation describes a novel class-balanced focal loss function (CBFL) for tackling the above-mentioned problem in DL-based PMPM. The CBFL and the long short-term memory network were employed for 3D exploration targeting by PMPM in the Wulong gold district (China). Bayesian optimization was used for automatic tuning the DL model's hyperparameters. The results of using the CBFL were contrasted with those of using popular focal loss and cross-entropy loss functions in DL-based PMPM. The outcomes reveal the superiority of the CBFL over the mentioned popular loss functions. The 3D exploration targets mapped in this work would benefit follow-up subsurface exploration in the the Wulong gold district.

Application of BEMD in Extracting Gravity Anomaly Components Showing Deep Ore-Forming Dynamic background of the Giant Jiaodong Gold Cluster Region

Yongqing Chen, Xu Zhu, Chen Chen
School of Earth Sciences and Resources, China University of Geosciences, Beijing, China

The Jiaodong gold cluster region with approximately gold reserve over 5000t, located at the southeastern edge of the North China Craton, is third largest gold cluster region in the globe. The BEMD (Bi-empirical mode decomposition) is applied in extracting gravity anomaly components showing the deep structures of the lithosphere at different depths. (a) At a depth of about 30 km, there are three tectonic units: the mantle uplift with gravity values ranging from 2 to 14 $\mu m/s^2$, a mantle depression with gravity values varying from 0 to -13 $\mu m/s^2$, and a mantle flat with gravity values ranging from -2 to 2 $\mu m/s^2$. All giant gold deposits are distributed within the mantle depression. It may illustrate the mantle uplifting triggered hot ore-forming fluids concentrating into the mantle depression to accumulate forming giant gold deposits. (b) At about 12.5 to 17.1 km, there are three tectonic units: Jiaolai-Jiaobei mantle uplift showing a strong positive gravity anomaly with gravity values ranging from 1.5 to 10 $\mu m/s^2$, the Sulu ultra-high pressure metamorphic block displaying a negative gravity anomaly with gravity field values ranging from -10 to -1.5 $\mu m/s^2$, and the Jiaoxibei gold cluster region exhibiting gravity background with gravity values varying from -1.5 to 1.5 $\mu m/s^2$. (c) At about 8.9 to 5.3 km, there is a series of positive and negative gravity anomalies. Most granites with low density display negative gravity anomalies, among which there are some negative anomalies with positive anomalous edges that contain gold deposits. This may illustrate that an ore-forming pattern of granite (negative gravity anomaly)- alteration(positive gravity anomaly). It has been illustrated that the geological architectures at different depths and the giant Jiaodong gold cluster region were formed by the asthenosphere rises triggered by NNW-ward subduction of the Izanagi Plate over a time period from about 200 to 100 Ma.

How do We Determine and Use Geochemical Background in Big-Data-Based Exploration?

Mason L. Dykstra, Thomas-Louis de Lophem
MinersAI Inc., Lafayette, CO, USA

Understanding how to determine and use a background geochemical signature in a region, a stratigraphic interval, or within a set of outcrops is critical to unraveling whether any particular geochemical excursion is of geological and economic significance in mineral exploration. But what, exactly, does 'background' mean in this context? Is the background determined geographically? Is it bounded by structure or determined by the watershed in which the sampling occurred? Is the geochemical background narrow or widely ranging, and how do you determine what is outside the background signal? Although the concept of a geochemical background has been used for a long time, it continues to be poorly defined, and thus the use of the concept is highly uneven in the literature and in exploration programs. In this contribution, we examine the importance of background geochemical data in a number of different contexts, and we demonstrate that results are critical to determine the relevance of signal excursions from a baseline. We examine how properly evaluating the 'background' signature is important for robust mineral exploration and targeting, and how spurious positives or negatives can be avoided by systematic use of a background geochemical baseline range. This work has implications for everything from local to regional studies, and from classic geochemical analysis to cutting-edge machine learning.

Enhancing Platinum Group Mineral Recovery from Flotation Tailings: A Process Mineralogy Approach for Alternative Pretreatment

Borbor Gibson[1, 5], Nonkuselo Madlakana[1], Glen Nwaila[1], Yousef Ghorbani[2], Eligiusz Gugala[3], Sehliselo Ndlovu[1], Jochen Petersen[4]
1. University of the Witwatersrand, Johannesburg, Gauteng, South Africa, 2. University of Lincoln, Lincoln, United Kingdom, 3. TESCAN ORSAY HOLDING, a.s., Brno, Czech Republic, 4. University of Cape Town, Cape Town, South Africa, 5. University of Liberia, Monrovia, Liberia

The shift toward a circular economy and green energy is driving increased interest in sustainable supplies of critical metals (CMs). In this geometallurgical study, platinum group mineral (PGM) flotation tailings from the highly weathered silicate reef in the Bushveld Complex were pretreated using acidic chemical media to optimise a unit process stage. This stage aimed to expose locked PGMs and base metals (BMs) not recovered in a prior flotation campaign for extraction in a downstream selective dissolution stage. We investigated the effects of sulphuric acid, nitric acid, and ammonium hydrogen sulphate as reagents, leaching times of 3, 24, and 96hrs, temperatures of 25°, 50°, and 70°C, and pulp densities of 20, 30, and 40w/v%.

Process mineralogy evaluation was conducted using a scanning electron microscopy (SEM)-based Tescan integrated mineral analyser (TIMA). The automated mineralogy (AM) investigation demonstrated significant alterations in the bulk mineralogy and surface morphology of the starting material. Acidic media, particularly sulphuric acid> nitric acid> ammonium hydrogen sulphate, effectively disintegrated the gangue minerals' crystal structure, exposing locked-in PGMs and BMs.

Postleaching TIMA analysis revealed that amorphous silica was the predominant product of the pretreatment, constituting ~30 to ~50wt%. While amphiboles and pyroxenes remained largely unaffected, serpentine, talc, chlorite, and other primary silicates, including feldspars and micas, were notably affected by the pretreatment, resulting in the formation of these postleached silica phases.

Optimised process conditions were identified as 3hrs leaching time, 20w/v% pulp density, 2M sulphuric acid, at 70°C in a two-stage process. However, AM results indicated that energy and economic efficiencies could be further optimised with 3hrs leaching time, 40w/v% pulp density, 4M sulphuric acid, at 50°C in a two-stage process. The chemical pretreatment was successfully validated on weathered flotation tailings, with future studies anticipated to extend the approach to low-grade run-of-mine stockpiles and other Bushveld orebody tailings.

SEG 2024 Conference: Sustainable Mineral Exploration and Development

Idaho's Proterozoic-Hosted Copper-Cobalt Belt: Characteristics, Challenges and New Geologic Insights

Virginia S. Gillerman[1], Reed S. Lewis[2], Darin Schwartz[3], Katharina Pfaff[4], Daniel Schmidt[4], George King[5], William Stone[6]
1. Idaho Geological Survey, Boise, ID, USA, 2. Idaho Geological Survey, Moscow, ID, USA, 3. Boise State University, Boise, ID, USA, 4. Colorado School of Mines, Golden, CO, USA, 5. Electra Battery Materials, Salmon, ID, USA, 6. Electra Battery Materials, Toronto, ON, Canada

The Idaho Cobalt belt (ICB) is a 60-km-long, northwest-trending belt of sedimentary rock-hosted copper-cobalt (Cu-Co) occurrences in the Lemhi subbasin of the Mesoproterozoic Belt-Purcell basin in Idaho. The ICB hosts the only primary cobalt resources in the US and is somewhat similar to the African Copper belt, though known ICB deposits are much smaller. Mineralization is strata-bound within siltites, quartzites, and argillites of the 1.4 Ga Apple Creek Formation. Cobaltite and chalcopyrite at the Blackbird Mine are associated with biotite-rich zones. To the southeast at Iron Creek, cobaltiferous pyrite is associated with magnetite, biotite, and chlorite. Genetic interpretations are complicated by three overprinting periods of intrusion, metamorphism, and deformation: bimodal magmatism at 1370 Ma, a poorly known "Grenville" event at ca. 1200 Ma, and widespread Cretaceous magmatism. The southeast end of the belt is less affected by Cretaceous overprinting. Slack suggested that chalcopyrite and cobaltite at Blackbird were variants of IOCG-type deposits and related to the 1370 Ma plutonism, compatible with U-Pb ages on xenotime. Ongoing geologic mapping by the Idaho Geological Survey is aided by new USGS-funded airborne geophysics. Recent work on the Iron Creek deposit includes LA-ICP-MS analyses of magnetite chemistry to assess its origin, hyperspectral scanning, and geochronology and petrographic studies. Hyperspectral drill core scanning of the fine-grained rocks suggests Co-Cu zones are associated with Fe-rich chlorite alteration. Magnetite and Co-bearing pyrite formed prior to the tight folding and stretching deformation and predate or are coeval with metamorphic biotite having a minimum age of 938 Ma ($^{40}Ar/^{39}Ar$). Magnetite compositions are compatible with an IOCG origin, though the metals may have been enriched earlier in the basin's evolution. The ICB is illustrative of the many challenges to critical mineral extraction, including economics and ESG issues.

Reuse of Granite Quarry Waste for REEs Recovery Along the Industrial Production Chain: Results from Preconcentration Tests

Silvia Gioiello[1], Licia Santoro[1], Alberto Cazzaniga[2]
1. Department of Earth Sciences, University of Torino, Torino, Italy, 2. Minerali Industriali Engineering S.r.l., Masserano (BI), Italy

Granite quarry waste from the "Graniti dei Laghi" plutons in Piedmont, Italy, underwent several lab-scale beneficiation tests to enhance the recovery of critical raw materials (CRMs) in the context of circular economy and sustainable mining. Preliminary analyses by ICP-MS, XRPD, SEM-EDS, and automated mineralogy highlighted that the magnetic waste from granite industrial reprocessing for feldspar production is notably enriched in rare earth elements (REEs), paving the way for simulating and refining a lab-scale treatment to optimize the concentration of REE-bearing minerals. Monazite-(Ce) was identified as the primary target mineral, predominantly locked within Fe-rich biotite (siderophyllite). The feed material was milled to various grain sizes and subjected to multiple magnetic separation stages to identify the optimal conditions for REE mineral beneficiation. Results indicate that approximately 90% of REEs are associated with paramagnetic fractions of 75-150 μm and 150-250 μm grain sizes. The 75-150 μm paramagnetic concentrate recovered at 1 A of current intensity shows the highest enrichment to ~1800 ppm REEs + Y and Sc compared to ~200 ppm in the raw granite. However, significant REE amounts remaining in the paramagnetic material at 0.4 A suggest possible incomplete liberation of monazite. Although grinding below 600 μm favours the rejection of diamagnetic gangue minerals from the monazite-bearing siderophyllite, SEM-EDS observations reveal liberation issues for monazite particles above 100 μm due to their smaller size compared to siderophyllite. Consequently, additional grinding below 100 μm is required to isolate monazite. However, dry electromagnetic separation on fine material demonstrated poor monazite recovery, indicating the need for wet magnetic separation and flotation for effective recovery from the fine-grained preconcentrate. This work underscores the potential of improving beneficiation techniques to maximise the recovery of CRMs as by-products along the industrial production chain, advocating the valorisation and sustainable management of mineral processing waste through its reuse.

Geometallurgical Evaluation of a Tailing Storage Facility to Address Environmental Concerns and Maximize the Extraction of Contained Resources

Petya Gutzmer[1], Kai Bachmann[2, 1], Raimon Tolosana-Delgado[2], Laura Tusa[3], Cecilia Contreras[3], Philipp Buettner[2], Jens Gutzmer[2, 1]
1. ERZLABOR Advanced Solutions GmbH, Freiberg, Germany, 2. Helmholtz-Zentrum Dresden-Rossendorf, Helmholtz Institute Freiberg for Resource Technology, Freiberg, Germany, 3. TheiaX GmbH, Freiberg, Germany

A comprehensive geometallurgical evaluation of tailings material from a high-sulfidation epithermal gold deposit stored in a historic tailings storage facility (TSF) was carried out with the main objectives to (a) constrain the presence and distribution of minerals and elements of environmental concern, in particular pyrrhotite, and (b) to provide a robust estimate of the resource potential (e.g., recoverable value) of the fine particulate material in the TSF. Resource potential was considered not only for the deportment and mineralogy of remaining gold, but also for possible use of the silicate fraction for industries such as ceramics, glass, and geopolymer production.

To achieve these objectives, a drill core campaign was initiated, strategically selecting six drill holes for optimal representativity based on expected sedimentation patterns in TSFs. Drill core material was thoroughly characterized using automated mineralogy, geochemical, and hyperspectral methods. Geometallurgical domains were then defined based on 64 geochemical assays, particle size measurements, and quantitative mineralogical assessments. A hierarchical Mahalanobis distance cluster analysis was used to distinguish domains, and predictions for these domains were extended to all hyperspectral imaging samples. This comprehensive approach resulted in the delineation of four distinct domains, along meaningful spatial continuity patterns, each domain characterized by distinct modal mineralogy and geochemistry. The exercise provided valuable insight into the heterogeneity of the tailings material, laying the groundwork for targeted interventions to address environmental concerns and maximize the extraction of resources contained in the studied TSF.

Muon Attenuation Radiography and Tomography in the Polymetallic St. Christoph Skarn Deposit, Ertzgebirge, SE Germany

Marko Holma[1, 2, 3], Pasi Kuusiniemi[1, 3], Gergely Surányi[1, 4, 3], Gábor Nyitrai[4, 3], László Balázs[1, 4, 3], Lars Starke[5], Eric Hohlfeld[5], Jens Metschurat[6]
1. Muon Solutions Oy Ltd, Pyhäjärvi, Finland, 2. Kerttu Saalasti Institute, University of Oulu, Oulu, Finland, 3. International Virtual Muography Institute, Tokyo, Japan, 4. HUN-REN Wigner Research Centre for Physics, Budapest, Hungary, 5. Saxore Bergbau GmbH, Freiberg, Germany, 6. Private, Freiberg, Germany

Attenuation muography, an innovative astroparticle geophysical technique, uses cosmic-ray muons to image the internal structures of objects based on the attenuation of muon intensity due to density. We applied it in the St. Christoph mine, a tourist attraction in western Erzgebirge, SE Germany. Mining of the local polymetallic ore containing Sn, Zn, Cu, Fe, In, and Ag began as early as 1558. Our measurements were conducted as part of the Horizon 2020-funded Mine.the.Gap project.

Muography was conducted at two depths of approximately 40 and 60 m within the mine using an MWPC-type muon telescope. At the first adit, the telescope was initially positioned vertically 2 m from the shaft axis, and subsequently tilted 30° toward the shaft at a second point 19 m away. Polar plot muon radiography images from both positions distinctly revealed the shaft. Furthermore, a two-point inversion tomography was developed for the rock volume around the shaft.

The third measurement was conducted in a complex underground gallery, producing a muon radiographic density image. Given the location's nature, we employed detailed laser scanning to account for known voids and detect potential unknown ones. The gallery also contained historical supports made of bricks and stones. Our results indicated that the extraction of the dense polymetallic ore horizon was incomplete during historical mining operations.

Our measurements demonstrate that muons can be successfully applied to scan rock densities for various purposes. These include the detection of drilling targets or additional resources for extraction. Muography can also be used to detect voids, like past mine infrastructures and collapsed adits and shafts. In conclusion, our study highlights the efficacy and potential of muon attenuation tomography in contributing to the broader aim of integrating cutting-edge science and innovative technologies across the mining industry supply chain.

A GIS and Remote Sensing Based Approach for the Classification of Mine Tailings in South Africa

Gaone Joseph
University of Witwatersrand, Johannesburg, South Africa

Innovative technology developments are key to sustainable management of mineral deposits and the environment. This study seeks to create a GIS database cataloguing mine tailings (waste material produced when processing ore of mineral deposits) across South Africa, both active and closed mines. This groundbreaking initiative in South African literature will create a publicly available database, paving the way for future research. With the escalating costs of deep-level mining, rising commodity prices, and advancements in ore processing, stakeholders are increasingly interested in reworking mine tailings for their economic potential. This study intends to identify mine tailings in the Witwatersrand Basin, Barberton Greenstone Belt and Northern Cape with p interest in gold, manganese and iron using Aster and Sentinel 2 sensors. In the methodology, we apply band ratios, mineral Indices, Principal Component Analysis and other image enhancement techniques for identification of mine tailings to the Aster and Sentinel 2 free, atmospherically corrected and, minimal-to-no-cloud cover images. Our approach enables larger area coverage and analysis of mine tailings, overcoming the cost and logistical limitations of traditional methods. Preliminary findings show potential in our approach to catalogue tailings and variations in gold mine tailings identification between Witwatersrand Basin mines and those in the Barberton Greenstone Belt, possibly attributed to different processing methods, potentially biological in the latter. Our study stresses the importance of monitoring, including identifying abandoned tailings, crucial for future environmental sustainability efforts. This research aims to compile mine tailings data into a geospatial database, guiding policymakers in sustainable mineral resource management and setting the stage for future South African research. Ultimately, it seeks to transform mine tailings from a liability into an asset.

Critical Mineral Exploration with Compiled Data: Joint Imputation and Clustering for Improved Analytics

Rebecca M. Montsion, Gerhard Visser, Carsten Laukamp
CSIRO, Kensington, WA, Australia

Exploration for critical minerals such as Li, Sn, W, and REEs presents a new challenge to geoscience as mineralizing processes and controls are often disconnected from current deposit settings. Smaller footprints, unique metallurgical requirements, and historically limited economic interest also contribute to this challenge when rapid discovery and production of critical resources influence the feasibility of global green energy targets. As exploration for these challenging targets leverages increasingly diverse data compilations, pitfalls related to variable precision, accuracy, and completeness of data become more pronounced and increase exploration risk. A new probabilistic imputation and clustering method was developed to overcome limitations from data gaps or lower precision vintages of data as well as extend the application of more comprehensive but less common techniques (e.g., hyperspectral). This method is demonstrated using lateritic material in the highly weathered southwest Yilgarn Craton in Western Australia, where a range of mineralization types have been reported.

Application of Machine Learning Algorithms to Predict Rock Types Using Geochemical Data: A Case Study from the Obuasi Gold District, Ghana

Abdullah B. Muhammed, Emmanuel D. Sunkari
University of Mines and Technology, Tarkwa, Ghana

Artificial Intelligence and Machine Learning (AI/ML) have increased precision and productivity in the current big data era. Machine learning has indicated its robustness in geosciences, particularly rock-type classification. The issue of lithological classification in the traditional way has raised critical concerns and the need to curb the limitations it breeds, such as time consumption and subjective results. The gold mineralisation occurrence is structurally controlled in the district. It exhibits complex patterns and relationships that may not be readily discernible through traditional methods, leading to geologists missing out on discovering new resources or potential exploration targets. Consequently, this research attempts to create a predictive model by exploring the best machine learning algorithms to predict rock types in the Obuasi Gold District using X-ray fluorescence (XRF) geochemical data. Comparative predictive modelling was established using four supervised classification algorithms: Gradient Boosting (GBoost), Adaptive Boosting (AdaBoost), Support Vector Machine (SVM), and Random Forest (RF). The acquired XRF data was integrated with the model using the Orange Data Mining software. The performance evaluation of the models indicated RF as the best algorithm for deployment with a Classification Accuracy (CA) of 0.952. Therefore, ML algorithms have been a great tool in rock-type classification, whereby RF emerged the best in the case of the Obuasi Gold District. However, it is encouraged to understand the geology of a particular area before employing the tool. The datasets must be balanced to yield good results and avoid model overfitting.

A Process Mineralogy Approach for Direct Leaching of Precious Metals from the UG2 Ore in the Bushveld Igneous Complex

Nonhlanhla Ndlovu
The University of the Witwatersrand, Johannesburg, South Africa

The Platinum Group Metal (PGM) industry currently relies on the traditional crush-mill-float-smelt and pressure leach extraction for ore processing. However, this approach is not always applicable to all ore types. Direct cyanide leaching has emerged as a cost-effective and environmentally friendly alternative to the traditional extraction method, particularly for low-grade or mineralogically complex PGE-bearing ore deposits. The effectiveness of direct cyanide leaching stems from cyanide's strong affinity for metals, enabling the formation of complexes with heavy metals at low concentrations and facilitating effective metal recovery.

Cyanide leaching offers several advantages over traditional methods in the extraction of PGMs. It ensures high recovery rates when the system is optimised. Additionally, direct leaching reduces treatment expenses by eliminating intermediate steps like flotation and smelting, resulting in overall cost savings. Direct cyanide leaching has a higher solid-to-water ratio, reducing the amount of water consumed, and making it more environmentally friendly. Furthermore, the flexibility of cyanide leaching enables mining companies to process mineralogically complex and low-grade ores.

This study explores the potential use of cyanide direct leaching for recovering PGMs from UG2 ore, building upon previous research and addressing the limitations by investigating the effect of elevated temperatures and the addition of oxidants to enhance leach kinetics and extraction levels. By assessing these parameters along with the mineralogy and texture of the ores, this research seeks to optimise cyanide leaching processes, improving both efficiency and sustainability in PGM extraction from challenging ore deposits.

Satellite-Based Mapping for Lithium Pegmatite Mineral Exploration in Namibia: A Criteria-Based Approach

Olivia Rhind, Christopher Lambert, Alfie Baines, Mark Broadley
CGG, Crawley, United Kingdom

The growing global demand for lithium (Li) requires innovative mineral exploration methods. Lithium pegmatites, a significant source due to their grade and conventional mining methods, represent a key target. In many cases, Li exploration methods involve observation of surface exposure or investigation through geophysical surveys. In this study we introduce targeting of Li pegmatites in Namibia using satellite-based mapping within a mineral system criteria framework. It integrates expert knowledge with remote sensing techniques to identify and prioritise exploration targets based on geological characteristics, such as source magma types, structural settings and host-rock metamorphic grade.

The study utilises a Bare Earth+ mosaic derived from the entire archive of ASTER satellite imagery. Every image is co-registered at the pixel level with atmospheric and vegetation suppression applied and the "barest" pixel chosen. The result is an enhanced representation of Earth's surface with optimal signal to noise ratio and minimum cover, allowing the most representative spectral signatures to be derived and targeted. The Bare Earth+ mosaic is the foundation for the regional multispectral lithological and alteration mapping, with processing tailored to the Li-bearing granitic pegmatite system and resulting mineral maps supporting target generation, spectral lithological mapping and prospectivity analysis.

Hyperspectral EMIT data are also integrated in the study, enabling enhanced mineral mapping and lithological discrimination. The additional spectral bands are used to further identify and map possible Li-rich pegmatites. Several key proxy clay minerals are used as vectors to highlight candidate pegmatite target areas. Variability in host-rock composition can be spectrally discriminated using both multi- and hyper-spectral data through careful use of Principal Component Analysis, and from the rocks and alteration styles that they highlight. This workflow facilitates rapid regional assessments of Li-pegmatite potential, enhancing target identification accuracy and reducing exploration risk.

Staged Pacha-Styled Mineral Prospectivity Mapping in NW Queensland

Jhon A. Amasifuen Leiva[1], Sebastián Y. Espinoza[2], Misael A. Robles Ancajima[1, 2], Katherine M. López[1, 3], Marcello Imaña[4], Stephanie Sykora[5], Ronald Yupa[6], Alejandro Caballero[7]
1. Pontificia Universidad Católica del Perú, Lima, Peru, 2. Universidad Católica del Norte, Antofagasta, Chile, 3. University of Queensland, Queensland, QLD, Australia, 4. Copperstone Resources, Svärdvägen, Sweden, 5. University of Tasmania, Tasmania, TAS, Australia, 6. The University of Western Australia, Perth, WA, Australia, 7. Universidad Nacional de Córdova, Córdova, Argentina

This study shows a novel approach to multicommodity exploration targeting in North-West Queensland, Australia. The approach involves the use of innovative tools to identify prospective areas using a guide for assigning importance, applicability, and confidence weighting values attributed to different factors related to the mineral system components. The dataset was chosen due to the need to target prospective areas with multicommodity lateral thinking and the creation of new vectors, which contribute as reliable tools to make a significant discovery that generates a positive impact on Australian society.

The key focus was data integration and innovation as tools for geological exploration. Therefore, our staged model integrated several datasets to create feasible new vectors concerning the mineral system components such as felsic and mafic anomalies, the use of MT Anomalies as vectors for producing igneous fertility at different depths, traps, and source of metals, reactivity/metal enrichment capacity index, the use of isotopic mantle ages (Terrane Definition Tool), application of concentration-number in multi-fractal model applied to the geochemistry and the calculation of an innovative factor termed "Litho-Structural Endowment". Likewise, the integration of environmental, social, and mining variables to create a final product named Staged Pacha-Styled Mineral Prospectivity Mapping which offers a holistic approach with scale of impact for multicommodity exploration targeting in North-West Queensland – Australia. The inclusion of environmental and social aspects as an overlay to this prospective framework has added an extra responsible, ethical, and sustainable approach to targeting. This applied methodology has allowed us to achieve a significant correlation with previously identified deposits, reaching a 90.50% correlation in all types of deposits and almost 98% excluding phosphate deposits.

This study was conducted by the Pacha Team in the context of the Frank Arnott - Next Generation Explorers Award 2024, which won the grand prize at the PDAC convention.

Tin and Zinc Stable Isotope Variations in Cassiterite and Sphalerite from the Bolivian Tin Belt

Evan C. Hirsh[1], Ryan Mathur[2], Adam C. Simon[1], Wayne G. Powell[3]
1. University of Michigan, Ann Arbor, MI, USA, 2. Juniata College, Huntingdon, PA, USA, 3. Brooklyn College, Brooklyn, NY, USA

The Bolivian tin belt is an arcuate northwest-south trending region of Sn-Ag-W-(Sb-Bi-Zn) mineralization that extends for more than 1000 km along the Andean Eastern Cordillera morphotectonic province. Tin mineralization occurs primarily as cassiterite as a result of magmatic-hydrothermal processes associated with reduced ilmenite-series, S-type granitic magmas emplaced within polydeformed pelitic basement rocks. Zinc mineralization occurs primarily as paragenetically later sphalerite and occupies distal positions along proximal-distal thermochemical gradients in many districts and deposits within the tin belt. Previous experimental and field-based work shows that tin and zinc isotopes fractionate as a consequence of the changing physicochemical conditions during the transport and precipitation of cassiterite and sphalerite from saline magmatic-hydrothermal fluid and vapor. In this study, tin and zinc isotopes were measured in ore samples from 10 Bolivian tin districts, which reflect distinct mineralization styles and metal transport-precipitation pathways. The tin isotope compositions reveal that cassiterite precipitated at the shallow porphyry level is isotopically heavier, which may be a consequence of the enhancement of fractionation by redox reactions and vapor phase separation. The residual intrusive-proximal fluids are depleted in heavy tin isotopes, and precipitate isotopically lighter cassiterite at distal positions along district-scale fluid pathways. The zinc isotope compositions reveal that the isotopic behavior of zinc may mirror that of tin during mineralization.

Hyperspectral Outcrop Surveys for 3-D Geological Mapping

Laura Tusa[1], Samuel Thiele[2], Erik Herrmann[1], Cecilia Contreras[1]
1. TheiaX GmbH, Freiberg, Germany, 2. Helmholtz Institute Freiberg for Resource Technology, Helmholtz Zentrum Dresden Rossendorf, Freiberg, Germany

Digital outcrop models based on photogrammetry, attributed with hyperspectral reflectance data, represent a 3-D environment for mineralogical mapping of natural cliffs or mine faces. With the collection of such spatially continuous datasets, the possibility arises for mapping extensive natural or human-made outcrops at high resolution and with a short turnaround time.

Such workflows rely on the integration of high resolution photogrammetric pointclouds and hyperspectral scans, acquired either from a tripod-based system or a drone. Once the hyperspectral data are projected onto a photogrammetric pointcloud, the derived hypercloud can be analysed using various techniques.

Endmember extraction methods allow for the identification of spectral endmembers characteristic of different mineral assemblages. Using this information, a selection of methods can be used for mapping the relative abundance of different spectrally diagnostic minerals, delivering relative abundance, alteration, or lithological maps.

With these extensive 2.5-D (i.e., surface) maps that are available to geologists on the ground just shortly after data acquisition, decision making can be improved for both exploration sampling as well as material handling during production. With the collection of samples and additional ground truth mineralogical or geochemical data, machine learning techniques can be employed to extend the relative abundance results to semi-quantitative abundances.

These obtained outcrop maps can be then integrated with drill-core data to improve the 3-D modeling of a specific deposit.

Precompetitive Support of Exploration and Development in the Critical Minerals Era

Richard K. Valenta
Sustainable Minerals Institute - The University of Queensland, Brisbane, QLD, Australia

Precompetitive support of exploration and development by government agencies has played a key role in driving exploration success for more than a century. Such support has traditionally taken the form of geoscientific datasets and value-added interpretations focused on the area selection end of the exploration and mining value chain. This focus has taken priority because the mining, processing, and downstream considerations of more traditionally sought commodities are well-understood, and companies and investors have been focused mainly on the technical challenges to attaining production. Several important developments have taken place over the past decade to change this situation.

- Exploration focus has moved more to critical and strategic minerals where sustainable processing pathway and supply chains are not as well understood;
- Mineral production opportunities are expanding to increasingly use exploration approaches applied to secondary sources such as tailings and other waste products;
- Sustainability performance at all stages of the mining value chain is becoming more important to stakeholders;
- Expectations have increased to begin at an early stages to identify potential challenges and opportunities associated with exploration and development projects in terms of waste disposal, circular economy opportunities, and closure considerations, all of which require early and different geoscientific data to be collected from the outset of a project; and
- Exploration targets continue to be more difficult to identify and test, and more effective means to provide a fuller spectrum of precompetitive data in a form that can allow explorers to recognise more subtle indications of haloes and expand the detection space is becoming increasingly important.

Sustainable Minerals has been collaborating with the Geological Survey of Queensland and other state and federal surveys over the past 8 years on over 40 projects aimed at addressing these challenges, with positive and tangible impacts in terms of exploration success and improved project development approaches.

Innovative Narrow Reef Mining (NRE) Technology for Ultrathin Subhorizontal Ore Deposits

Sibila Borojevic Sostaric[1], Ema Vokic[1], Vjecislav Bohanek[1], Luka Petro[2], Paulo Plese[2]
1. University of Zagreb - Faculty of Mining, Geology and Petroleum Engineering, Zagreb, Croatia, 2. DOK-ING, Zagreb, Croatia

Traditional underground mining methods, utilizing standard LP equipment (low profile; 2.2 m), when applied to low-grade, thin, subhorizontal orebodies, result in significant dilution of the ore due to the extraction of large amounts of waste rock. To address this issue, ultralow-profile mining equipment (0.9–1.7 m; the NRE fleet) has been developed by Croatian DOK-ING company. The NRE fleet includes the NRE dozer, drill rig, and support rig. Each machine operates within a width range of 0.9 to 1.7 m at a slope of 22° (up to 25° for the dozer), and they are all remote-controlled, fully automatic, equipped with tracked undercarriages, and electrical.

To estimate the market potential of the NRE fleet, basic geological and mineralogical features of over 200 subhorizontal ore deposits/prospects/mineralizations have been analyzed:

1. PGE reef deposits within layered mafic-ultramafic complexes are NRE-suitable. Bushveld complex hosts 25 NRE-suitable PGE underground mines, Great Dyke of Zimbabwe hosts 10 suitable mines. Mafic-ultramafic intrusions at Stillwater, Muskox, and Greenland are NRE-suitable. Other European, Asian, and Australian deposits (Penikat&Portimo, Kemi, Fedorov Tundra, Monchegorsk Complex, Rinco del Tigre) are tectonically highly disturbed or a productive zone is too thick.
2. Witwatersrand gold reef deposits are NRE-suitable (productive zone 0.35 – 3 m, dipping 10–22°)
3. Magmatic Ni-Cu-PGE sulfide deposits within:
    1. Mafic-ultramafic rocks of the Noril'sk–Talnakh intrusion are NRE-suitable. Other Asian massifs (Kluane, North Rock, Uitkomst, Jinchuan, Kola, Ural, Kondyor, Kempirsay, Khabarninsky, Nuralinsky) are not suitable.
    2. Gabbro-norite are generally not suitable (too thick productive zone or steeply dipping deposits). A single NRE-suitable project in Sudbery basin is detected. Analyzed Duluth and Lac des Iles complex are not suitable.
    3. Komatiite are generally not suitable. Only two deposits/projects out of 20 analysed from Australia (Kambalda, Lake Johnston), Europe (Central Lapland, Suomussalmi), and North America (Abitibi, Capa Smith) are suitable.
4. European Cu-kupfershiefer deposits are highly suitable.

Modeling Field Strength and RQD to Interpret Potentially Poor Ground Conditions in Southwest Wau Graben, Papua New Guinea

Farzi Yusufali, Daniel Mogilny, Yush Rajcoomar
StratumAI, Richmond Hill, ON, Canada

Machine learning is creating value within multiple facets of the mining industry, especially when utilizing larger data sets to augment predictions of parameters that are characterized by small or incomplete data. The authors will introduce deep learning tools like convolution neural networks, how datasets of disparate size and characteristics are combined together as inputs into neural networks, and the results of these methods applied to geotechnical modeling. The authors present the results from a vein-stockwork gold-silver epithermal deposit in the Morobe Granodiorite of the Wau Graben in Papua New Guinea. They demonstrate how statistically significant non-linear correlated elements from assays and lithology are used as pathfinders for field strength (FS) and RQD to categorize each block as potentially poor ground conditions (PPGC). The PPGC threshold for both parameters were RQD <25% or field strength = soil, 0, 1 designations. KNN baseline models for RQD and field strength were created to (1) understand the complexity of the modeling task and (2) to create a performance benchmark against which the neural network methods would be evaluated. The results of this baseline demonstrated that geostatistic-esque interpolation method resulted in an 81% error rate in FS correctly diagnosing PPGC while RQD had a 79% false positive rate of indicating PPGC. Modeling each parameter separately and using the resulting block models to assign a PPGC classification to each block was a more accurate result than a singular block model predicting a PPGC value directly. The block model evaluation method, due to the sparse dataset, was to spatially split the data where 10% of the dataset was kept aside as a blind test set. It is demonstrated that alteration logging (propylitic (PR), potassic (K), sericitic (SE)) with RQD and FS as inputs identified over 55% of PPGC events, far exceeding the KNN baseline.

Application of Anomaly Components of Multi-Element Combination Extracted from Stream Sediments Data by BEMD in Mapping Ore Prospectivity Within the JGCR, Eastern China

Aoyue Zheng
School of Earth Sciences and Resources, China University of Geosciences, Beijing, China

The significant decline in the number of newly discovered mineral deposits and the dramatic increase for the global demand for mineral resources in recent years have led exploration geologists to search for more efficient and innovative methods to explore recognized and concealed mineral resources. However, the multi-stages of metallogenesis have made the distribution of ore-forming elements exhibit complexly overlapped characteristics within geological units. We have been tasked with finding an innovative technology capable of recognizing ore footprints from the complexly overlapped phenomena created by the multi-stages of metallogenetic processes. In this study we establish a hybrid method by integrating Bi-dimensional empirical mode decomposition (BEMD) with Principal Component Analysis (PCA) to extract local and regional anomaly components of multi-element combination from stream sediment concentration data, which can be used for recognizing ore footprints. The proposed model is applied to spatial analysis of the mineralization in the giant Jiaodong Gold polymetallic Cluster Region (JGCR), Eastern China.

The research results have shown as follows: the local anomaly components can identify Au-Ag-Cd and Be-La-Mo-Nb-Th-U-Y ore targets, and the regional anomaly components can recognize their high background zone favorable mineralization in the JGCR, Eastern China. Combined with the spatial distribution of both the local anomaly components and the granitic intrusions, it is inferred that the Au-Ag-Cd mineralization be related to the Linglong granitic intrusion and the Guojialing granodioritic intrusion, and the Be-La-Mo-Nb-Th-U-Y to the Guojialing granodioritic intrusion and the Weideshan monzonitic intrusions. The Guojialing granodioritic intrusion may be dual mineralization characteristics of both precious metals and critical metals. With the insights acquired from this study, the proposed hybrid method by integrating BEMD with PCA can serve as an effective tool in the exploration information system to trace the footprints left behind by magma hydrothermal fluid flow for directing prospecting concealed ore bodies.

The Prediction of Uranium Occurrence Using GIS Based Geostatistical Algorithms in Erongo Region

Silas T. David
EDS Namibia, Windhoek, Namibia

Mineral resources are pivotal for Namibia's socioeconomic progress and play a crucial role in its energy transition and on a global scale. The Erongo region boasts significant uranium deposits, fostering the establishment of uranium mines, with potential for further discoveries. Consequently, mapping uranium occurrences in this region is imperative for effective mineral resource management. Interpolation techniques provide a means to accurately forecast mineral values at unsampled locations, generating a continuous dataset of spatial distributions. This study employs spatial and geostatistical analytical tools to compare various interpolation methods concerning uranium occurrence patterns in the Erongo region, utilizing ArcGIS 10.6 software. Seventy uranium occurrences were selected to construct prediction models. Deterministic interpolation methods like Radial Basic Function and Inverse Distance Weighting, alongside geostatistical techniques such as kriging, including Empirical Bayesian Kriging, were utilized in both spatial and geostatistical analyses. The interpolation method exhibiting the lowest average standard error, root mean square standardized error, root mean square error, and mean square error was deemed the most suitable for capturing spatial variations in mineral occurrences. Geostatistical approaches, particularly Empirical Bayesian Kriging with a linear semivariogram model, demonstrated superior performance compared to Universal Kriging in spatial analysis. In conclusion, geostatistical interpolation emerges as the preferred method over deterministic interpolation for accurately mapping and predicting potential uranium occurrences in the Erongo region.

EIS: Exploration Information System – Linking of Mineral Systems and Mineral Prospectivity Mapping

Andreas Knobloch[1], Vesa Nykänen[2], Hafsa A. Munia[3], Tobias E. Bauer[4], Guillaume Betrand[5], Juha Kaija[3], Joy Cremesty[6], John M. Carranza[7]
1. Beak Consultants GmbH, Freiberg, Germany, 2. Geological Survey of Finland, Rovaniemi, Finland, 3. Geological Survey of Finland, Espoo, Finland, 4. Luleå University of Technology, Lulea, Sweden, 5. Bureau de Recherches Géologiques et Minières, Orleans, France, 6. LGI – Sustainable Innovation, Paris, France, 7. University of the Free State, Bloemfontein, South Africa

The Exploration Information System (EIS) project, funded by the European Union (EU) Horizon Europe research and innovation-funding programme, aims to discover new sources of critical primary raw materials within the EU by combining mineral systems modelling and mineral prospectivity analysis methods. As exploration data continues to emerge, the need for efficient data analysis becomes crucial, considering the high cost of data acquisition. Efficient analysis of expensive exploration data has become crucial and recent advancements in artificial intelligence and machine-learning algorithms within geographic information system (GIS) platforms have enabled the integration of geological knowledge and exploration data into complex mathematical models for predicting the existence of new mineral occurrences.

The EIS project promotes a hybrid approach that utilizes mineral systems modelling as the foundation for mineral prospectivity modelling. A crucial component of EIS is a library of geological fingerprints representing diverse types of mineral systems. These fingerprints are used to identify the most relevant mappable geoscientific features essential for successful prospectivity analysis. The project focuses on three mineral systems as case studies: cobalt minerals in volcanogenic massive sulfide (VMS) systems, lithium-tin-tantalum-tungsten minerals in granite/pegmatite-related systems, and rare earths-cobalt minerals in iron oxide copper-gold (IOCG) systems. Selected mineral deposits within the partner countries of the EU serve as study or test sites, and reference sites in South Africa and Brazil, provide additional insights.

The presentation will showcase the newly developed EIS Toolkit, which includes among others preprocessing tools and validation tools, but also new data analysis methods by incorporating artificial intelligence with machine learning and deep learning. In addition, it will be highlighted how geo-models and mineral systems modelling is being integrated into mineral prospectivity mapping. Finally, the first beta release of the open source EIS QGIS Plugin consist of a collection of software tools for semi-automated exploration targeting in QGIS.

Mineral Exploration Mapping Tools Using Machine Learning and Drone Geophysics

Andreas Knobloch[1], Sami Niemi[2], Jochen Kamm[3], Rene Forsberg[4], Ari Saartenoja[5]
1. Beak Consultants GmbH, Freiberg, Germany, 2. Geological Survey of Finland, Kuopio, Finland, 3. Geological Survey of Finland, Espoo, Finland, 4. Technical University of Denmark, Lyngby, Denmark, 5. Radai OY, Oulu, Finland

Remote sensing methods and applied geophysics techniques gather information about subsurface structures without disturbance. Airborne surveys are preferred due to their speed, safety, and cost-effectiveness. Drone-based geophysical surveys are versatile, fast, economical, and ecologically friendly. However, drone payloads have limitations. Some geophysical methods are unsuitable for drones, and some may be prohibited by law. Drones have new opportunities in mining operations, especially for local scale support for ore and top cover volume estimations by repeated photogrammetry and LIDAR, constructions, and a general understanding of geological bodies in large-scale pit mining. Magnetic measurements are the most mature technology, while electromagnetic sounding and gravimetry are still developing.

In DroneSOM, Radai is developing a drone-based geophysical electromagnetic (EM) survey system called Louhi. It has two modes of operation – one uses a large loop on the ground as an EM source, while the other uses a small portable loop EM transmitter. Both systems utilize a lightweight three-component EM receiver towed by a drone.

Second, DTU Space is developing a drone-based gravity system with a moving platform. The goal is to make the operation feasible by exploring new sensor systems. A new configuration called the strap-down configuration of an inertial measurement unit (IMU) has many advantages over traditional gravity sensors. The IMU is physically fixed to the chassis of the aircraft, making it simpler, smaller, lighter, and requiring less power.

Third, self organizing maps (SOM) method have proven in the past to be applicable for geoscientific data integration and mineral prospectivity analysis. In DroneSOM, software tools for implementation of SOM were developed by Geological Survey of Finland and Beak Consultants. One developed tool is the open source standalone application GisSOM for mineral predictive mapping (MPM) in 2D, while the other currently developed option is advangeo® Prediction, which will allow MPM in 3D using SOM.

Mapping Nigeria's Mineral Frontier: AI-Driven Prospecting for Investment Opportunities

Luis Pizano[1], Abba Usman[2], Abdulrazaq Garba[2], Pablo Martin-Cocher[1], Andreas Barth[1], Andreas Brosig[1], Andreas Knobloch[1]
1. Beak Consultants GmbH, Freiberg, Germany, 2. Nigerian Geological Service Agency, Abuja, Nigeria

Increasing demand for mineral resources, exploring previously overlooked areas has become essential. This presents a unique opportunity for African countries endowed with mineral-rich geological settings to present their mineral resource potentials for investment promotion with the use of scientifically verifiable artificial intelligence algorithm. Nigeria, traditionally focused on oil production has been actively promoting its mining industry over the past five years. Recent geoscientific projects have provided valuable data, confirming the country's potential for key commodities.

The Mining Investment Facilitation Mission played a crucial role in consolidating recent data, identifying promising target areas for various commodities, including Gold (Au), Tin (Sn) – Niobium (Nb), Lithium (Li) – Tantalum (Ta) Pegmatites, and Lead (Pb) – Zinc (Zn). The methodology required the definition of a new metallogenic scheme for Nigeria, categorizing regions, and utilizing machine learning algorithms to generate favourability and prospective maps. Derivation of information from existing data by understanding the respective mineral systems was crucial to obtain applicable findings. The results allowed identification of several undocumented considerable mining areas as a validation control.

Outcomes, compiled in prospectuses with relevant information, revealed at least eight main targets per commodity, indicating favourable conditions for large deposits, often only mined by artisanal miners. These findings have facilitated engagement with investors and provided guidance for future exploration efforts by the Nigerian Ministry of Solid minerals development and Geological Survey agency.

SEG 2024 Conference: Sustainable Mineral Exploration and Development

Exploring for Cu-Au Alkalic Porphyry Deposits from the Fringes

Fabrizzio Abello[1], Farhad Bouzari[1], Shaun Barker[1], Devin Wade[2]
1. MDRU, Vancouver, BC, Canada, 2. New Gold, Vancouver, BC, Canada

The Iron Mask Batholith (IMB) from the late Triassic to early Jurassic period has many mineralized centers such as New Afton, Copper King, and Ajax, which feature alkalic porphyry-type deposits and calc-alkaline stocks like the Nora molybdenum porphyry mineralization. Exploring the IMB district is difficult because of the diverse magma composition and deposit characteristics. However, it provides a unique opportunity to analyze and contrast the properties of alkalic and calc-alkalic porphyry deposits. The research effort seeks to analyze district-scale intrusions, factors influencing orebody placement, and distant alteration by detailed fieldwork, mineralogical analysis, and geochemical techniques. Initial results show similar footprints for New Afton and Copper King with an epidote, chlorite, prehnite, white mica chalcopyrite-rich inner zone surrounded by a pyrite, chlorite, epidote-calcite outer zone. Element distribution in New Afton shows a concentric zonation grading outward from a depleted Zn center to anomalous Te-Cu-Ag and finally Sb and As. On the other hand, the calc-alkalic Nora deposit shows a more extensive footprint with an inner white mica zone, grading outward to a chlorite zone and, finally, an extensive epidote-calcite outer zone. Although Nora presents a much broader footprint, element zonation is similar, with a depleted Zn inner halo, to a Mo center, grading to Cu, and finally an extensive Sb halo. Mineral chemistry in the distal Ajax samples showed encouraging results for chlorite and epidote chemistry in the area, characterizing areas interpreted in other studies as distal and proximal to porphyry centers. Future work on vertical zonation, mineral chemistry, and vein characterization is expected to help define the footprints of the studied deposits.

Unlocking Gold Potential: Remote Sensing Insights and Applied Analysis of the Qena-Safaga Area in Egypt's Eastern Desert

Mohamed Badawi
Miskolc Egyetem, Miskolc, Hungary

The eastern desert of Egypt has undergone prolonged deformation as a result of the cratonization of the juvenile crust known as the Arabian Nubian Shield (ANS). This geological phenomenon has raised prospects for potential gold discoveries associated with the corresponding tectonic events. Within this context, the Qena-Safaga zone, located on the border of Egypt's Northern and Central Eastern Deserts, stands out for its metavolcanic rocks, which have been subjected to significant twisting due to the movement along the Qena-Safaga Shear Zone (QSSZ) relative to the Barud magmatic body.

In our recent study, we conducted a thorough examination of numerous deformed post-orogenic granitic intrusions and felsite dikes within the Qena-Safaga area. Our investigation aimed to identify promising mineralization sites, with a particular focus on orogenic gold deposits. Applying a combined approach involving detailed field surveys, advanced analysis of remote sensing data—including band combinations from ASTER and Sentinel 2—and interpretation of aeromagnetic enhanced maps, we systematically distinguished various rock types, structural features, and hydrothermal alterations across the study area.

By integrating insights from multi-sorce data, including the development of remote sensing band combinations and aeromagnetic maps, we were able to validate previously hypothesized mineralized zones within the Qena and Safaga areas, while identifying additional areas with similar mineralization potential. Notably, our analysis, which included sophisticated methods such as the utilization of ASTER and Sentinel 2 data alongside hydrothermal alteration and orientation entropy heat maps, revealed a strong correlation between mineral-rich regions and major structures associated with late-stage deformation of the QSSZ.

Our study stands out for its applied analysis, which provided valuable insights into the geological processes at the Egyptian eastern desert. By elucidating the spatial distribution of mineralization and its relationship with structural features through advanced analytical techniques, our findings offer important guidance for future exploration efforts in this promising area.

Trace Element Systematics of Auriferous Sulfides and Its Implications on the Gold Mineralization Within the Vempalle Dolostone of the Cuddapah Basin, India

Sahendra Singh[1], Vijay K. K S[1], Ramesh C. Behera[1], Vinod K. Tirkey[2], Shalivahan S[3]
1. Indian Institute of technology (Indian School of Mines), Dhanbad, JH, India, 2. St.Xavier's College, Ranchi, Ranchi, JH, India, 3. Indian Institute of Petroleum & Energy, Vishakhapatnam, Vishakhapatnam, AN, India

Tummalapalle is well known for the uranium mineralization hosted within the Vempalle dolostone of Cuddapah Supergroup, southern India. The present study deals with the pyrite hosted invisible gold mineralization within the uraniferous dolostone. Petrographic studies reveals the presence of various sulfide phases i.e. pyrite, arsenian pyrites and chalcopyrite. On the basis of textural characteristics and trace elemental distribution (Co/Ni values), three generations of pyrites (Py1, Py2, and Py3) are identified. Py1 is fine grained and early to syn-diagenetic to the uranium minerals. Py2 is late diagenetic and show overgrowth texture. Py3 is euhedral in shape and is of hydrothermal origin. Py3 shows higher gold concentration among all the three pyrite generation. The sub-microscopic gold is present in association with the other trace elements i.e., As, Sb, Hg, Tl, and Cu. The Au/As ratio of all generations of pyrite indicate that, gold within the Vempalle dolostone occur as solid solution. Gold within the Py3 precipitated as solid solution from Au-As rich hydrothermal fluid, which is evident by the well-developed As rich zones over Py2. This study reveals that pyrites hosted gold mineralization within the Vempalle dolostone of Tummalapalle area was of sedimentary origin during the initial stages, but latter on remobilized and enriched by the hydrothermal fluid. It is inferred that the nanoscale gold mineralisation in the host sulfide is triggered by the combined effects of dissolution, transportation, precipitation, and biological activities, which is evident from the preservation of the stromatolitic cluster in auriferous sulfide bearing dolostone. This study is the first report on the pyrite hosted invisible gold occurrence within the uraniferous Vempalle dolostone of the Cuddapah sedimentary basin in and around Tummalapalle area.

Rare Earths and Other Minor Elements in Hungarian Coal Formations

Csilla Balassa[1], Mohamed Elbalawy[1, 2]
1. University of Miskolc, Miskolc, Hungary, 2. Ain Shams University, Cairo, Egypt

During the 1960s, several sampling campaigns were carried out to determine the trace element content of domestic hard coals, brown coals, and lignites, which were major resources of Hungary's energy production. The results of these surveys were not of any commercial interest in those times. As part of the 2012, 2018, and 2024 work programs, these records were revisited because many of the elements were ranked as of critical or strategic importance.

The available historic data are sporadic and of poor quality by today's standards, but sufficient to highlight and compare different critical element enrichments. The highest concentrations (reaching over 600 ppm total REE average values) are found in the Mecsek Coal Basin hard coals, claystones, and carbonaceous clays of Jurassic age. There are also high Sc, Be, Hf, Ge, Ta, Nb, Ga, Hf, and Zr anomalies. Anomalous concentrations of vanadium (with maximums over 1,600 ppm V) are characteristic of the Cretaceous Ajka sub-bituminous coals. Enrichments of Ge were recorded in the Miocene Borsod basin coals, with maximums over 300 ppm Ge. A similar sampling program is now being carried out in the Pliocene coals, the Visonta and Bükkábrány lignite mines. The data also point to the untested potential of coal residues and disposed fly ash.

These poorly known enrichments may mean radically new exploration objectives in the coal deposits, which recently faced closures due to the restrictions imposed on fossil fuel based energy production. In the present global geopolitical turmoil, it is increasingly important to get reliable information about the geology, resources, and availability of critical raw materials within the borders. The approach that has been offered in our study may be a serious contribution in this regard.

Crustal Architecture of the EMARK Oceanic Core Complex

Olivia Filson, Melissa O. Anderson, Jonathan Umbsaar
University of Toronto, Toronto, ON, Canada

Oceanic core complexes are domal structures uplifted and exposed at the sea floor by long-lived slip along deep-seated, low-angle, detachment faults. This is an important process of crustal accretion, exposing large sections lower crust and upper mantle material at the surface, as well as for the formation of hydrothermal ore deposits. Oceanic core complexes have been observed in ultraslow and slow-spreading mid-ocean ridges, especially along the northern Mid-Atlantic Ridge, however most studies of oceanic core complexes and their associated hydrothermal activity focus on mature complexes in late to end stage development. Research herein examines a young, early-stage oceanic core complex called the EMARK (East Mid-Atlantic Ridge Kane) at approximately 23°N, just south of the Kane Fracture zone on the Mid-Atlantic Ridge and east of the mature Kane oceanic core complex.

In collaboration with Schmidt Ocean Institute, high-resolution bathymetry, remotely operated vehicle dive observations, and sampling at EMARK have uncovered this large, in-situ crustal section in detail. Results from remote predictive geologic mapping illustrate the crustal architecture of the EMARK oceanic core complex from the uppermost paleoseafloor basaltic flows to sheeted dyke complexes to gabbroic sills and mainly gabbroic dome to serpentinites at the exposed fault surface. Active black smoker hydrothermal chimneys were also discovered on an adjacent off shoot of the spreading axis. The abundance of gabbro sampled in addition to local high-temperature hydrothermal venting, suggest robust magmatism during the development of the oceanic core complex, in contrast to the general model of detachment faulting during periods of relatively low magmatism. Furthermore, these findings indicate that oceanic core complex development may increase potential for hydrothermal ore formation providing fluid pathways, even in early stages. This work has implications for the understanding of oceanic core complex evolution, controls on the formation of hydrothermal ore deposits and future exploration of these deposits.

The Role of Magmas in Forming a Big Skarn System: A Case Study from La Colorada Deposit (Zacatecas, Mexico)

Pedro Francisco[1], Zhaoshan Chang[1], Shiqiang Huang[1], Chris Emerson[2], Dante Juarez[2], Patricio Perez[2], Robert A. Creaser[3]
1. Colorado School of Mines, Golden, CO, USA, 2. Pan American Silver, Durango, DUR, Mexico, 3. University of Alberta, Edmonton, AB, Canada

The La Colorada deposit in Mexico is a large skarn system, containing 299 Moz Ag (34g/t), 7.4 Mt Zn (2.67%), and 3.4 Mt Pb (1.21%; 12/18/2023), and is among the world's top three Zn-Pb skarns. It features skarn, porphyry, and carbonate replacement (CRD) mineralization in deeper carbonate-rich sedimentary sequences, and intermediate sulfidation epithermal mineralization in both overlying volcanic rocks and sedimentary sequence. Exploration is ongoing and the deposit remains open.

Multiple intrusions of andesitic, dacitic and rhyodacitic compositions occur in this deposit as stocks, dikes, and/or sills. Some of them are identified as syn-skarn based on zoning patterns of exoskarns next to the intrusions that include A) garnet color becomes darker; B) pyroxene color changes from brown to green; and C) garnet and pyroxene Mn contents decrease (garnet, from 1.9 to 0.1; pyroxene, from 15.5 to 0.2 MnO wt%), towards the causative intrusions. Additional evidence includes the presence of red garnet endoskarn, with or without porphyry-style alteration/mineralization. Other intrusions are either pre- or post-skarn.

Syn-skarn intrusions include andesite, dacite, and rhyodacite porphyries. Cross-cutting relationships suggest that fertile magmas evolved from intermediate to more felsic compositions, with the latter responsible for more skarn development. Syn-skarn porphyry molybdenite Re-Os (n=3) and zircon U-Pb (20) dating, together with skarn garnet U-Pb (10) dating show a mineralization age of ca. 62 Ma in the east, and ca. 63 Ma in the western side of the deposit.

The plurality of mineralizing intrusions, intruded at close but different locations over ~1 million years, is believed to be critical in the formation of this large system. The intrusions may have stemmed from a deeper batholith and were emplaced in contact with carbonate-rich sequences. Early andesite/dacite intrusions produced only narrow skarns but enhanced wall-rock permeability. Later rhyodacite phases formed wider skarns and significantly increased the deposits' metal endowment.

Geology, Mineral Alteration and Mineralization of the Santa Cecilia Porphyry Au-Cu Deposit, Maricunga Belt, Northern Chile

Jose Franco[1], Antonio Arribas[1], Luciano Bocanegra[2], Piotr Paleczek[2]
1. The University of Texas at El Paso, El Paso, TX, USA, 2. Torq Resources, Vancouver, BC, Canada

The Santa Cecilia porphyry gold-copper deposit is located within the Maricunga gold-silver-copper belt in northern Chile. Santa Cecilia presents a complete porphyry copper system alteration column, from potassic alteration at depth to shallow advanced argillic and silicic assemblages near the surface, and it has not been clear where the highest ore grade is located. Therefore, part of this study aims to narrow down this information and identify the specific location within the alterations. Adjacent to Santa Cecilia is the Caspiche porphyry Au-Cu deposit (proven and probable mineral reserve of 1,091 million tonnes (Mt) averaging 0.55 g/t Au), separated by a 250-m-wide valley. The deposit formed in the latest Oligocene (~24 Ma) during the first of two volcanic and corresponding metallogenic epochs that define the Maricunga belt. The main objectives are: (1) identify the mineral assemblages and timing of each of the magmatic-hydrothermal alteration and veining stages present in the deposit; (2) evaluate the grade of Cu and Au mineralization within different alteration assemblages or stages; and (3) study the relationship between the different alteration and mineralization centers existing within the Santa Cecilia property. These objectives will be addressed through petrographic, mineralogical, geochemical, and isotopic techniques, including XRD, micro-XRF and radiometric dating. Preliminary results indicate that this deposit has different alteration zones: (1) potassic alteration consisting of biotite-magnetite (at ~1,400 m); (2) chlorite-sericite alteration with chlorite-muscovite-albite (at ~550 m), (3) sericitc (white mica) alteration with quartz-muscovite-albite (at ~1,100 m), and in the lithocap (4) quartz-pyrophyllite alteration is present with advanced argillic alteration which comprises quartz-alunite with alunite ledges and vuggy residual quartz. Finally, further work will involve generating a model of the deposit and understanding the genesis of the mineralization with an emphasis on the magmatic and structural setting and evolution of this sector of the Maricunga Belt.

Chlorite Chemistry of the New Afton Alkalic Porphyry Cu-Au-(Pd) Deposit, Canada

Andres Gonzalez[1], Adam Simon[1], Devin Wade[2], Catherine Ryan[2]
1. University of Michigan, Ann Arbor, MI, USA, 2. New Gold Inc, Kamloops, BC, Canada

This study delves into the chemical and textural characteristics of chlorite, serving as an initial step toward its potential application as a vectoring tool in alkalic porphyry deposits. Chlorite is ubiquitous, both in the proximal and peripheral alteration zones of the New Afton orebody, occurring in two distinct domains: Chlorite-R, resulting from chloritization of biotite, actinolite, and ferromagnesian phenocrysts, characterized by numerous inclusions, predominantly Ti oxides; and Chlorite-V, found as veinlet infillings, often associated with late-stage chalcopyrite, bornite, specularite, and Co-Ni-rich pyrite. Analysis of major and trace elements from Chlorite-R reveals that its composition mirrors that of precursor minerals, with partial contamination by inclusions, particularly tiny rutile grains contributing to high Ti content. Conversely, Chlorite-V compositions exhibit a distinct and systematic depletion of chalcophile elements, consistent with co-precipitation with sulfides. Moreover, temperature estimations range from ~210° to ~260°C for Chlorite-R and from ~250° to ~330°C for Chlorite-V. These constraints on temperature formation align with a Cu-Au-Pd mineralization pulse in New Afton, resembling other alkalic porphyry deposits in the region. Furthermore, previous studies have proposed that the presence of Co-Ni-rich pyrite serves as an indicator of PGE enrichment in New Afton. Similarly, Chlorite-R compositions exhibit significant enrichment in Ni and Co relative to chlorite from other porphyry deposits, whereas Chlorite-V is depleted in these elements, reflecting its co-precipitation with copper sulfides. Nonetheless, the presence of Co and Ni in chlorite holds promise for exploring PGE-bearing porphyry systems. The coexistence of both chlorite domains in several samples highlights the necessity for comprehensive textural characterization before delving into trace element analysis, thus accentuating the complexity of mineralization processes in alkalic porphyry deposits. Although the prevalence of multiple chlorite generations in these deposits offers potential for temperature estimation, exercising caution is essential to ensure accurate geological interpretations.

Retrograde, Shear-Related Copper Mineralization at the Onganja Mining District, Namibia

Matthew P. Hales, Judith A. Kinnaird
University of the Witwatersrand, Johannesburg, South Africa

The Onganja mining district is located on an antiformal dome in the southern zone of the Pan-African Damara orogenic belt. The district comprises several surface and underground workings that targeted Cu-mineralised breccias and massive sulfide ore bodies, which were hosted in a sequence of amphibolites, biotite-plagioclase schists, and pelitic schists. The breccias are associated with a N- to NE-trending quartz-albite vein system but are only mineralised where black-mica schists form the wall rock to the veins. The massive sulfides are parallel to the schistosity within the black-mica schists. They are dominated by chalcopyrite and secondary chalcocite with subordinate pyrite, magnetite, and hematite and an enrichment Au-U-Mo±REE. Albite and biotite are the dominant alteration phases associated with Cu mineralisation.

The black-mica schists comprise typically well-aligned greasy black phlogopite that defines the strong schistosity. Veins and lenses of albite and quartz may be aligned to this schistosity. The black-mica schists are interpreted as shears in the host amphibolites with differing mineral assemblages representing the varying degrees of hydrothermal alteration and mineralisation of the wall rocks. Although similar mica-rich, coarse-grained schists with a strong schistosity occur throughout the district, the amphibolites and subsequent black-mica schists appear to have acted as a redox boundary, whereas the higher Fe content favoured the deposition of the Cu sulfides and associated elements.

The ore and alteration assemblages as well as the subvertical orientation of the breccias bear some similarity to the iron oxide-copper-gold (IOCG) mineralisation style, however, field observations in conjunction with whole-rock and mineral chemistry data suggest that the mineralisation at the district is unlikely to be representative of an IOCG. Rather, the mineralisation at the district is comparable to other retrograde, shear-related Cu-Au-U-Mo mineralisation in the Damara-Lufilian orogenic belts such as at Omitiomire, Lumwana, and Kansanshi.

High Grade Cu-Au-Ag Overprint at the Filo del Sol Porphyry Deposit, Argentina-Chile

Jack Halloran, Jeffrey Hedenquist, Keiko Hattori, Jose Perello
University of Ottawa, Ottawa, ON, Canada

The Filo del Sol porphyry deposit is associated with Miocene porphyritic diorite dikes along the Argentina-Chile border. The deposit contains high copper and gold grades due to early porphyry-style mineralization overprinted by high sulfidation-state minerals. This overprint bringing high gold and silver extends to depths greater than 1000 m. The early event formed mushketovite and biotite plus pyrite with subsequent anhydrite intergrown with chalcopyrite and bornite. Co-existing pairs of anhydrite and chalcopyrite yield sulfur isotope equilibrium temperatures between 550° and 450°C at sulfide-dominant conditions. Oxygen isotope compositions of anhydrite support a high-temperature magmatic-hydrothermal origin of the early mineralization. Later hydrothermal activity caused a deep overprint of residual quartz and quartz-alunite on the early mineralization. This later stage of hydrothermal activity was relatively oxidized and cool. It produced an assemblage of enargite, digenite, covellite, chalcocite, and pyrite. The sulfide minerals are accompanied by native gold grains, and silver sulphosalts and selenides. The cool temperature of the late mineralization is confirmed by intergrown K-rich alunite and enargite, which yield sulfur isotope equilibrium temperatures of ~260°±30°C. Pyrite is common and crystallized throughout hydrothermal activity, with diverse crystal habits and textures and variable sulfur isotope compositions.

High-grade Cu-Au porphyry deposits with high sulfidation-state mineral assemblages occur around the world. This study shows that detailed petrographic and mineralogic studies combined with stable isotope analyses are effective in unraveling the evolution of the mineralization stages of magmatic hydrothermal activity.

Hitting the the Right Frequency - Not All Conductors are Created Equal

Vessela Hobson
Axiom Exploration Group, Saskatoon, SK, Canada

Airborne time domain electromagnetic (EM) surveys are indispensable tools for subsurface exploration, particularly in resolving conductors at depth. Their depth of investigation depends in part on transmitter frequency and moment, as well as the noise levels of the receiver.

In this study, we present a comparison of conductor detection and resolution capabilities of the XCite time-domain EM system operating at frequencies of 15 Hz and 30 Hz. Through detailed modelling of data collected over a study area, we investigate the effect of these frequencies on the system's ability to detect and resolve conductors. Methodologically, we describe the parameters and settings used for both frequencies and outline the data analysis techniques used.

Our results reveal significant differences in conductor detection and resolution between the two frequencies. The practical implications of our study extend to improvements in survey design, suggesting that the choice of frequency can significantly impact survey outcomes. Moreover, our findings underscore the importance of further research to enhance understanding and optimization of airborne EM survey methodologies.

Nickel-Copper Sulfide Mineralization in the Maibele-North Ni Sulfide Prospect, Central-Eastern Botswana

Jacob Kaavera[1], Thatayaone Oletile[1], Elvis Mosweu[2], Thembiso O. Basupi[1], Elisha Shemang[1]
1. BIUST, PALAPYE, Botswana, 2. Si6 Metals, GABORONE, Botswana

The Maibele Ni-Cu (PGE) prospect in central-eastern Botswana represents partially metamorphosed mafic-ultramafic intrusions. In this study we focus on the detailed description of sulfide mineralization and petrologic characterization of the partially metamorphosed mafic-ultramafic host rocks aiming at better constraining the geodynamic context of the prospect. The Maibele prospect occurs within the Magogaphate shear zone, which marks the boundary between the Zimbabwe craton and the central zone of the Limpopo Complex. Host rocks in the Maibele prospect can be divided in two: (1) the amphibolites herein referred to as meta-gabbros; (2) ultramafic zone (UZ) characterized by pyroxenite cumulates intruded into granitic gneisses. Petrographic observations reveal a crystallization sequence for the UZ comprising olivine, orthopyroxene, plagioclase and minor clinopyroxene. The Ni-Cu sulfides occur as disseminated and massive mineralization within the UZ and the meta-gabbros. The main sulfide assemblage comprises pyrrhotite, pentlandite, chalcopyrite and pyrite. The mineralized pyroxenites feature variable forsterite (Fo) contents of olivines ranging from Fo 82.4 to 88.2 suggestive of a primitive parental magma source composition. The olivines are Ni depleted and exhibit sulfide fractionation trends suggesting that magmas from which the sulfides at the Maibele prospect precipitated underwent sulfide extraction, resulting in Ni-depleted olivines. S/Se ratios extend beyond typical mantle compositions. Ratios of trace elements including $(Th/Yb)_{PM}$ and $(Nb/Th)_{PM}$ are unlike those of typical mantle rocks and interpreted to indicate significant crustal throughput by the magmas. Together with the S/Se ratios, the trace element ratios likely suggest that crustal sulfur played a role in the sulfide saturation at the Maibele prospect.

Unravelling Manganese Mineralization Through Intergration of Remote Sensing and Geophysical Techniques in Mansa Area of Luapula Province of Zambia

Chanda Katotobwe, Kalela Banda
University of Zambia Student Chapter, Lusaka, Zambia

Recent interest in manganese occurrences in Zambia's Luapula and Central provinces and the resurgence of manganese mining indicates the potential for the region to be considered a Tier 1 manganese mining district, akin to the Central African Copperbelt as the understanding of geology in the area increases. Manganese is classified as a critical mineral defined by the European Union (risk list, 2023) due to its pivotal role in green energy components. This also aligns with Zambia's strategy to diversify mineral portfolios and boost the nation's GDP. Historical data from the Rhodesian Selection Trust and Anglo-American Corporation (1992-2000) regarding mineral occurrences in Zambia is the earliest mention of manganese deposits in Zambia. Using remote sensing and airborne geophysics the study aims to identify manganese mineralisation by locating regional and local faults which are controls of mineralisation. Analysis of geoscientific and remote sensing data will try and identify surface and subsurface manganese anomalies associated with preferential rock formations. Manganese mineralisation is hosted within Mesoproterozoic metasediments of the Muva Supergroup of the Irumide belt and Bangweulu block. Manganese deposits are observed to form in deformed supracrustal rocks, including pelites and sandstones, with manganese occurring as veins, nodules, and pockets within fracture zones. Faulting and fracturing serve as fluid pathways for manganese-bearing hydrothermal fluids. Economic deposits, with reserves up to 40 Mt, have been identified in Chipili district, Luapula province, and areas such as Kabwe through to Nakonde exhibit significant manganese potential, due to the extensive coverage of the Irumide belt in Central and Northern Zambia. This study aims to identify lineaments or faults that would function as sites for manganese deposition, informing exploration and extraction strategies in the study area. It also contributes to increased geoscientific knowledge and socioeconomic development in the associated communities.

Unidirectional Solidification Textures and Magmatic Silica Caps; What Do They Tell Us?

Douglas J. Kirwin[1], Axel Müller[2]
1. Consultant, Makati, Philippines, 2. Natural History Museum, University of Oslo, Oslo, Norway

Unidirectional Solidification Textures (USTs) are primary magmatic textures which comprise parallel to sub-parallel bands of crystals, commonly quartz with oriented c-axis terminations. The UST zones are products of rhythmic precipitation of quartz and quartz plus feldspar (or other rock-forming minerals) during periods of fluctuating over pressuring of volatiles and fluids within relatively small cupolas. They occur in a wide variety of intrusion-related mineral deposits and display several textural types. The exsolution of a magmatic vapor phase accompanying or following UST formation may serve as an important marker for magmatic-hydrothermal transition processes and the beginning of peak ore precipitation. Magmatic silica caps comprising massive milky to cloudy mottled quartz have been observed in a number of cupolas associated with mineralized felsic intrusions. Furthermore, those silica-enriched zones may transition to an underlying comb quartz UST horizon reflecting magma quench processes due to undercooling and oscillatory crystallization of excess silica during the magmatic to hydrothermal transition. Silica caps normally occur at the cupola contact with the host wall rocks and may vary considerably in terms of lateral extent and thickness from several meters to hundreds of meters. In many cases, they are well mineralized, particularly in highly evolved porphyry stocks hosting copper-gold-molybdenum mineralization, subject to the timing and conditions of the magmatic to hydrothermal transition. There are also many barren examples of USTs and magmatic silica caps which, at a district scale, represent pre-mineral magmatic accumulations, and as such, are vectors for proximal fertile mineralized intrusion-related ore deposits.

Petrogenesis of the PGE Mineralized Mafic-Ultramafic Sills on Uitloop Farm, Northern Limb

Mabatho Mapiloko[1], Marina Yudovskaya[1, 2], Paul Nex[1]
1. CIMERA, University of the Witwatersrand, Johannesburg, South Africa, 2. IGEM RAS, Moscow, Russia

The concept of Platreef-style mineralization has reshaped exploration perspectives, highlighting PGE and sulfide precipitation enrichment beyond discrete stratabound layers in the northern limb of the Bushveld Complex. Recent borehole investigations in the western Uitloop area unveiled PGE-enriched mafic-ultramafic sills nestled within the Transvaal Supergroup metasediments, alongside the Uitloop II intrusion of lower zone affinity. The complex interfingering contacts and composite structure of the sills, encompassing ultramafic subunits, present a significant challenge in assigning them to either the Platreef/critical zone offshoots or marginal sills. Their differentiated and locally layered nature favours the former option, supported by occurrences of uneven lenses and seams of chromitites, characteristic of the Platreef body.

Geochemical data and the elevated proportion of interstitial plagioclase further indicate a CZ/Platreef affinity for the mafic-ultramafic sills. Correlation arrays of Cr# (0.6–0.83) vs. Mg# (0.12–0.39) for chromites from the Uitloop sills and the main Platreef body, suggest a similar type of contaminated magma, akin to that parental to the Platreef chromitites. The relatively moderate PGE-enrichment of Uitloop chromitites and marginal-style mineralization align with the characteristics of the lower, contaminated portions of the main Platreef body. Multiple S isotope compositions of sulphides from magmatic and contact metamorphosed sedimentary rocks indicate the involvement of crustal S in the formation of Ni- Cu-PGE mineralization in the LZ and Platreef sills at Uitloop. The elevated $\delta^{34}S$ values (+3.1 to +12.9‰) and mostly near-zero $\Delta^{33}S$ values of Uitloop LZ and Platreef magmatic sulphides on Uitloop suggest S derivation from Paleoproterozoic sedimentary sulphates of the upper Duitschland Formation rather than from Archaean dolomites and shales, consistent with earlier conclusions based on the multiple S isotope data for neighbouring areas.

Controls on the Rarity of Porphyry Cu Deposits in the Archean

Xuyang Meng
China University of Geosciences (Beijing), Beijing, China

Porphyry Cu deposits are mainly formed in the Phanerozoic and are rare in the Archean. They are closely associated with oxidized S-rich magmas originating mainly from asthenospheric mantle metasomatized by slab-derived fluids in subduction zones. Because of extremely low seawater sulfate concentration and relatively reduced seawater-altered oceanic seafloor, prevention of oxidation of the mantle is hypothesized to have precluded porphyry Cu deposit in the Archean.

To test this hypothesis, we carried out systematic electron probe and *μ*-XANES analyses on zircon-hosted apatite inclusions from sodium-rich TTG and volcanic rocks and late potassium-rich intrusive rocks formed at 2,750 to 2,670 Ma in southeastern Superior Province, Canada. The main findings include: (1) primary apatite grains from > 2700 Ma pre-modern-subduction volcanic rocks yielded exceptionally low S concentrations except for a few samples yielding high apatite S concentrations but low $S^{6+}/\Sigma S$ ratios. The host volcanic samples yielded elevated La/Yb ratios and zircon Eu/Eu* ratios, suggesting that the operation of immature subduction at >2,700 Ma cannot form oxidized S-rich magmas. (2) Melt S content estimated using the compositions of the primary apatite grains from the intrusive rocks increased sharply upon the onset of the subduction event at ~2,700 Ma, but the apatite $S^{6+}/\Sigma S$ ratios and the magmatic fO_2 values for igneous rocks slightly decrease with increasing zircon $\delta^{18}O$ value (reflecting the increase in crustal assimilation in the source area) and crustal thickness.

These results suggest that the asthenospheric mantle and the derivative magmas may have been oxidized at the onset of the ~2.7 Ga persistent subduction in Superior province, Canada. We conclude that the scarcity of porphyry Cu deposits in Archean cratons may be attributed to (1) the limited operation of persistent subduction at >2.7 Ga and (2) subsequent erosional loss since the surface uplift associated with continental formation.

Late Cretaceous Metallogeny of the Turkish Eastern Pontides and the Georgian Bolnisi District: Evidence for a ~700 km-Long Epithermal-Porphyry and VMS Belt Along Eurasia

Robert Moritz, Şafak U. Sönmez
University of Geneva, Geneva, Switzerland

Late Cretaceous magmatic rocks of the Eastern Pontides in Turkey and the Bolnisi district of the northernmost Lesser Caucasus in Georgia have recorded comparable evolutions and host abundant base and precious metal ore deposits and prospects with a similar age range. Together, these two segments of the Tethyan orogenic belt document the presence of a highly fertile ~700 km-long eastwest-oriented metallogenic belt, which has been formed along the Eurasian margin during north-verging subduction of the northern Neotethys. The Late Cretaceous magmatic arc hosts abundant volcanogenic massive sulfide (VMS), porphyry and epithermal systems. The total metal potential (Cu, Au, Zn, Pb, Mo, Re, etc.) of this belt remains unknown, since the Late Cretaceous magmatic arc is partly concealed by Oligocene and Quaternary volcanic and sedimentary rocks, and its eastern extension into Azerbaijan still needs to be understood. The Late Cretaceous fertile metallogenic event is related to reactivation of magmatism along the southern Eurasian convergent margin after a magmatic lull of about 40 m.y. at the Early to Late Cretaceous transition. In the Eastern Pontides, bimodal Turonian to Santonian tholeiitic to calc-alkaline magmatic rocks host major VMS deposits (e.g., Murgul, Tunca, Hod Maden). A major magmatic and tectonic switch at about 83 Ma marks the onset of Campanian calc-alkaline to shoshonitic magmatism coinciding with the uplift of the arc, and resulting in the formation of epithermal (e.g., Yanıklı, Sisorta) and porphyry systems (e.g., Elbeyli, Emeksen). In the Georgian Bolnisi district, epithermal systems (e.g., Madneuli, Sakdrisi, Beqtakari) are coeval with late Santonian to early Campanian magmatism, dominated by felsic rocks ranging between low-K and high-K calc-alkaline compositions. Cenozoic post-ore tectonic, hydrothermal and magmatic events have overprinted the Eastern Pontides and its ore deposits. By contrast, such overprints are minor in the Bolnisi district, except local Eocene magmatism, therefore preserving its Late Cretaceous settings.

Understanding of Ground Gravity Responses Over the Kiala Deposit, Western Foreland, Democratic Republic of Congo

Melanie J. Postman, Roger Salumu
Ivanhoe Mines Exploration, Lubumbashi, Democratic Republic of the Congo (Kinshasa)

The Western Foreland (WF) of the Democratic Republic of Congo is host to the Kamoa-Kakula copper complex, and the Makoko and Kiala Deposits. Extensive airborne geophysics has been instrumental in developing the geological models used for exploration on the WF which led to the more recent Makoko and Kiala discoveries. Airborne gravity was flown in conjunction with magnetics and provided clear gravity low anomalies over the Kakula and Kamoa domes defined by Roan Group clastic rocks in the core; however, for smaller features the resolution was too coarse. For this reason, an extensive program of ground gravity was carried out to improve resolution.

At the Kiala deposit, mineralisation is hosted mainly in siltstones deposited in sub-grabens which accommodate thicker sequences of Nguba stratigraphy, dominated by diamictite. Ground gravity collected in the area highlighted two distinct north-south linear features that were assumed to be associated with higher density ferricrete/laterite within the lower density saprolite. Forward modelling of the higher resolution ground gravity data was carried out using the density of the rock types generated by routine measurement on core, to determine the cause of these anomalies. It was determined that the anomalies were not associated with ferricrete/laterite, but rather a thicker Nguba sediment sequence and mafic intrusive.

It is difficult to distinguish structures from the airborne magnetics over Kiala due to low magnetic susceptibilities. The ground gravity results highlighted low gravity responses in the same orientation as the sub-grabens. These correlated to the location of faults and lower density Nguba sediments adjacent to the faults, potentially caused by increased weathering along structures or associated alteration of faults. Thus, the ground gravity proved useful for delineating structures where the magnetic survey was ambiguous.

Unidirectional Solidification Textures from Two Andean Intrusion-Related Cu-Au Prospects

Misael Robles[1, 2], Lisard Torró[1], Antonio Arribas[3], José Franco Moraga[3], Douglas Kirwin[4]
1. Pontificia Universidad Católica del Perú, Lima, Peru, 2. Universidad Católica del Norte, Antofagasta, Chile, 3. University of Texas at El Paso, El Paso, TX, USA, 4. International Consultant Geologist, Makati, Philippines

Unidirectional Solidification Textures (USTs) are rhythmic bands of magmatic quartz formed in felsic cupolas where there are over-pressured volatiles and fluids. Two Andean examples are Pijilí, a newly-discovered Oligocene calc-alkaline Cu-Au-(Mo) porphyry system close to the Miocene Chaucha deposit in southern Ecuador, and San Lorenzo, an intrusion-related Cu prospect located near La Serena in the Cretaceous Coastal Belt in Central Chile. At Pijilí, USTs were found in artisanal mining stopes and host the highest grades (> 1% Cu-eq). In San Lorenzo, USTs are associated with unmineralized monzodiorite cupolas.

USTs from Pijilí show a heterogranular rhythmic distribution of cm-scale contorted quartz- and feldspar-rich bands and atoll textures. Thin (<1 cm) sinuous-straight quartz veins overprint the USTs and represent the magmatic to hydrothermal transition. Chalcopyrite and pyrite mineralization is more abundant in quartz bands and veins. These textural observations suggest syn- and post-sulfide mineralization with respect to UST formation.

USTs from San Lorenzo are mainly defined by irregularly-stacked, heterogranular, sub-parallel, contorted multilayers of quartz, K-feldspar, and magnetite. Magnetite defines narrow bands hanging from quartz layers into K-feldspar-rich layers indicating oxidized magmatic conditions. Individual magnetite crystals are 1 to 2 mm in size and show oscillatory compositional zoning in terms of Ti and V. Sulfide mineralization while present in sheeted fracture zones, has not been detected in the UST samples from San Lorenzo.

Timing and Characteristics of Skarn-Hosted Cu-Au Mineralization at Blue Copper, Montana, USA

Erin Shervey[1], Eric LeLacheur[2], Zhaoshan Chang[1], Shiqiang Huang[1], Mathias Burisch[1]
1. Colorado School of Mines, Golden, CO, USA, 2. Falcon Copper Corp, Deer Lodge, MT, USA

The Blue Copper Cu-Au prospect, located in the Ophir District on the northern border of the Boulder Batholith in Montana, USA, hosts several mineralized skarn zones. These zones are developed along the contact between the Blackfoot City granodiorite intrusion and carbonate rocks of the Mississippian Madison Group, Devonian Jefferson Formation, and the Cambrian Meagher and Pilgrim Formations. Here we present preliminary results of geological mapping, petrography and geochronology of skarn Cu-Au mineralization at Blue Copper. Outcropping skarn zones vary from 2 to 15 m in thickness and can be traced along several hundreds of meters of strike length. Their vertical extent is yet to be constrained.

Proximal prograde exoskarn comprises mainly dark red garnet overprinting local pyroxene skarn. Distal/intermediate skarn is characterized by green garnet, green pyroxene, and vesuvianite. Prograde skarn varieties are cross-cut by retrograde epidote-calcite-quartz veins. Endoskarn is subordinate, instead most of the intrusion in contact with skarn is overprinted by propylitic alteration. Locally, magnesian skarn alteration with abundant magnetite and serpentine is recognized, which shows elevated gold concentrations compared to calcic skarn. Copper-gold mineralization is intimately associated with skarn alteration. Hypogene sulfides are completely weathered to secondary copper minerals.

Four samples of garnet across the district were dated by LA-ICP-MS U-Pb geochronology, yielding ages ranging from 82 ± 3 Ma to 85 ± 2 Ma. The ages indicate that skarn alteration at Blue Copper is the oldest yet recognized hydrothermal mineralization event within the Boulder Batholith. We are planning to carry out zircon U-Pb geochronology of samples from the Blackfoot City granodiorite and place our age data in the regional context to update the metallogenetic understanding of the area.

Eolian Organic and Pyritiferous Siltstones as Redox Traps for Sediment-Hosted Ore Deposits

Leandro Silva, Peir Pufahl
Queen's University, Kingston, ON, Canada

Stratigraphic and spatial controls on organic- and pyrite-rich fine-grained sediments are critical for base metal exploration because they are both redox traps for metal-bearing fluids and a potential sulfur source for sulfide precipitation. In the modern ocean, windblown silt and clay accumulates on the seafloor together with organic matter downwind from active eolian systems adjacent to deserts and recently deglaciated landscapes. Bioessential P, Fe, and other micronutrients derived from eolian dust delivered to the surface ocean increases primary productivity and export of organic carbon to the seafloor. Authigenic pyrite may form as bacterial sulfate reduction degrades accumulating sedimentary organic matter. Such eolian derived siltstones are mineralogically immature, punctuate the stratigraphic record, and are thickest in the Neoproterozoic. When placed in a paleogeographic and sequence stratigraphic context, the deposition of Neoproterozoic organic-rich sediments peaked in the arid horse latitudes (30°N and 30°S) during interglacial periods between the low-latitude Sturtian (ca. 717-660 Ma) and Marinoan (ca. 650-635 Ma) snowball glaciations. The relationship between an extreme icehouse climate, paleogeography, black shales, graphite deposits, hydrocarbons, and thick siltstone successions is compelling evidence for nutrient mobilization to the Neoproterozoic ocean via eolian processes. The occurrence of framboidal pyrite assists with preconditioning these organic-rich sediments to reduce more effectively oxidizing metal-bearing fluids and provide a sulfur source for sulfides forming sediment-hosted ore bodies. Thus, understanding the stratigraphic and spatial distribution of wind-derived, organic-rich siltstones through time is an important exploration tool for identifying areally extensive redox traps for base metals that are correlative across and between basins.

Hydrothermal Alteration and Cu-(Co) Mineralization of the Fishtie Deposit: Implication for Metal Source and Timing of Mineralization in the Zambian Copperbelt Deposits

Mukuka Simusokwe[1], Yasushi Watanabe[2], Takuya Echigo[2]
1. University of Zambia, Lusaka, Zambia, 2. Akita University, Akita Prefecture, Japan

This study presents lithologic, petrographic, mineralogic, and sulfur isotopic data for the recently discovered Fishtie deposit to discuss controls of Cu-(Co) mineralization, and shed light on the contentious issues of metal sources and timing of mineralization in Zambian Copperbelt sediment-hosted deposits. Cu-(Co) mineralization occurs in stratified sandstone and conglomerate of the Grand Conglomerate formation, as well as the overlying folded dolomite and dolomitic siltstone of the Kakontwe Limestone Formation. The host rocks were affected by two hydrothermal events,resulting in the formation of; (i) magnesium (Mg-) alteration assemblage occurring as patchy to pervasive replacement of calcite and plagioclase in conglomerate matrix and clasts by dolomite and clinochlore, and (ii) iron (Fe-) alteration assemblage of chamosite, biotite and siderite that overprints the Mg-alteration assemblage. Cu mineralization occurs as chalcopyrite in host rock and veins, with minor bornite and chalcocite in Mg- and Fe-alteration assemblages, whereas Co mineralization occurs as cobaltite in chalcocite zones and is limited to Fe-alteration assemblages.

Sulfur isotopic results of sulfide minerals show strong lithologic control, as evidenced by the highest $\delta^{34}S$ values (from ~+15‰ to +27.2‰) recorded in sulfides disseminated in dolomitic siltstone, while the lowest $\delta^{34}S$ values (from ~-0.07‰ to +7‰) were recorded in vein-hosted hosted sulfides in dolomite and dolomitic siltstone. Intermediate $\delta^{34}S$ values (+9.2‰, +11.2‰, +13‰) are observed in sulfides disseminated in sandstone and conglomerate unit. Variations in $\delta^{34}S$ values suggests mixing of hydrothermal sulfur and reduced sulfur from evaporitic sulfate. Trapped hydrocarbons in dolomitic siltstone, sandstone and conglomerate acted as a reductant.

We propose a metamorphic model for Cu-(Co) mineralization of the Fishtie deposit that involves circulation of oxidizing hydrothermal fluids capable of remobilizing Cu and Co from basement mafic rocks. Reaction of metal-bearing hydrothermal fluids with reduced sulfur led to deposition of Cu-(Co) mineralization in fault-controlled veins and folded dolomite and dolomitic siltstone.

Physicochemical Conditions of Ore-Forming Fluids at Filo del Sol Deposit, NW Argentina: New Insights from Laser Ablation Split Stream S Isotope and Trace Element Chemistry

M. Camila Sojo Aguero[1], Hervé Rezeau[1], Michael Pribil[2]
1. University of Arizona, Department of Geosciences, Tucson, AZ, USA, 2. U.S. Geological Survey, Denver, CO, USA

Filo del Sol (FDS) is a high-sulfidation Cu-Au-Ag epithermal system superimposed on porphyry Cu-Au mineralization in the recently defined Vicuña metallogenic belt, between the world-class Maricunga and El Indio Cu-Au-Ag belts in the Central Andes. The physicochemical conditions of the ore-forming fluids at FDS remain unknown. We present new petrographic observations combined with in situ S isotope and trace element chemistry collected simultaneously by laser ablation multicollector inductively coupled plasma mass spectrometry (MC-ICP-MS) and quadrupole (Q)-ICP-MS split-stream analyses in sulfides, sulfosalts, and sulfates. In FDS, pyrite is ubiquitous and exhibits a diversity of textures and trace-metal compositions that correspond with paragenesis. Although the porphyry environment is dominated by a first generation of nonporous Cu-rich pyrite, the epithermal comprises porous and colloform Au-Ag-rich pyrites that post-date a second generation of nonporous Cu-Au-Ag-poor pyrite. In contrast, several generations of enargite are only identified using combined in situ S isotope and trace element analysis. Sulfide/sulfosalts-sulfate isotopic pairs associated with porphyry mineralization indicate high temperature of ~500°C and H_2S-dominated fluid, whereas mineral pairs from the epithermal stage revealed more complex physicochemical conditions at 300° to 400°C. The progressive shift toward δ^{34}S-depleted compositions from nonporous Cu-Au-Ag-poor pyrite (avg δ^{34}S of –7.1‰) to porous and colloform Au-Ag rich pyrites (avg δ^{34}S from –8.5 to –10‰) suggest an ore-forming fluid path toward lower temperatures and/or more oxidizing conditions. The occurrence of cinnabar associated with billingsleyite (Ag sulfosalt) at shallower levels yielding extreme δ^{34}S-depleted compositions from –42 to –23‰, combined with alunite-pyrite and alunite-enargite pairs in isotopic disequilibrium, argue for supergene processes at low temperatures and oxidizing conditions. Collectively, we show the power of coupled in situ S isotope and trace element analyses in high-spatial investigations to resolve geochemical changes of mineral assemblages, while preserving information about microscale crosscutting relationships and the relative timing of cryptic ore-forming events.

The Geological Characteristics of the Newly Outlined Cu-Au Deposits at the Sharlo and Krasta Prospects, Northeast of the Chelopech Mine, Bulgaria

Anastasiya D. Stoyanova, Beihan Chaush, Betina Petkova, Georgi Ivanov, Hristomir Stanev, Ilian Atanasov, Ivan Petrov, Kuncho Kunchev, Martin Dobrev, Miroslav Kavarimov, Ross Overall, Strahil Terziyski, Viktor Stoychev, Aneta Aleksieva, Mariana Trifonova, Miglena Grigorova, Stefan Metodiev, Paul Ivascanu, Istvan Marton
Dundee Precious Metals Chelopech, Chelopech, Bulgaria

The Sharlo Dere and Krasta prospects, located on the northeast flank of the Chelopech hydrothermal system, host newly defined zones of high-sulfidation (HS) Cu-Au mineralisation, that represent future mineral resource inventories at the Chelopech mine. The deposit is part of a large, Late Cretaceous magmatic complex formed in a back-arc basin.

The high-sulfidation (HS) mineralisation formed within a transition from a structurally influenced shallow intrusive related epithermal environment to an underwater hydrothermal vent system with sulfide- and sulfosalt-rich replacement zones. The mineralization is associated with a zoned advanced argillic alteration footprint and is hosted by a multi-phase 91.9±0.2Ma old dioritic intrusive system, which produced a variety of peperitic, mingling and fluidized features on the contacts. The Cu-Au (+Ag-Zn-Pb) mineralisation is manifested within a variety of settings and brecciation events, including structurally controlled subvertical phreatomagmatic feeders, stratabound breccia injections into the epiclastic formations, hydro-magmatic vent breccia flows, syn-sedimentary exhalative sulfide precipitates and reworked mineralised clasts in avalanche and debris flow deposits that were subsequently tilted according to the syn-formal basin architecture.

Mineralisation consists of pyrite, marcasite, sulfosalts, chalcopyrite, bornite, covellite, sphalerite, galena, with occasional presence of tellurides and other Cu sulfides (digenite, mawsonite, chalcocite, colusite) associated with quartz and barite. Two chalcophile associations can be distinguished by the geochemical analysis, including the economically more interesting Cu-Au-As-Sb-Bi-rich (sulfosalt-rich) and the Au-Mo-Tl-rich (pyrite-dominated) sub-domains, which shares similarities with the main ore bodies at the Chelopech deposit.

Two types of sulfosalt assemblages have been distinguished: enargite-luzonite and tennantite-tetrahedrite. The first one is characterized by high gold contents and is inherent to the more central parts of the ore bodies, the second occupies the outer reaches of the mineralized zones, being closely associated with the chalcopyrite-bornite paragenesis. An increased intensity of polymetallic minerals such as sphalerite, galena, and chalcopyrite is observed in the peripheral areas.

Hydrothermal Alteration Footprint of the Chelopech High-Sulfidation Cu-Au Epithermal Deposit, Bulgaria

Mariana G. Trifonova, Anastasiya Stoyanova, Aneta Aleksieva, Beyhan Chaush, Betina Petkova, Georgi Ivanov, Hristomir Stanev, Ilian Atanasov, Ivan Petrov, Kuncho Kunchev, Miroslav Kavarimov, Ross Overall, Strahil Terziyski, Viktor Stoychev, Miglena Grigorova, Stefan Metodiev, Paul Ivascanu, Istvan Marton
Dundee Precious Metals Chelopech, Chelopech, Bulgaria

The Chelopech high-sulfidation (HS) Cu-Au epithermal deposit is located in Bulgaria, SE Europe, 75 km east of the capital Sofia and is part of Apuseni-Banat-Timok-Srednogorie Metallogenic Belt. The HS mineralisation is related to a Late Cretaceous (91.9±0.2 Ma) age magmatic complex which is represented as transition from shallow intrusive related phreatomagmatic environment to an underwater hydrothermal vent system, characterized by sulfide and sulphosalts-rich replacement zones associated with well-developed alteration and mineralisation zonation.

The alteration footprint has been identified and modeled, based on short-wave infrared (SWIR) analysis, whole rock geochemistry and drill logging data, systematically captured during the recent exploration and mining history of the deposit.

Based on SWIR data the core of the ore bodies is characterized by intense acid-sulfate alteration, including vuggy silica and APS minerals (K and Na alunite, svanbergite, woodhouseite), which laterally grades into silica-dickite-kaolinite alteration associated with Cu-Au-As-Ag-Te-Bi mineralisation assemblages. Low-crystallinity kaolinite forms a large diffuse envelope surrounding high sulfidation mineralisation, transiting at depth into a diaspore-pyrophyllite dominated facies that marks the lower extent of Cu mineralisation. There is a large phyllic halo, surrounding the Cu-Au orebodies, extending laterally and below, with relative enrichment of Pb-Zn-Mn-Tl-Ag. Detailed analysis of the SWIR spectral features provides vectoring within a wide range of illite-sericite mineral assemblages. Shifting of absorption at the 2200 nm wavelength, caused by aluminum hydroxyl vectors acidity and the sharpness of water or hydroxyl absorption feature at 1900 nm indicates degree of crystallinity, which increases at higher temperatures. The distal parts are presented by phengite-chlorite-smectite dominated mineral assemblage. The use of multielement whole-rock interpolants provides additional vector within this alteration footprint by quantifying feldspar destruction with Sr-Ca-Na depletion, tracing distal parts of the system with changes in Mg/Fe ratios, and identification of illite and APS minerals alteration with relative enrichment of K-Rb and Sr-K-Ca, respectively.

3D Lithological and Structural Geo-Modelling: A Case Study of The Mudilandima Deposit, Tenke Fungurume Mining District, Lualaba Province, Democratic Republic of Congo

Robert K. Tshilumba
University of Zambia, Lusaka, Zambia

1. Introduction

- Background

The Mudilandima deposit or Mudi is a complex structured deposit located within the northern apex of the larger Central African Copperbelt, primarily extending from Zambia to the Democratic Republic of Congo. Despite the existence of a production geological model, the excavation of the copper-cobalt ore still faces challenges, leading to suboptimal ore recovery, loss, and dilution. Integrated approaches have significantly enhanced the high-resolution, accurate, and detailed representations of complex structured mineral deposits and have improved their understanding. Therefore, this research will attempt to construct a 3D lithological and structural model to mitigate the challenges induced by the high variability of the Mudilandima subsurface.

- Objectives

The main objective of this research is to produce an integrated three-dimensional model, based on a pit geological map and RC drill hole data, focusing on the lithological, alteration, and structural aspects in the ore delineation of the Mudilandima deposit. Specific objectives include: (i) to explore the lithologies and the alteration footprint; (ii) to explore the geologic structures (faults, folds, caverns); (iii) to investigate the influences on the ore mineral redistribution and the geometry of the ore.

2. Methodology

A field trip has been conducted on the Mudilandima deposit to investigate the lithologies, alteration, and geological structures from outcrops at the faces of the pit. It has resulted in a 2D geological map. Additionally, an RC chip tray logging campaign was undertaken to establish the spatial variation of lithologies in the 3rd dimension. Additional data has been collected from the 3D exploration geological model provided by Tenke Fungurume Mining SA. From the above 3 sources of data will derive The Integrated Geological Model. Complementary information will be sourced from the literature, thin section observations, and data acquired through XRD, XRF, and SEM analyses conducted on samples.

The Skarn to Porphyry Transition: Establishing Geochronological and Geochemical Links Between Skarn and Porphyry-Like Mineralization at Craigmont, British Columbia

Warren Wegener, Shaun Barker, Farhad Bouzari, Matthew J. Manor
The University of British Columbia, Mineral Deposit Research Unit, Vancouver, BC, Canada

The Craigmont Cu-Fe skarn deposit is located in the Quesnel terrane of the Canadian Cordillera in southern British Columbia. Craigmont was originally mined from 1958-1982, when it produced 36.75 Mt of ore averaging 1.28% Cu, and is one of the highest-grade copper mines in British Columbia to date. Craigmont is adjacent to the Late Triassic Guichon Creek Batholith, which hosts calc-alkalic Cu-Mo porphyry deposits of the Highland Valley District. The Cu-Fe skarn mineralization is restricted to Upper Triassic Nicola Group volcaniclastic rocks and marine sedimentary rocks in actinolite-epidote-magnetite-chalcopyrite skarn and brecciated specular hematite-chalcopyrite±K-feldspar. Early academic work suggested the Guichon Creek Batholith only acted as a heat source to skarn mineralization from fluid infiltration into Fe-rich Nicola Group host rocks. Recent drilling, however, has confirmed porphyry-style alteration assemblages in the Guichon Creek Border Phase (diorite to quartz-diorite) immediately north of the historic Craigmont open pit mine. This study aims to identify the link between the skarn and porphyry-style mineralization at Craigmont and characterize porphyry-style mineralization and alteration in order to develop vectors for porphyry Cu exploration. Field and microXRF petrographic studies on selected Border Phase intrusive rocks show K-feldspar-biotite, epidote-chlorite, and sericite-quartz±chlorite alteration. Vein types include K-feldspar-chalcopyrite-pyrite-(±molybdenite), locally with bornite, epidote-chlorite, and quartz. Chalcopyrite, bornite, and molybdenite are typically associated with K-feldspar-biotite alteration. Field mapping, core logging, and deployment of modern analytical techniques, such as microXRF analysis, Re-Os molybdenite dating, short-wave infrared (SWIR) spectroscopy, U-Pb zircon dating, and whole-rock lithogeochemistry will be applied to characterize both skarn and porphyry type mineralization and establish spatial and temporal links between two different mineralizing systems. Results are expected to define mineralization and alteration footprints and help exploration targeting in the porphyry belts of southern British Columbia and similar orogenic belts globally.

An Investigation of Helium Liberation During Olivine Serpentinization as a Potential Future Experimental Method for Research into Natural Subsurface Hydrogen Production

Tori Weston
University of Oxford, Oxfordshire, United Kingdom

This study investigates the experimental viability of measuring helium liberated from olivine crystal breakdown during a serpentinisation reaction and using it as a quantitative proxy for natural hydrogen production. If proven, such a methodology would significantly simplify future investigations into natural subsurface hydrogen production from this reaction.

Olivine samples were sourced from Aheim, Norway. Gas streams produced from dry olivine samples heated to 300°C, and wet olivine samples, which had been serpentinised at 300°C and 1,000 bar for six weeks were collected, cryogenically separated, and measured using mass spectrometry to quantify helium release. Account is taken of the natural diffusion of helium out of olivine crystals at elevated temperatures independent of any serpentinisation.

The control experiments in this study were the first to investigate helium diffusion in natural olivines and demonstrated a linear relationship between sample grain size and the diffusion rate coefficient of proportionality.

For an olivine sample with grain size of 20–30 μm in the presence of 4 molal NaCl, the results confirm that measuring helium liberation during serpentinisation is a viable proxy experimental method for future investigations of natural, subsurface hydrogen production.

Other samples tested did not produce reliable data. Further research is needed to definitively prove the viability of this method, to investigate its experimental limits and to test a range of pressure and temperature reaction conditions, including those simulating specific areas of the crust.

This study showed that measuring helium release from olivine can be successful as a method of investigating serpentinisation and ultimately for use in future research of natural subsurface hydrogen production.

PGE Behavior During Serpentinization of Ultramafic Sequences in the Bushveld Large Igneous Province

Marina Yudovskaya[1, 2], Bruno Bartolomeu[1], Judith Kinnaird[1], Roger Key[3], Andy Moore[3], Ian McGeorge[3], Mabatho Mapiloko[1]
1. CIMERA, School of Geosciences, University of Witwatersrand, Johannesburg, South Africa, 2. IGEM RAS, Moscow, Russia, 3. Kalahari Key Mineral Exploration Company, Gaborone, Botswana

Behavior of magmatic sulfides and associated platinum-group minerals (PGMs) during serpentinization can be highly variable depending on the PT conditions of the process. The S isotope composition of interstitial sulfides in the ultramafic zones of the Bushveld northern limb and Molopo Farms Complex indicates crustal S contribution to the parental S undersaturated siliceous komatiitic magmas. Primary PGE distribution in all mineralized ultramafic rocks is characterized by a typical Bushveld-type pattern albeit with a pronounced Pt negative anomaly and a lower level of PGE enrichment roughly proportional to a reduced proportion of interstitial liquid in adcumulates. In contrast, chromite-rich rocks have a distinct flatter PGE pattern thought to result from preferential crystallization of Os, Ir and Ru compounds with respect to Pt, Pd and Rh phases within the zone of massive chromite precipitation. Three different types of PGE and base metal redistribution are observed in the studied serpentinized varieties. In chromite-poor rocks from Uitloop and Grasvally remobilization is limited to microscale replacement. In Grasvally chromite-rich rocks, a primary pyrrhotite-pentlandite association is mostly dissolved and replaced by secondary heazlewoodite and pentlandite, whereas only Ir-Os-Rh-Pt sulfarsenides are mainly preserved among the PGM relics in secondary sulfides. In Molopo Farms serpentinites, a secondary Ni sulfide-alloy-arsenide assemblage replaces primary sulfides with liberated PGE stabilized as solid solutions in secondary Ni arsenides. Secondary mineralization of all three types is depleted in Cu and enriched in Ni, which is derived from dissolution of both sulfides and silicates, and is mostly concentrated in secondary ore minerals. The PGE capture by arsenides, preventing their dissipation, is facilitated by a superimposed relatively high-temperature hydrothermal event as seen in the Molopo Farms mineral assemblages, whereas serpentinization on its own results in loss of noble metals while gaining Ni.

Embedding Good Governance in Mining Operations and Value Chains

Dennis O. Achoki[1, 2]
1. RWTH Aachen University, Kisumu, Kenya, 2. Shanta Gold Kenya Ltd, Kisumu, Kenya

The core focus of a mining business should be to create value not volume growth. Africa is home to 30% of the world's minerals. Minerals crucial to the energy transition including copper, lithium, cobalt, titanium, nickel and manganese are found in the continent. Africa is also rich in precious minerals including gold, silver and gemstones. In particular, the geology of the countries in the East African Community (Kenya, Tanzania, Uganda, DRC, Rwanda, Burundi and South Sudan) has been demonstrated by a number of exploration activities to be rich in all these minerals. It is therefore imperative to strengthen the exploration and mining industry as a backbone for development. This work aims to tackle the two biggest bottlenecks to the success of the extractives industry: governance and value chains.

An effective way to tackle the bottlenecks is to embrace technology. Application of the right technology would ensure transparency and collaboration within the mining industry. One such technology is blockchain. This is a distributed and decentralized ledger system in which blocks on the chain hold information such as contracts and management transactions that no one person or party can change retroactively. It is similar to a shared database where information can be stored and accessed by designated individuals or groups. No one person oversees the data and as such it is secure, transparent, consistent, and reliable. Blockchain technology is the key to tackling a ubiquitous data problem in the mining industry, increasing efficiency in operations and transactions, and ultimately lowering costs across the board. Within the core mining business, blockchain technology would be applied to monitor mine stockpiles and reserves and ensuring appropriate assessment, disclosing information on sales, revenues, taxes, expenditure, profits and royalties due, as a trading platform and, allow for third party reviews.

Partitioning and Mobility of Deleterious Elements from Sulphide Oxidation in the Historical Gold Mine Tailings

Steve J. Chingwaru, Bjorn Von der Heyden, Margreth Tadie
Stellenbosch University, Stellenbosch, South Africa

The Witwatersrand Goldfields, responsible for 27% of global gold production, face environmental challenges due to over a century of extraction, resulting in 6 billion tons of historical tailings. These tailings directly contribute to acid mine drainage and release deleterious elements from unrecovered sulphides. Iron oxyhydroxide alteration products, formed during sulphide oxidation, intricately regulate the mobility of deleterious elements through absorption and incorporation. Despite extensive research on the immobilization properties of iron oxyhydroxides, little attention has been given to understanding the mobility and distribution of deleterious elements during sulphide oxidation, especially regarding the sulphide core and secondary mineral alteration layers, and their impact on environmental contamination. This study extensively characterizes the 70-year-old Klerksdorp tailings dump within the Witwatersrand, focusing on deleterious element partitioning and mobility between sulphide grains and iron oxyhydroxide alteration rims. Using a multi-method approach, including laser ablation inductively coupled plasma mass spectrometry, automated mineralogy, and wet chemistry data integration, metal release and retention across phases were quantified. Results indicate that under net acid-producing conditions, iron oxyhydroxide alteration rims retain a significant proportion (3–15%) of As, Ni, Cu, and Zn (>250 ppm), whereas Co, Au, and Pb exhibit limited compatibility with iron oxyhydroxide alteration (<3% deportment). The majority of remaining deleterious elements reside in mobile, unstable phases, exceeding 88%. Projections highlight a significant partitioning of elements (As, Ni, Cu, Zn, and Pb) into the environment, exceeding 9000 tons, underscoring the urgency of addressing these challenges. Remedial measures for these historical Witwatersrand tailings dumps are imperative, potentially involving metallurgical reprocessing flowsheets targeting sulphides and effective mechanisms for capturing deleterious elements. Such comprehensive approaches are vital for mitigating risks to human health, aquatic systems, and local ecosystems. Additionally, capturing metals such as Cu, Co, and Ni could serve as secondary sources to support the green energy transition.

Assessing Opportunities for Treating Acid Mine Drainages and Environmental Degradation of Kettara Mines in Morocco

Loubna Elamraoui[1], Abdellatif Elghali[1], Mostafa Benzaazoua[1], Olutoyin A. Fashae[2]
1. University Mohammed VI Polytechnic (UM6P), Benguerir, Morocco, 2. Department of Geography, University of Ibadan, Ibadan, Nigeria

Mining mineral ores have unintended adverse environmental and health effects on surrounding communities. This is a part because mining produces massive tonnes of waste, which could be solid, liquid, or gaseous. Among others, solid and liquid waste in abandoned mines poses the most significant concern and requires sustainable solutions. This research simulated the ex-situ neutralisation of acid mine drainage (AMD) from the Kettara mines in Morocco. Batch tests were conducted to assess the extent to which phosphate limestone waste rocks (PLW) can mitigate AMD acidity by using continued contact to represent the potential of PLW in neutralising AMD from the abandoned Kettara mines of Morocco. The results indicate that Kettara AMD is very acidic, with a pH of 2.5 and an electrical conductivity of 11,8 mS/cm. The Inductively Coupled Plasma Optical Emission Spectroscopy (ICP-OES) analysis showed a high sulphate concentration of 3668,83 mg/L, displaying the existence of some metals, notably Copper, Aluminum, and Iron. The neutralisation process of Kettara acid mine drainage using PLW in oxic conditions was highlighted by the variation of pH from the time my water was in contact with PLW. The pH rose from 2.5 to 7 while electrical conductivity decreased from 11,8 to 7.37. During the treatment of Kettara AMD with PLW, the effluent's sulphate concentration was reduced to 35%. In addition, Iron and Aluminium were significantly removed from Kettara AMD with a percentage of 99% in the leachate. Therefore, these results indicate that neutralising AMD using the passive treatment is effective and may serve as a cost-effective mitigation for AMD.

Sustainable Mining Through Bioleaching in the Central African Copperbelt (CACB)

Liseli Kunda
The University of Zambia, Lusaka, Zambia

Africa is home to several notable ore districts. One example is the Central African Copperbelt (CACB), spanning Zambia and the Democratic Republic of Congo with similar geology on both sides. It stretches about 50 miles wide and 140 miles long and is renowned for copper and cobalt deposits. Copper mining has been happening in this region for centuries, but it became more focused in the late 19th century with the discovery of copper deposits. As mining operations grow and new exploration techniques are developed, they provide valuable information about the type and amount of minerals present to the miner and the geologist. Although the phrase "copper no longer crops out" got on the lips of old hands to describe each depletion of the lucrative surface deposits, there are always new discoveries to be made. And thus, after 77 years of discovery, attainment, and depletion, the story of copper in the two countries closed one chapter and opened another. As easily exploitable high-grade deposits are becoming increasingly depleted, there is a need for new technologies to improve exploration and mining strategies. Taking a microbial perspective in mining is an innovative approach enhancing the recovery of valuable copper resources while minimizing environmental impact. Through the introduction of bacteria and fungi, minerals are broken down, releasing copper ions that can be easily extracted. This environmentally friendly method reduces reliance on harsh chemicals and energy-intensive processes, making it a promising solution for the mining industry in the CACB. The application of bioleaching showcases the potential for sustainable mining practices that balance resource extraction with environmental stewardship. Exploration from the microbial perspective can provide insights into the formation and enrichment of copper deposits while microbial remediation techniques can aid in addressing copper contamination and ability to tolerate or even thrive in environments within the Copperbelt.

United Nations Resource Management Systems as a Guide for Sustainable Resource Management: Case Example from Copperbelt Province, Zambia

Michael M. Musialike[1], Eva Marquis[3], Gabriel Ziwa[1], Mubanga Mutale[2], Karen Hudson-Edwards[3], Frances Wall[3]
1. The Copperbelt University, Kitwe, Zambia, 2. Geological Survey Department of Zambia, Lusaka, Zambia, 3. Camborne School of Mines, University of Exeter, Penryn, United Kingdom

In this contribution, we present the results of an initial scoping study of the application of the United Nations Resource Management System (UNRMS) to the critical raw materials (CRM) sector in the Copperbelt and Northwestern provinces of Zambia. As the global economy transitions to use of clean energy technologies, the demand for CRM needed to develop these technologies is rising. The surge in demand for CRM presents real economic benefits to producer countries, especially those that are resource rich and developing. To sustainably benefit from these resources, producers should build legislative and policy frameworks that ensure that operations across the value chain in the sector leave minimal climate footprints and have high environmental standards and quality social circumstances. Zambia is a major producer of CRMs such as copper and cobalt, with major nickel projects underway. Like the rest of the world, Zambia has realised the need to adopt a more sustainable way of managing its mineral resources in order to deliver the socio-economic transformation necessary for improved well-being for its people and the planet. As the country is in the process of reforming its mining policy to align it with the sustainable development agenda, it is timely to investigate how sustainable management of the CRM sector can support this objective. Published in 2019, the UNRMS aims to provide a comprehensive framework for sustainable resource management by government, industry, investors and the civil society. It links to the African Mining Vision, African Minerals and Energy Resource Classification (AMREC) and the Sustainable Development Goals.

Navigating Namibia's Energy Crossroads: Balancing Renewable Ambitions and Fossil Fuel Opportunities

Shepherd M. Nyambe
University of Namibia, Windhoek, Namibia

Namibia finds itself at a crucial crossroads, balancing aspirations for renewable energy supremacy with the newfound promise of oil and gas reserves. This paper explores the nation's trajectory, emphasizing its fervent pursuit of becoming a renewable energy hub, notably through initiatives like green hydrogen. Concurrently, it delves into the seismic shift caused by the revelation of substantial oil and gas discoveries by prominent energy corporations.

The discourse unfolds around the challenging dilemma confronting Namibia: whether to prioritize sustainable energy goals or embrace economic opportunities presented by fossil fuels. Highlighting the country's proactive measures in renewable energy, such as the Hyphen and DAURES Green Hydrogen village, it underscores the employment opportunities and capacity-building initiatives already underway.

To navigate this multifaceted landscape, Namibia requires a delicate balancing act. Strategies encompass synergistic utilization of both renewable and fossil fuel resources while ensuring sustainable growth and environmental stewardship. The paper advocates for meticulous governance, collaboration between sectors, and robust policies to harness the potential of both industries without compromising long-term sustainability.

The analysis acknowledges the critical role of political dynamics in shaping Namibia's energy future, particularly as a transition in leadership looms. The conclusion underscores the pivotal importance of balancing these industries to secure Namibia's economic, environmental, and energy independence trajectories.

Resilience and Governance in Artisanal Gold Mining: Addressing Challenges and Opportunities in Western Kenya

Christine A. Omuombo[1], Joy A. Owidhi[2]

1. The Technical University of Kenya, Nairobi, Kenya, 2. Independent Researcher, Nairobi, Kenya

The artisanal and small-scale gold mining (ASGM) sector plays a critical role in global supply chains. Kakamega experienced growth during the 'colonial gold rush' in the 1930s but faced decline after the closure of the Rosterman mine in the 1940s. Since 2019, there has been renewed interest in the Artisanal Gold Mining Sector in Western Kenya due to significant deposit discoveries by mining companies. This surge has resulted in increased engagement in ASM activities by local communities and small-scale investors, leading to notable challenges in environmental and social governance. This study focuses on the role of the local Kakamega ASM committee's resilience team in promoting responsible mining practices and addressing the sector's opportunities and challenges. Kenya is currently in the process of formalizing the ASM sector. Key challenges include a high dependence on artisanal mining and quarrying for livelihoods, a low uptake of modern sustainable resource utilization technologies, and a general lack of awareness about natural resource technologies. Key areas of support identified include access to engineering solutions, environmental data, and geoscience information, which provide a foundation for future mining activities. Additionally, the organization of miners into cooperatives presents an opportunity for the local community to benefit from available financial resources for user-needs-based activities. Miners find these benefits appealing as they seek to mitigate physical and regulatory risks associated with their operations. Moving towards a formalized and resilient ASGM sector requires capacity building and technical assistance from both governmental and non-governmental sector players. This study highlights the pivotal role of the ASM committee's resilience team in facilitating sustainable practices and governance improvements within the ASGM sector in Western Kenya.

Local Consideration for Blast-Induced Ground Vibration in Mining Communities

Harriet N. Tetteh[1], Wenfeng Li[2], Virginia T. Mclemore[3], Alex Rinehart[1]
1. New Mexico Institute of Mining and Technology, Socorro, NM, USA, 2. Los Alamos National Laboratory, Los Alamos, NM, USA, 3. New Mexico Bureau of Geology and Mineral Resources, Socorro, NM, USA

The mining industry relies on drill and blast operations to liberate rocks for metallurgical processing. These operations if not mitigated can drive mines into serious social license and environmental consequences in communities sensitive to blast-induced ground vibration.

The $83 billion mineral industry is fast growing with demand for energy transition minerals -- recording 2.78% production increase since June 2022 (Trading Economics). With this, there is rising expectation of the mining industry for green and sustainable metals. The minerals education coalition in 2021 estimated an average of 40,630 pounds of mineral product consumption per person in the United States alone. This demand re-emphasizes the need for mining operations to meet growing demands by increasing production targets even in hard and difficult grounds or those in close proximity to communities.

From pre-feasibility stages into the life of a mine, social license to operate is key to a successful mine, and issues of blast-induced ground vibration arise. By understanding the behavior of rocks to blast-induced waves, civil and geotechnical mitigation strategies could be adopted by considering rock structure and behavior under dynamic loading conditions.

This study considers meta-sedimentary and igneous rocks under field-representative loading and unloading, deformation, damage accumulation, and active ultrasonic wave propagation to suggest advancements in building designs and sustainable material solutions for local structures to remain resilient throughout the life of the mine.

Minimising the Environmental and Social Impacts of Exploration: Lessons from the GREENPEG Project

Kate Smith[1], Frances Wall[1], Robert Pell[2]
1. Camborne School of Mines, University of Exeter, Penryn, United Kingdom, 2. Minviro, London, United Kingdom

Mineral exploration is essential to provide the metals that we need for a low-carbon future. This exploration must be environmentally and socially responsible.

We have applied life cycle assessment (LCA), a powerful technique to quantify environmental impact, to exploration techniques developed and refined within the GREENPEG project. These assessments have highlighted ways to reduce the impact of exploration, many of which involve practical adjustments that are easy to implement. We have investigated satellite remote sensing, ground-based, drone- and helicopter-borne geophysics and geochemical techniques at province, district and small prospect scale. During fieldwork, travel choices are clearly very important, demonstrating the need to consider local hiring of staff, car-sharing, use of electric vehicles when viable and minimising travel during fieldwork by using accommodation as close to site as possible. LCA also allows us to demonstrate the great value in making data publicly available, for example the GREENPEG spectral library aids spectral identification of outcropping pegmatites with a much reduce environmental impact.

We made qualitative studies of social impacts so we could explore local community and individual perspectives in depth. GREENPEG interviews carried out in Norway and Ireland and surveys in Portugal and the UK highlight the importance of building company-community relationships as early as possible, and of understanding social context and history and the value in trying to look at an exploration project through a community lens. In our case studies, the exploration activities themselves were less of a worry than concern about possible future mining. The growth and spread of this anxiety and fear in a community, even from an early project stage, is the main social impact of exploration that needs recognition and empathy. Relationship building and community participation in projects may be key to tackling these worries, which otherwise may stall or halt the progress of a project.

Circular Economy Can Start from the First Stages of Exploration Geology

Frances Wall[1,4], Eva Marquis[1,4], Karen Hudson-Edwards[1,4], Markus Zils[5], Gawen Jenkin[2], Phil Bird[2], Carol Pettit[3], Vicky Smyth[3]
1. Camborne School of Mines, University of Exeter, Penryn, United Kingdom, 2. School of Geography, Geology and the Environment, University of Leicester, Leicester, United Kingdom, 3. Exeter Innovation, University of Exeter, Penryn, United Kingdom, 4. Environment and Sustainability Institute, University of Exeter, Penryn, United Kingdom, 5. University of Exeter Business School, Exeter, United Kingdom

Creating a more circular economy is important to achieve materials sustainability and lower carbon emissions. The term ‘circular economy’ (CE) is often used to describe the reverse loops in the materials cycle (e.g. recycling, re-use and repair) but CE thinking needs to cover the whole value chain right from the beginning. Important actions to ensure circularity take place in the production and manufacturing ‘inbound’ steps of the materials cycle – and the first of these inbound steps are geology and mining. The Ellen MacArthur Foundation defines three key CE principles: ‘designing out waste and pollution’, ‘keeping products and materials in use (at highest value)’, and ‘regenerating natural systems’. All of these are applicable to geological exploration and mining (https://geoscientist.online/sections/features/the-circular-economy-a-view-from-the-front/), which are core value chain steps and significant contributors to the ‘emission-rucksack of materials’. CE thinking during the exploration phase includes, for example, early consideration of by- and co-products as well as the main commodity of interest; designing resource efficient mining and extraction processes; using techniques such as life-cycle assessment to test mine design and extraction technology; re-use of mine waste and capture of ‘other’ value streams; biodiversity and ecology regeneration and enhancement; joining up primary production with recyclers as ‘materials solutions’ providers; and taking part in activities to ensure onwards good materials stewardship and participation into future CE value capture through each step in the value chain. In our UKRI Interdisciplinary Circular Economy Centre for Technology Metals (Met4Tech.org), we are carrying out research across the full value chains for the specialist technology metals (rare earths, lithium, cobalt, tin and tungsten) defined as critical minerals in many countries, and vital for the strategic technologies needed for the energy transition and net zero aims. We will also set this CE thinking in the context of the new UN Resource Management System.

The Kuboos-Bremen Line in Namibia and South Africa: New Genetic Insights and its Potential as a Hub for REE Enrichment Research

Jorge C. Arthuzzi[1, 2], Robert J. Giebel[3, 4], Jochen Kolb[1, 2], Aratz Beranoaguirre[5], Michael Marks[6], Paul H. Macey[7], David L. Reid[7], Lorenz Kemmler[3], Benjamin F. Walter[6]
1. Karlsruher Institute of Technology, Karlsruhe, Germany, 2. Laboratory of Environmental and Raw Material Analyses (LERA), Karlsruhe, Germany, 3. Technische Universität Berlin, Berlin, Germany, 4. University of the Free State, Bloemfontein, South Africa, 5. Goethe-Universität Frankfurt, Frankfurt, Germany, 6. Universität Tübingen, Tübingen, Germany, 7. University of Cape Town, Cape Town, South Africa

Carbonatites and associated alkaline igneous rocks are usually enriched in critical metals, such as the Rare Earth Elements (REEs), and they are a vital REE source in the current mineral and energy/technology industries, with 40% of REEs exploration projects targeting such rocks. However, there is still a gap in the understanding about which processes and mechanisms govern REE enrichment.

The Kuboos-Bremen Line (KBL; South Africa and Namibia) is defined by aligned alkaline igneous intrusions that contain carbonatites that have REE-rich occurrences. The KBL extends for about 250 km with both the varying intrusions level (increase in altitude by about 1.2 km) and a suggested younging of the intrusions, occurring from SW to NE. It hosts a large variety of rocks, including granite and syenite, foid-syenite, carbonatite, and a carbonate-bearing diatreme field in the west, central, and the eastern portions, respectively.

Zircon, titanite, and apatite U-Pb ages provide a range from 530 to 457 Ma for the KBL. The older zircon and titanite ages indicate the onset of magmatism with apatite ages being around 30 Ma younger in individual complexes. The total REE content in apatite from different complexes varies from ~0.02 to 0.68 apfu and correlates positively with Si, which varies from ~0.01 to 0.73 apfu (britholite substitution).

The obtained ages are scattered having no clear relationship with earlier interpretations of a systematic younging of the intrusions toward NE. Moreover, the apatite's lower closure temperature for U-Pb, compared to zircon and titanite, reveals a varying cooling history for the individual complexes.

Enrichment of REEs with increasing Si in apatite does not show apparent connection with compositional and age evolution or emplacement depth of the individual complexes, but rather is related to magma contamination and wall rock interaction, indicating that these processes are critical for REE enrichment to economic levels.

Nb-Ta Primary Oxides and U Secondary Minerals from Pegmatites in Central Portugal

Santos Barrios Sánchez[1], Juan G. Barreiro[1], Julio G. García[1], Vitor d. Maia[2], Wilck G. de Campos[3], Ricardo Augusto S. Cipriano[3]
1. University of Salamanca, Salamanca, Spain, 2. Universidade Federal Minas Geraes, Belo Horizonte, MG, Brazil, 3. Universidade Federal Ouro Preto, Ouro Preto, MG, Brazil

The central Portugal stands out for the presence of important pegmatites with different fractionation and enriched in elements such as Nb-Ta-Be-P-Li. They are embedded in variscan granites (syn and mainly post kinematic) belonging to the Schist and Greywacke Complex of the Central Iberian Zone, which would have been produced by crustal melting with the change from compressional to extensional conditions (between 320 and 290 Ma).

In this work, we have analyzed Nb-Ta and U minerals from less known Be-P pegmatites, using a SEM-EDS (Department of Geology at UFOP). They occur in veins or groups of veins up to 1 m thick (e.g.: Chãs de Tavares, Rãs) or in pockets of tens of meters (e.g.: Real), with coarse to very coarse grain size, evident zoning and quartz core. In general, the Be-P pegmatites show a lower fractionation than the Li-enriched ones in the same area. The objective is to study their evolution degree and their application to the prospection of critical elements, comparing with previous data and those obtained by the analysis of new samples collected in well studied portuguese pegmatites (e.:g: Senhora de Assunção).

BSE images show homogeneous Nb-Ta crystals. EDS chemical characterization shown mainly columbite-(Fe) and in minor proportion columbite-(Mn). In some crystals Ti and Sn partially substitutes for the Nb-Ta site. The Nb-Ta x Fe-Mn diagram allows us to observe a differentiation trend. Secondary U minerals are torbernite and autunite and occur in association with primary oxides.

Authors thank FAPEMIG-APQ 00764-23 and GREENest project.

The Control of Biotite on the Lithium, Fluorine and REE Budget of Crustal Silicate Magmas

Charles D. Beard[1, 2, 4], Vincent J. van Hinsberg[2], John Stix[2], Madeleine C. Humphreys[3], Owen M. Weller[4]
1. Utrecht University, Utrecht, Netherlands, 2. McGill University, Montreal, QC, Canada, 3. Durham University, Durham, United Kingdom, 4. University of Cambridge, Cambridge, United Kingdom

Of the common rock-forming minerals, biotite contains the highest concentration of lithium and fluorine in both alkaline and calc-alkaline magmatic suites globally. Its control on the distribution of minor and trace-elements, including Li and F, and its impact on metal enrichment, however, is not well understood. Here, we use experiments to quantify that control in sodic (per)alkaline H_2O-saturated magmas with variable F-content at 650–800 °C, 200 MPa total pressure, and log fO_2 ~FMQ +1. Biotite-glass pairs from (tephri)phonolite fall deposits from Tenerife, Canary Islands complement our experimental data set. Our results, alongside a literature compilation covering a wide compositional range constrain empirical models that: (1) describe the exchange of F and OH between the silicate melt and the biotite W-site; (2) predict the partitioning of 1+ cations Li–Cs between silicate melt and the biotite A- and M-sites. The models use the major-element composition of biotite and melt, pressure and temperature as input. Lithium partition coefficients vary from 0.06 to 33 and are lowest in Al-rich, Si-poor biotite formed at high temperature. Residual enrichment of Li should therefore be most efficient during crystallisation of high Al/Si magmas, a finding consistent with the association between peraluminous granites and lithium-caesium-tin pegmatites. Further, biotite-melt $Kd_{F/OH}$ exchange coefficients are 1.2–51, with minima for Al-rich, Mg-poor biotite. Where $Kd_{F/OH}$ values are high, crystallisation of even small amounts of biotite will strip fluorine from the melt. In silica undersaturated peralkaline magmatic systems, biotite $Kd_{F/OH}$ values are low and fluorine can be residually enriched during crystallisation, complexing with REE in the melt and facilitating their enrichment too. Our partitioning models can be used to interrogate biotite from natural systems to determine the composition of their source melt, or to forward model the metal budgets of mafic to evolved igneous systems at upper-mantle to crustal pressures.

Constraining the Magmatic-Hydrothermal Evolution of a Gneiss Dome: Insights from U-Pb Dating and Trace Element Analysis of Sn-W Mineralization

Daniel Bermejo-López[1], Lorena Ortega[1], Pedro Castiñeiras[1], Lorenzo Tavazzani[2], Santos Barrios[3], Juan Gómez-Barreiro[3], Elena Crespo[1], Cyril Chelle-Michou[2]
1. Complutense University of Madrid, Madrid, Spain, 2. ETH Zürich, Zürich, Switzerland, 3. University of Salamanca, Salamanca, Spain

The Martinamor extensional gneiss dome (Salamanca, Spain) represents a late-Variscan geologic feature resulting from the gravitational collapse of the Variscan orogen. This dome formed during the E2 extensional stage (~315–300 Ma), and it is characterized by a wide extensional shear zone leading to tectonic denudation and thermal attenuation of the crust. The footwall exhumation also led to isothermal decompression and heat transfer through the detachment zone, inducing syn-kinematic low-pressure metamorphism and ultimately reaching melting conditions in the dome core. Partial melting ultimately gave rise to S-type magmatism locally associated with Sn and W mineralizations.

This study focuses on Sn-bearing pegmatites, and cassiterite and wolframite vein-type mineralizations, which occur in different structural levels, both in the shear zone hanging wall and footwall. Cassiterite and wolframite in situ U-Pb dating by laser ablation-inductively coupled-mass spectrometry (LA-ICP-MS) indicate a prolonged mineralizing period spanning nearly 25 m.y. (324–301 Ma). This interval initiated with the emplacement of Sn-bearing pegmatites at 324 Ma and W-veins at 323 Ma, followed by the development of Sn veins between 320 and 301 Ma. Partial melting and crystallization + hydrothermal activity took place within uncertainty of the obtained U-Pb dates, implying an earlier onset of the Martinamor gneiss dome exhumation, during the E1 deformation stage (330–315 Ma).

Trace element composition of cassiterites, when used as thermo indicator (i.e., Zr-Hf and Nb-Ta ratios), reveals a cooling pattern from older (324 Ma) to younger (305 Ma) cassiterite, followed by an inverse, reheating trend from 305 to 301 Ma. This suggests a protracted cooling of the magmatic-hydrothermal system during dome exhumation until a heating event, which nature is still to determine, occurred around ~305 Ma, elevating cassiterite temperatures in the Sn vein-type mineralization.

Detailed Mineralogical Study of the Bentonitic Basin of Northwestern Sardinia (Italy): the S'Aliderru and Sa Pigada Bianca Deposits

Sebastiano Coticelli[1], Isabel Abad[3], Giuseppina Balassone[1], Piergiulio Cappelletti[1], Concetta Rispoli[1], Fabio Granitzio[2], Mauro Farenzena[4], Nicola Mondillo[1]
1. Dipartimento di Scienze della Terra, dell'Ambiente e delle Risorse, Università degli Studi di Napoli Federico II, Napoli, Italy, 2. Clariant Prodotti (Italia) Spa, Milano, Italy, 3. Departamento de Geología and CEACTierra, Universidad de Jaén, Jaén, Spain, 4. Società Sarda di Bentonite Srl, Cagliari, Italy

In recent years, bentonite has been gaining increasing importance from an industrial viewpoint due to its many properties that make it a valuable product, especially in the oil and steel industries. The objective of this study is the detailed mineralogical investigation of mineralized samples from the S'Aliderru and Sa Pigada Bianca deposits, the two most important deposits in the bentonite basin of northwestern Sardinia (Italy). The samples collected represent the different varieties found at the two sites and differ in color and the nature of parent rock. The study was based on XRPD analysis of the samples, which were subjected to various processes (AD, EG, DMSO, heating) to identify the occurring mineralogical phases to understand the geological processes that led to the formation of the deposits and possible genetic differences at the basin scale. The analyses showed a very complex mineralogical assemblage. In detail, the S'Aliderru bentonites are composed mainly of Ca-Montmorillonite, with the accessory presence of kaolinite, chlorite, muscovite, and biotite. The Sa Pigada Bianca samples are composed of Ca-Montmorillonite, Muscovite, Biotite, Kaolinite, and Zeolite. In the case of S'Aliderru, the new mineralogical information, combined with information from previous studies, indicates that the volcanic material that gave rise to the mineralization was probably deposited in a marine-transitional environment and was then affected by at least one hydrothermal stage. About Sa Pigada bianca, from the mineralogical assemblage and the limited extent of mineralization, a continental-type genesis can be inferred, driven by the faults affecting the area, with fluid evolution over time resulting in the present mineralogy. In the near future, it is planned to conduct analyses of other deposits in the area to formulate a genetic model at the basin.

The Black Angel Zn-Pb district, West Greenland: Germanium Mineralization and Syn-metamorphic Ore Formation

Michael Eigler[1], Benjamin F. Walter[2], Jochen Kolb[1]
1. Karlsruhe Institute of Technology, Karlsruhe, Germany, 2. University of Tübingen, Tübingen, Germany

The Black Angel district in central West Greenland hosts several Paleoproterozoic marble-hosted Zn-Pb occurrences, including the closed Black Angel mine (11.2 Mt at 12.3% Zn, 4.1% Pb, and 29 ppm Ag). The sphalerite-pyrite-galena dominated massive sulfides in the district are host to briartite, a rare Ge sulfide mineral. The origin of mineralization remains ambiguous, as both Mississippi Valley-type and Kipushi-type deposit models have been suggested. Detailed petrographic studies, electron microprobe, and LA-ICP-MS analyses were applied to breccia-hosted massive sulfide from the South Lakes Glacier prospect to study briartite formation and ore genesis. Graphite thermometry, biotite thermometry, and phengite barometry indicate peak metamorphic conditions of 500° ± 40°C and 4.5 ± 1.5 kbar. Macro- and microstructures, such as isoclinally folded marble clasts in the sulfide matrix, poikilitic pyrite, and chalcopyrite trails parallel to S_2 foliation, indicate syn-peak metamorphic and syn-tectonic ore formation during regional D_2 deformation, supported by available Re-Os ages of 1828 Ma. The mineralization was affected by progressive solid-state and dissolution-precipitation sulfide remobilization and is structurally controlled by reactivated D_1 thrusts and D_2 reverse faults and shear zones. Arsenopyrite thermometry (390° ± 30°C) indicates that mineralization continued during retrograde cooling. Briartite, occurring as clusters of µm-sized grains as well as anhedral grains of up to 3 mm in diameter, formed by precipitation from Cu-Ge-(Sn)-bearing prograde metamorphic fluids coeval with Ge-bearing chalcopyrite. Whole rock data show Ge grades of 120 ppm and Ga grades of 130 ppm. The widespread occurrence of briartite, the syn-tectonic timing of ore formation, and syn-metamorphic fluid with intermediate sulfidation suggest a Kipushi-type deposit model. The structurally controlled, syn-metamorphic style of Zn-Pb-Ge mineralization highlights structures in metamorphosed carbonate platforms of foreland fold-thrust belts as exploration targets for critical metals.

Recovery appraisal of Ag, Bi, In, Sb, Se, and Te in processing samples from the Ruwai Pb-Zn skarn deposit, Central Kalimantan, Indonesia

Shelly M. Faizy[1], Alkiviadis Kontonikas-Charos[2], Mathias Burisch[3], Doreen Ebert[1], Robert Möckel[1], Bradley M. Guy[1], Joachim Krause[1], Valby van Schijndel[4], Arifudin Idrus[5], Ernowo Ernowo[6], Anggara Widyastanto[7], Agata V. Kindangen[8], Max Frenzel[1]
1. Helmholtz-Zentrum Dresden-Rossendorf, Helmholtz Institute Freiberg for Resource and Technology, Freiberg, Germany, 2. School of Earth and Atmospheric Sciences, Queensland University of Technology, Brisbane, QLD, Australia, 3. Department of Geology and Geological Engineering, Colorado School of Mines, Golden, CO, USA, 4. GFZ German Research Center for Geosciences, Potsdam, Germany, 5. Department of Geological Engineering, Universitas Gadjah Mada, Yogyakarta, Indonesia, 6. National Research and Innovation Agency, Bandung, Indonesia, 7. PT. Kapuas Prima Coal, Lamandau, Indonesia, 8. Center for Mineral, Coal and Geothermal Resources, Geological Agency, Jakarta, Indonesia

Processing samples from the Ruwai skarn deposit, the largest polymetallic skarn deposit in Indonesia, were analyzed to assess mass balances, recovery rates, and deportments of selected critical elements in order to understand their distribution and behaviour within the deposit. These samples included feed ore, Pb scavenger and final Pb concentrates, Zn scavenger and final Zn concentrates, and final tailings. They underwent bulk geochemical, mineralogical, and mineral chemical analyses.

The analyses revealed that Ag and Bi are the predominant critical elements in the samples, with highest recoveries occurring in the Pb concentrate at 75-90% and 78-90%, respectively. This observation is consistent with the results of the Ag and Bi deportments, as these elements are mainly hosted in solid solution within galena (PbS), whereas particulate Ag- and Bi-bearing minerals, such as acanthite (Ag_2S), freibergite ($(Ag,Cu,Fe)_{12}(Sb,As)_4S_{13}$), pyrargyrite ($Ag_3SbS_3$), bismuthinite ($Bi_2S_3$), native bismuth (Bi), and cosalite ($Pb_2Bi_2S_5$) are commonly associated with galena.

The majority of In reported to the Zn concentrate (61-84%), suggesting an association with sphalerite. However, a notable portion, approximately 13-33%, was lost to the final tailings. A significant proportion of Sb (44-62%) was lost to the final tailings, while only 33-52% reported to the Pb concentrate. Recovery rates of Se in the Pb and Zn concentrate reach 30-61% and 32-63%, respectively. Meanwhile, 56-77% of Te was recovered in the Pb concentrate, and 16-34% was lost to the final tailings. Sphalerite is expected to be the primary carrier of In, whereas freibergite and pyrargyrite have been identified as carriers of Sb, and pilsenite/tsumoite (Bi_4Te_3/BiTe), hessite (Ag_2Te) and BiTeAg host Te. Additionally, it is possible that some gangue minerals are responsible for carrying some of these elements to the final tailings

Hydrothermal Rare Earth Element (REE)-Enrichment in an Iron-Oxide-Apatite (IOA) Type Deposit at Kwyjibo, Québec, Canada: A Fluid Inclusion and Stable Isotope Study

Gary Fung, Anthony E. Williams-Jones
McGill University, Montreal, QC, Canada

The Kwyjibo deposit, Québec (Canada) is an unusual rare earth element (REE)-enriched iron-oxide-apatite (IOA)-type deposit. Hydrothermal processes led to the mobilization of the REE from fluorapatite in a magnetite-apatite host to britholite and andradite, followed by the replacement of andradite by allanite. The physio-chemical conditions of the REE mobilization, however, are poorly understood. Microthermometric measurements identified four distinctly different types of aqueous fluid inclusions in minerals coeval with the mineralization, representing a NaCl-KCl-H_2O-dominated (NaCl) fluid, an intermediate (NaCl-$CaCl_2$)-H_2O fluid, a CaCl2-NaCl-H_2O-dominated ($CaCl_2$) fluid, and a CO_2-NaCl-H_2O (CO_2-bearing) fluid. The NaCl-$CaCl_2$-H_2O fluid inclusions, hosted by fluorapatite, have a higher Na/Ca ratio than the $CaCl_2$-NaCl-H_2O fluid inclusions (quantified by laser ablation inductively coupled plasma mass spectrometry), and some of them are anhydrite-bearing (identified by Raman spectroscopy). The inclusions of medium-salinity (18-24% NaCl eqv.) sulphate-bearing NaCl-$CaCl_2$ composition homogenize to liquid at 185-290 °C. The NaCl-KCl-H_2O fluid inclusions, hosted dominantly by andradite, and the CO2-bearing fluid inclusions, hosted by quartz and fluorite, have low salinity (9-16% NaCl equiv). However, whereas the aqueous NaCl-KCl-H_2O inclusions homogenize to liquid at 290 to 365°C, the CO_2-bearing inclusions decrepitated upon heating. The exclusively secondary $CaCl_2$-NaCl-H_2O inclusions have high salinity (17-27% NaCl equiv) and homogenize at low temperatures (110-170°C). The $\delta^{13}C$ and $\delta^{18}O$ values of calcite, which was crystallized in the REE deposition stage, range from -4.61‰ to -7.6‰ and from +8.86‰ to +17.32‰, respectively. We interpret a magmatic-origin fluid unmixing event occurred, from NaCl-$CaCl_2$ fluid to NaCl and CO_2-bearing fluids, and $CaCl_2$ fluid. The transport of REE were attributed to the predominant chloride and sulfate complexes, given the elevated sulphate activity in the NaCl-$CaCl_2$-H_2O fluid. The deposition was facilitated by the precipitation of fluoride, phosphate, and carbonate ions, reflected by the spatial proximity of fluorite, tertiary fluorapatite, and calcite with respect to the REE-bearing minerals.

Remote Sensing Aided Geological Investigation of Rare Earth Element Mineralization in Sangu-Ikola Carbonatites, Southwestern Tanzania

Japhet N. Fungo, Nelson Boniface, Elisante Mshiu
University of Dar es Salaam, Dar es Salaam, Tanzania, Tanzania

Carbonatites are unique igneous rocks that are rich in carbonate minerals. These rocks are known to host the world's largest and richest deposits of rare earth elements (REEs). REEs have become increasingly important in modern technologies such as electric vehicles, renewable energy (wind turbines), aerospace applications, and telecommunication (smartphones). This has brought great attention to research and exploration for carbonatites, as they are the principal sources of REEs and niobium.

This study presents a remote sensing aided geological investigation of carbonatites from Sangu-Ikola, with a focus on understanding the mineralization processes that have led to the concentration of REEs. The study examines the mineralogy, geochemistry, and textural features of carbonatites using an optical microscope, an analytical spectra device, and X-ray diffraction. Petrographic analysis revealed that the carbonatite rocks were composed mainly of calcite and dolomite, with minor amounts of quartz, apatite, pyrochlore, magnetite, hematite, sulfides, and mica. Remote sensing approaches were able to map tectonic blocks, structures, and alteration minerals, which are the key targets for exploration.

These results improved our understanding of the geological processes that control the formation and distribution of REE-bearing carbonatites. By unlocking the economic potential of carbonatites and exploring new targets, the study contributes to the sustainable development of critical mineral resources. This chains the high-tech economy and meets the growing demand for these critical elements.

Sokli – A Multimetal Carbonatite Deposit Abounds with Critical Elements

Tuomas Gråsten, Teo Lehto, Pasi Heino, Mikko Numminen
Sokli Oy, Helsinki, Finland

Sokli, one of the largest carbonatite complexes in the world, is a world-class multimetal deposit located in northeastern Finland. In the center of the oval-shaped complex is a magmatic core composed mostly of carbonatites and phoscorites, which transition into metasomatized carbonatites and ultramafic rocks toward the outer zone of the complex. They are surrounded by an extensive fenite aureole, and the whole complex is cut by lamprophyre dykes of varying ages as well as late REE-bearing carbonatite dykes. On top of the bedrock are layers of weathered rock, laterite, and glacial till, where the laterite constitutes a major resource of P, REE, Nb, and Fe.

The lateritic P-deposits are mainly composed of primary fluorapatite, micas, and magnetite, with secondary francolite, crandallite, and iron and manganese oxides. In some areas, the laterites can be enriched in REE and Nb, where REE are mainly incorporated in fluorapatite and monazite/rhabdophane, whereas the bulk of Nb is in pyrochlore. Laterite units enriched in Fe, Mn, and vermiculite are also found.

The most promising hard rock deposits in Sokli are related to phoscorites and late REE-bearing carbonatite dykes. Phoscorites are potential targets for several different elements that are incorporated mainly in fluorapatite (P, REE), pyrochlore (Nb, Ta, REE, U, Th), and magnetite (Fe). The late REE-bearing carbonatite dykes comprise a highly potential exploration target, with REE incorporated largely in kukharenkoite as well as ancylite and monazite. Additionally, the whole carbonatite massif can be considered as a giant low-grade P-resource with approximately 3.5 wt % of P_2O_5.

Chiefly due to challenging location, complexity of the ores, and political decisions, the deposits in Sokli are unexploited to this day. Elements such as P, REE, Nb, and Mn are in high demand, which emphasizes Sokli's position as one of the most important critical element deposits within Europe.

Cassiterite LA-ICP-MS U-Pb Geochronology of Greisen- and Vein-Hosted Sn Mineralization in the Eastern Erzgebirge (Germany/Czech Republic)

Marie Guilcher[1], Mathias Burisch[2], Richard Albert[3], Axel Gerdes[3], Jan Černý[1], Sam Thiele[1], Uwe Lehmann[4], Henrik Kaufmann[4], Jens Gutzmer[1]
1. Helmholtz-Zentrum Dresden-Rossendorf, Helmholtz Institute Freiberg for Resource Technology, Freiberg, Germany, 2. Mineral Systems Analysis Group, Department of Geology and Geological Engineering, Colorado School of Mines, Golden, CO, USA, 3. Goethe University Frankfurt, Frankfurt Isotope and Element Research Center (FIERCE), Frankfurt am Main, Germany, 4. Landesamt für Umwelt, Landwirtschaft und Geologie (LfULG), Freiberg, Germany

Various types of polymetallic magmatic-hydrothermal deposits are present in the eastern part of the Erzgebirge, located across the eastern part of Germany and the northwestern part of Czech Republic. Most of these deposits are greisen- and vein-type deposits that are well-endowed in Sn, Li, and W and spatially associated with the Altenberg-Teplice caldera. However, only a few of those deposits have been reliably age dated (e.g., Sadisdorf district), leaving the timing of hydrothermal mineralization on the regional scale widely unconstrained.

Here, we report new U-Pb LA-ICP-MS ages of cassiterite from Li-Sn-(W) mineralization at Zinnwald, Altenberg, Niederpöbel, Schmiedeberg, Bärenfels, Lauenstein and Krupka. The new ages of the different localities span between 317.7±3.8 and 306.6±3.5 Ma, with a majority of the ages comprised between 315 and 309 Ma. Greisen- and vein-hosted cassiterite ages constrain the timing of hydrothermal mineralization, on a regional scale, to a narrow time window of ~10 Ma years and are significantly younger than previously proposed ages (325-318 Ma). The new ages are consistent with recent zircon ages of (sub-)volcanic rhyolite units (315-313 Ma), which are the host rocks of some of the Li-Sn-(W) granites. Greisen formation and associated cassiterite crystallization thus temporally coincides with the formation of the 315 to 310 Ma ring dikes linked to the collapse of the Altenberg-Teplice caldera.

From Eboundja Nepheline Syenite (Southern Cameroon) to Ion Adsorption REEs Deposit: Implications for REE Deposits Genesis

Kenneth K. Kwanang[1], Licia Santoro[2], Louise Marie Bidjeck[1], Paul-Désiré Ndjigui[1]
1. Department of Earth Sciences, University of Yaounde I, Yaounde, Cameroon, 2. Earth Science Department, University of Turin, Turin, Italy

REEs are essential metals for green technologies and high-tech industries. REE deposits are found in specific rocks (e.g., nepheline syenite) and regolith (ion adsorption deposits). Among these deposits, ion-adsorption REE deposits present the advantage of a low-cost extraction process. However, most of these deposits are found in southern China, and few have been discovered in other parts of the world. This work focuses on a newly discovered ion adsorption REE deposit in southern Cameroon. In this study, we compare the geochemical, mineralogical, and geological characteristics of two regolith profiles developed over the Neoproterozoic nepheline syenite of Eboundja (southern Cameroon) to assess the key factors controlling the formation of weathering-related REE deposits. The study area is characterized by a tropical rainforest climate with four seasons; and annual rainfall of about 2,800 mm. The two profiles, here referred to as EB2 and EB6, are located on two slopes and have similar morphological features (from base to top: source rock, saprolite, mottled zone, ferruginous zone, loose clay zone, and organo-mineral horizon). Both source rocks are metaluminous nepheline syenites with different mineralogical content. The two regolith profiles have similar secondary mineralogical assemblages including kaolinite, muscovite, halloysite, gibbsite, goethite, quartz, and hematite. The main findings of this work highlight that: (1) within Eboundja nepheline syenite, allanite, zircon, and pyrochlore fractionation during magmatic differentiation have particularly influenced the REE-ratio in the primary rock; (2) hydrothermal alteration in the presence of high salinity fluorine and carbonate-rich aqueous fluid may have favored re-mobilization of REEs and thus overprinted the REEs initial enrichment; (3) weak to strong lateritization, and strong kaolinization processes are highly effective for REE concentration at the Eboundja weathering profiles; and (4) the proportion and distribution of relict primary and secondary minerals, combined with pH, strongly influenced REE enrichment and distribution in weathering profiles.

Rare Earth Element Mobilization and Enrichment in Weathered Carbonatite of the Dong Pao Deposit, Vietnam

Martin Li[1], Martin Smith[1], Mei-Fu Zhou[2]
1. University of Brighton, Brighton, United Kingdom, 2. Institute of Geochemistry, Guiyang, China

The Eocene Dong Pao rare earth element (REE) deposit in Vietnam is one of the world's largest, with reserves of 0.55 Mt REO at an average grade of 3.6 wt %. The REE mineralisation is associated with a carbonatite-syenite complex and the related laterite contains 0.9 Mt REO at an average grade of 4 wt %. An ~100 m thick profile from the laterite to the carbonatite bedrock has been characterised to investigate the REE mobilisation and enrichment processes during weathering.

The carbonatite bedrock is mainly composed of calcite and has an accessory mineral assemblage of tainiolite, apatite, pyrochlore-group minerals, barite, celestine, bastnäsite-(Ce), synchysite-(Ce), fluorite, monazite, strontianite, quartz, and sulphides. In the laterite residual barite and monazite occur, but residual bastnäsite-(Ce) is only observed in the saprolite. Secondary rhabdophane-(La,Ce), florencite-(Ce), cerianite, Fe-Mn oxides, and kaolinite occur throughout the laterite. Acicular rhabdophane-(La,Ce) occurs filling fractures, while fibrous florencite-(Ce) fills cavities. Cerianite rims the edges of corroded monazite and voids in florencite-(Ce) and occurs as veinlets cutting rhabdophane-(La,Ce).

The REE concentration of the bedrock is highly variable, ranging from 170 to 57,000 ppm total REE with $(La/Yb)_N$ ratios from 13 to 3,000, and occasionally with positive Y anomalies and high SiO_2 contents. In the laterite, the REE patterns are consistently highly LREE-enriched and without Y anomaly. Positive Ce anomalies are observed in the uppermost laterite, with negative Ce anomalies lower in the profile. REE concentration shows a depletion-enrichment pattern with depth, from ~5,000 ppm at the top of the profile to ~30,000 ppm in the saprolite and decreasing to ~10,000 ppm in the saprock. Geochemical correlation indicates phosphates are the main REE hosts, with niobates deeper in the saprolite. Substantial REE re-distribution has taken place during weathering to generate an REE-rich ore horizon and homogenise the variable REE compositions of the bedrock.

The Petrography and Paragenetic Sequence of U-REE-Cu-Au Skarn Mineralisation in NW Queensland

Christina Loidolt, John Mavrogenes, Ross Chandler, Antony Burnham
Australian National University, Canberra, ACT, Australia

The Mary Kathleen deposit is the best-known U-REE-Cu-Au skarn deposit in the Mount Isa Inlier, NW Queensland, Australia. Similarities in mineralisation style and host rocks connect the Mary Kathleen deposit to several Cu-Au ± U-REE prospects in the region indicating the existence of an unrecognised large U-REE-Cu-Au mineralising system. However, the deposits are complex and the processes controlling trace element enrichments are contentious. Understanding the paragenetic sequence and links to regional tectonic, metamorphic, magmatic, alteration, and mineralisation events will define their provenance and assist with exploration for similar systems in Australia and worldwide.

Regionally extensive clinopyroxene skarn--overprinted by an extensive network of garnet skarn along faults and aplite dikes followed by scapolite-feldspar dominated alteration and U-REE mineralisation--extends westward from the Burstall granite to the Mary Kathleen shear zone (MKSZ). Known U-REE-Cu-Au skarn deposits and prospects in the region occur in the Corella Formation within or proximal to the MKSZ west of the ca. 1730 Ma Burstall granite. Mineralisation is hosted in garnet- or clinopyroxene-dominated, texturally and mineralogically diverse skarn lithologies. U-REE mineralisation occurs as uraninite-apatite-allanite assemblages primarily associated with garnet-dominated skarn. Cu-Au mineralisation is hosted in a texturally diverse sulfide assemblage associated with titanite. Complex overprinting relationships of skarn lithologies and mineralisation indicate multiple mineralisation episodes and significant changes in mineralisation style. We propose a multi-episode mineralisation sequence spanning ~280 Ma from ca. 1750 to 1470 Ma. Multi-stage skarn formation and initial U-REE mineralisation is associated with syndeformation 1750 to 1700 Ma granitic magmatism. Cu-Au mineralisation and U-REE remobilisation occur synchronous with extensive post-orogenic A-type magmatism, Na-Ca alteration, and diverse Cu-Au ± U-REE mineralisation in the eastern Mount Isa Inlier. Similarities in fluid composition and vein mineralogy with IOCGs and Cu-Au ± U-REE mineralisation indicate a genetic link with the ca. 1550 to 1500 Ma skarn-hosted Cu-Au mineralisation.

Insights Into U-REE-Cu-Au Skarn Occurrences in the Eastern Mount Isa Inlier from Garnet Geochemistry and Geochronology

Christina Loidolt, John Mavrogenes
Australian National University, Canberra, ACT, Australia

Complex U-REE-Cu-Au skarn occurrences in the Mount Isa Inlier, Australia, are suggestive of a larger mineralising system. U-REE mineralisation hosted in uraninite-apatite-allanite assemblages are intimately associated with skarn garnet, but the nature of that relationship is ambiguous. Although understanding of individual skarns has improved, processes controlling trace element enrichments are contentious. Previous and ongoing research on garnets related to mineralisation in other deposits demonstrates that garnet records information that is key to understanding the evolution of those deposits.

Detailed studies of garnet mineral geochemistry and U-Pb geochronology using EPMA, LA-ICP-MS, and Synchrotron XFM indicate two main events of garnet formation associated with U-REE mineralisation between ca. 1750 and 1500 Ma. Initial skarn formation and U-REE mineralisation is associated with intrusion of the Burstall granite during E-W-directed compression ca. 1750 to 1700 Ma. Garnet-dominated skarn (Group A) overprinted early, regional clinopyroxene-dominated skarn and was itself partially replaced by late-stage U-REE mineralisation. Although garnet U-Pb geochronology suggests one phase of garnet growth associated with Cu-Au mineralisation and U-REE remobilisation between ca. 1550 and 1500 Ma, garnet mineral geochemistry reveals two phases of garnet growth during that period – a premineralisation group (B) and a synmineralisation group (C).

Group A garnet generally forms inclusion- and fracture-rich crystals with well-developed concentric zonation of Fe, Ti, Y, and Zr. Group B garnet mainly forms concentrically zoned (Fe, Ti, Zr, Y) rims around vugs in older garnet skarn. Group C garnets are characterized by higher Fe, Mn, Ti, Nd, Y, Zr, U, Th, and Rb than group B garnet, and are associated with Cu-Au-U-REE mineralisation hosted in allanite-sulfide-feldspar-epidote assemblages. This assemblage occurs in vugs rimmed by group B garnet as well as in veins cross-cutting older skarn lithologies. This study illustrates the ability of garnet to record important stages of mineralisation in complex multi-element skarn deposits.

Mineral Chemistry, Fluid Inclusion, and Isotopic Characteristics of the Enargite Orebodies at Lepanto, Mankayan, Philippines

Pearlyn Manalo[1], Ryohei Takahashi[1], Manuel Nopeia[1], Kalin Kouzmanov[2], Nigel Blamey[3], Andrea Agangi[1], Akira Imai[4], Leo Subang[5], Marlina Elburg[6], Henriette Ueckermann[6], Linda Iaccheri[7]
1. Akita University, Akita, Japan, 2. University of Geneva, Geneva, Switzerland, 3. Western University, London, ON, Canada, 4. Kyushu University, Fukuoka, Japan, 5. Lepanto Consolidated Mining Co., Quezon City, Philippines, 6. University of Johannesburg, Johannesburg, South Africa, 7. University of the Witwatersrand, Johannesburg, South Africa

Enargite and luzonite are major ore minerals in high-sulfidation epithermal Cu-Au deposits. These minerals have also recently gained interest due to their enrichment in some critical metals, such as Sb, Te, and Ge. In this study, we report new results on mineral chemistry, fluid inclusion microthermometry, sulfur and lead isotope signatures of enargite from the Lepanto high-sulfidation epithermal Cu-Au deposit in the Philippines, and their implications for deciphering the ore-forming processes in the deposit.

Elemental mapping shows that enargite hosts native tellurium, goldfieldite and Au-Ag tellurides inclusions. Enargite and luzonite contain up to 5 wt % Sb and up to 5,000 ppm Te. Variations in Sb and Te contents generally follow the crystal growth planes. Near-infrared microscopic observations revealed that enargite hosts two-phase liquid-vapor fluid inclusions that homogenize to liquid between 205 to 240°C. Fluid salinity ranges from 2.5 to 5.7 wt % NaCl equiv. Bulk gas analysis indicates that the fluid inclusions dominantly contain H_2O with 0.1 to 3 mol % CO_2 and 0.0006 to 0.02 mol % H_2S. N_2/Ar ratio widely varies between 2 and 120. Sulfur isotopic ratios ($\delta^{34}S_{CDT}$) of enargite range from -6.0 to -3.0 ‰, typical of sulfide minerals that were formed from H_2S produced from magmatic SO_2 disproportionation at magmatic-hydrothermal temperatures. Lead isotope ratios of enargite range from 38.48 to 38.53 for $^{208}Pb/^{204}Pb$, 15.57 to 15.59 for $^{207}Pb/^{204}Pb$, and 18.41 to 18.43 for $^{206}Pb/^{204}Pb$, which are similar to the Pb isotopic composition of the Pleistocene dacite in northern Luzon.

Hydrogeochemical Modelling of Indium in a Mine Waste Environment: Opportunities for Recovery from Aqueous Sources

Olivia Mejías[1], Thomas Poulet[2], Pouria Behnoudfar[2], Laura Jackson[1], Anita Parbhakar-Fox[1]
1. Sustainable Minerals Institute, The University of Queensland, Brisbane, QLD, Australia, 2. CSIRO Mineral Resources, Perth, WA, Australia

Indium (In) plays a crucial role in the energy transition, particularly in copper indium gallium selenide (CIGS) solar panels, where its demand is projected to increase by approximately 200% by 2050. This surge has driven intensive exploration into various resources, including mining waste. However, the extraction of indium from mine aqueous sources as an alternative to traditional hard rock ore mining remains poorly understood. To address this issue, we present a hydrogeochemical model of indium aimed at developing a reactive transport model to simulate advection, diffusion, and fluid-rock interactions within the mine waste material in a multi-component environment. In this study, we used surface water samples from the abandoned Baal Gammon mine in Queensland, Australia, and the surrounding Jamie creek. Due to insufficient thermodynamic indium data, we compiled and created a database from literature. Using the geochemical code PHREEQC, we calculated speciation, complexation, mineral saturation indices (SI), and reactions to derive the stability regimes and kinetics of indium within the mine waste system. The results show that indium is distributed mainly as In^{3+}, $InSO_4^{+}$, $In(SO_4)_2^{-}$, and $InOH^{2+}$ in acid mine drainage (AMD). This was confirmed by simulating Pourbaix diagrams with different compositions of solutions. SI indicate that jarosite and schwertmannite are expected to precipitate and may control indium transport. We then couple the geochemistry with advection and diffusion in a machine learning approach to investigate how much indium could potentially be released into the environment if no treatment is applied. This work sheds light on the environmental impacts and recovery opportunities of indium from aqueous sources as an alternative in supply chains. Further research is needed to address the technological challenge of developing processes for the selective extraction of targeted indium dissolved constituents from AMD. This entails elucidating sorption/desorption reaction mechanisms, specifically in waters containing competing constituents.

Multi-stage Evolution and REE Mineralization of Wigu Hill Carbonatite, Uluguru Mountains, Tanzania

Mary C. Moshi[1, 2], Yasushi Watanabe[1], Nelson Boniface[2], Tatsuki Tsujimori[3], Carmela Tupaz[1], Daisuke Araoka[4], Shogo Aoki[1], Elisante E. Mshiu[2]
1. Akita University, Akita, Japan, 2. University of Dar es Salaam, Dar es Salaam, Tanzania, 3. Tohoku University, Sendai, Japan, 4. Geological Survey of Japan, National Institute of Industrial Science and Technology (AIST)), Tsukuba Ibaraki, Japan

The Wigu Hill Carbonatite (WHC) in Uluguru Mountains, Tanzania, is among the REE-endowed carbonatites in Africa. The carbonatite comprises large, fine-grained apatite magnesiocarbonatite that has been locally brecciated and intruded by late-stage carbonatite dike swarms. This work presents the first detailed petro-geochemical and mineralogical characteristics of WHC.

REE mineralization at Wigu Hill is concentrated within late-stage carbonatite dikes, which occur as magnesiocarbonatite to ferrocarbonatites that host pegmatitic, well-preserved REE-bearing hexagonal pseudomorphs after burbankite. Mineralogy of the pseudomorphs varies in space, reflecting the dissolution of primary burbankite through reaction with evolving residual carbothermal fluids. Two major burbankite alteration phases were identified: (1) early alterations are observed in dikes with yellow-colored pseudomorphs typified by an association of synchysite-(Ce), barite, Ca-strontianite, calcite, quartz ± bastnaesite-(Ce), and high synchysite $(La/Ce)_N$ - $(La/Nd)_N$ ratios. (2) The second stage alteration is evident by an assemblage of synchysite-(Ce), barite, Ca-strontianite, fluorite, calcite, quartz, apatite, monazite, Al-REE-phosphates, and low synchysite $(La/Ce)_N$ - $(La/Nd)_N$ ratios. This alteration is recognized in dikes with pink-colored pseudomorphs and has been interpreted as a product of interaction of primary burbankite with highly evolved fluids of elevated HREE.

Stable C, O, and Mg isotopes dolomite signatures across Wigu Hill are of pristine mantle ($\delta^{13}C_{VPDB}$ -4.1‰ to -6.2‰; $\delta^{18}O_{VSMOW}$ +6.5‰ to +7.31‰ and $\delta^{26}Mg$ -0.44 to +0.19‰) and indicate magmatic fractionation is responsible for REE enrichment in the late-stage dikes. $\delta^{26}Mg$ isotopes further suggest that WHC magma may have differentiated by liquid immiscibility from a parental alkaline-rich magma. Our textural, geochemical, and stable isotope findings picture a polygenetic evolution of WHC. The main REE mineralization phase occurred during the end of the magmatic phase, but subsequent reworking by carbothermal fluids has resulted in outstanding REE enrichment.

Thermal Separation of Silicate-Fluorosilicate Melts: Trace Element Partitioning and Implications for Granite-Hosted REE Deposits

Wil K. Munro, Neil R. Bennett
University of Toronto, Toronto, ON, Canada

Demand for the rare earth elements (REE) and other high field strength elements (HFSE), driven by the green energy transition, motivates continued exploration for REE deposits. Several A-type granite-hosted HFSE-rich ore deposits have been identified globally with local REE concentrations up to 40 wt %. Silicate-fluorosilicate melt immiscibility is proposed to play a key role in HFSE enrichment, however, formation of the observed compositions is hard to explain with existing experimental partitioning data. Experimental determination of two-liquid trace element partitioning is often complicated by incomplete separation of immiscible liquids and quench effects that lead to complex run-product textures. Some studies have attempted to mitigate these issues using high temperature centrifuges. However, such equipment is uncommon, limited to moderate temperatures, and does not guarantee complete phase separation.

We are developing a novel technique to fully separate immiscible melts via thermodiffusion, using the small thermal gradients (<10°C/mm) intrinsic to commonplace vertical tube furnaces. Preliminary results have successfully separated REE-doped immiscible silicate-fluorosilicate melts, with the fluorosilicate phase hosting liquidus fluorite and concentrating at the hot end of the charge. Importantly, the small thermal gradients used do not induce significant major-element gradients internal to each melt. Work is ongoing to determine trace element compositions of each phase, confirm the timescales required to achieve steady-state, and extend the range of studied melt compositions.

Silicate-fluorosilicate REE partition coefficients obtained from well separated melts, where cross-contamination during analysis can be avoided, will provide new constraints on genetic models of A-type granite REE deposits, such as Strange Lake, QC, Canada. Well-constrained genetic models are necessary to develop novel and effective exploration tools. Although our focus is REE-enriched A-type granite deposits, the thermodiffusion methods developed can be applied to other immiscible melt pairs, providing a new tool to aid the study of these challenging systems.

Rare Elements Granite-pegmatites in Kivu Region (D.R.Congo)

Rub'son H. N'nahano[1], Moise Luemba[2], Huan Li[1], Mohammed Abdallah E. Ibrahim[1]
1. Central South University, School of Geosciences and Info-Physics, Department of Geological Engineering and Geological Resources, Changsha, Hunan, China, 2. China University of Petroleum, School of Geosciences, Qingdao, China

In Democratic Republic of Congo, the Kibaran orogenic and metamorphic belts host many syn- or late tectonic granite-pegmatites, which are genetically linked to rare metals endowments. This contribution investigated four major granite-pegmatites in Kivu region worldly known for their Nb-Ta-Be-Sn-W producing mineralization in central Africa. The selected granite-pegmatites fields are Lemera, Numbi, Kalehe (South-Kivu province), and Mokama (Maniema province). Major- and trace-elements data from previous published data of the mentioned pegmatitic fields were used in this contribution with the objective of further clarifying their petrogenesis, genetic relationship, and geodynamic emplacement. The pegmatites investigated were shown to be S-types formed in a very active continental margin during the orogenic process associated with the almagamation of the Rodinia megacontinent. The petrogenetic processes involved in the emplacement of these pegmatitic fields were controlled by a magmatic evolution through a strong fractional crystallization. Lemera pegmatites are geochemically the less differentiated pegmatites in the Kivu region, a result consistent with its Sn value <5 ppm while the pegmatites of Numbi, Kalehe, Mokama depicted a higher Sn value ranging between 12 and 2,451 ppm, confirming their strong geochemical differentiation signature. The pegmatites of Kivu region are likely to be originated by muscovite-dehydration melting environment of dominantly metagraywackes and clay-poor sources. However, Numbi, Kalehe, and Lemera are petrogenetically derived from sediment melts, whereas some of Mokama pegmatites are proving to have a slab-derived fluids signature, which might be supported by the Cu endowment noticed in the previous research.

Trace Element Characterization of Scheelite in the Polyphase Intrusive W-(Sn-Ta) deposit of Santa Comba, NW Spain

Candela Pita Díaz[1, 2], Iñigo Borrajo Maza[1], Elena Crespo Feo[2], Fernando Tornos Arroyo[1]
1. IGEO-CSIC, Madrid, Spain, 2. Complutense University of Madrid (UCM), Madrid, Spain

The study analyzes the two distinct types of W mineralization found in the Santa Comba deposit through scheelite geochemistry. The mine is part of a polyintrusive complex that aligns with the Exogranite-Stockscheider-Endogranite model. Santa Comba is a W-(Sn) deposit characterized by the presence of cassiterite (SnO_2), wolframite (Fe, $MnWO_4$), and scheelite ($CaWO_4$). All ores are found disseminated in the endogranite associated with the typical alteration sequences in porphyry deposits and concentrated in a late quartz-vein system.

Scheelite, found in both types of deposit mineralization, is a better environment indicator than wolframite due to its ability to incorporate different elements. Discriminant analysis using elements like Mo, Sr, and Nb, along with specific ratios, can distinguish scheelite from different types of mineralization within the same deposit.

Scheelite disseminated in the endogranite (Sch I) shows low Mo and positive Eu, suggesting a relationship with reduced hydrothermal fluids. Scheelite associated with quartz veins, sulfides, and chlorite (Sch II), exhibits lower rare earth elements (REEs), high Mo, and a negative Eu anomaly. Mo concentration displays restricted ranges within each mineralization type, indicating its usefulness as a redox-sensitive element for discriminating scheelite formed in different geologic settings.

Sch II is impoverished in light REE (LREE) compared to Sch I, which allows for easy differentiation of the generations. Scheelite tends to incorporate LREE over heavy REE due to ionic radii differences. The low LREE content in Sch II may be due to hydrothermal apatite precipitation, which can remove LREE from the system.

The trace element signatures of the different types of scheelite are very distinctive and showcase important information about the fluids and genetic context of a deposit. By deepening our understanding, scheelite could become an effective tool in identifying specific types of mineral deposits in potential mineralization areas.

Mineral Forms of Thallium and the Genetic Model of the Vorontsovskoe Gold Deposit (North Urals, Russia)

Fedor D. Sandalov[1], Mikhail A. Rassomakhin[1], Sergey Y. Stepanov[1], Elena V. Belogub[1, 2]
1. Institute of Mineralogy, South Urals Federal Research Center of Mineralogy and Geoecology UB RAS, Miass, Russia, 2. Saint Petersburg State University, Saint Petersburg, Russia

The Vorontsovskoe gold deposit is located in Sverdlovsk region in Northern Urals (Russia), about 310 km north of the Ekaterinburg city. The Vorontsovskoe deposit was discovered in 1985 and actively mined for gold during 1999–2020. Total of the gold content of the deposit was initially estimated *ca.* 101 t (with 7 g/t Au on average). Two main primary ore types were exploited: disseminated pyrite ore with low Au grade and more rich veinlet ore presented by carbonate breccias with orpiment-pyrite-realgar cement. Gold in breccias associates with extremely specific Tl-Hg-Mn-As-Sb-S mineralization, which makes Vorontsovskoe a unique gold deposit both in Russia and the world.

Thallium is an indicator element of renowned Carlin-type Au–Tl–As–Hg–Sb deposits such as Carlin (US), Allchar (North Makedonia) or Lanmuchang (China). Other crucial localities with specific minerals of Tl are related to metamorphic carbonate rocks: the Lengenbach quarry (Switzerland) or Monte Arsiccio mine (Tuscany, Italy). The thallium mineralization of the Vorontsovskoe deposit includes 33 mineral species, which characterizes it as the second rank locality after Lengenbach. Thallium minerals of the Vorontsovskoe deposit are predominantly sulfosalts, seven of them were discovered as new mineral species in the past decade. The presence of Mn is a distinctive feature of low-temperature thallium mineralization of the deposit. The source for Tl remains uncertain.

According to previous comprehensive structural, mineralogical, and geochemical studies of the Vorontsovskoe ores, two genetic models are proposed: Carlin-type and «jasperoid»-related. Thus, the true nature of Vorontsovskoe's origin is still debatable.

On the basis of detailed analyses of previous works and new mineralogical data we propose the new view for an origin of the deposit as an epithermal object relating to a late stage of formation of the Auerbach gabbro-diorite-granite complex.

Dolocrete as a Tool for Rare Earth Elements Exploration Undercover – A Case Study of Mt Weld Carbonatite, Western Australia

Anderson Santos[1], Artur Deditius[1], Hans Oskierski[1], Arthur Oliveira[2], Denis Fougerouse[2], Pete Kinny[2], Ganesh Bhat[3], Malcolm Roberts[4]
1. Murdoch University, Perth, WA, Australia, 2. Curtin University, Perth, WA, Australia, 3. Lynas Rare Earths Ltd, Perth, WA, Australia, 4. University of Western Australia, Perth, WA, Australia

The 2.06 Ga Mt Weld carbonatite in Western Australia is renowned for its rich, rare earth elements (REE) deposit, comprising 54.3Mt at 5.2% total RE oxide and 37.7Mt at 1.07% Nb_2O_5. It consists of a calcite carbonatite outer rim (~4km) and a ferroan dolomite carbonatite core (FDC) (~1.5km) hosting REE mineralisation. A 20-40m thick layer of sedimentary rocks covers the carbonatite and poses significant challenges for exploration. A lenticular layer of dolocrete (máx. 25m) occurs below the transported sedimentary layer and above the REE-rich saprolite derived from the FDC. Dolocrete has three distinct units: (a) Upper White Dolocrete (UWD), thickness ~8m, characterised by minor palygorskite and traces of quartz, REE-aluminophosphates, rutile, ilmenite, monazite, and pyrite; (b) Middle Pisolite Dolocrete (MPD), 1-10 m thick, containing Fe-oxide nodules allogenic quartz grains, monazite, and Ce-florencite. The pisolites (10μm-4cm) contain iron oxide bands and fragments from the underlying units; (c) Basal Breccia Dolocrete (BBD), <6m thick, comprising clays and ironstone clasts with fine grains of Nb-bearing ilmenite, rutile, quartz, baddeleyite, and zircon. Except for dolomite, the mineralogical composition of these units is consistent with the underlying regolith. Stoichiometric dolomite with a Ca/Mg ratio of 1.01-1.10 is the main constituent of dolocrete. Veinlets of silica crosscut the MPD and BBD, which share some textural features such as coarse regolith-transported grains and porosity. The concentrations of FeO+MnO (in wt.%) in dolomite increase with depth: UWD (0.14%), MPD (0.65%), and BBD (0.89%). Whole-rock analysis revealed a declining $(La/Lu)_N$ ratio from FDC (165.42) upwards: BBD (97.98), MPD (16.31), and UWD (6.62), indicating light REE enrichment in regolith due to carbonatite weathering. The isotopic signature of dolocrete ($\delta^{18}O_{VSMOW}$=29.1 to 30.7‰, $\delta^{13}C_{VPDB}$=−6.6 to −5.9‰) is consistent with low-temperature alteration. The REE and Nb mineralisation preserved in dolocrete above the carbonatite ore body makes it a promising tool for REE exploration.

Hunting the Magmatic Processes that Lead to Granite-Related W Mineralization

Angela I. Santos Costa[1], Coralie Siégel[2], Rosa Didonna[2]
1. University of Tasmania - CODES, Hobart, TAS, Australia, 2. Discovery Program, Mineral Resources, Commonwealth Scientific and Industrial Research Organisation, Perth, WA, Australia

Tungsten (W) is globally considered a critical metal due to its high economic importance and limited supply. Major W resources form within magmatic-hydrothermal systems, such as greisen, skarn, and porphyry. It is broadly accepted that W mineralization associated with granites is derived from evolved volatile-rich melts that are expelled from differentiating granitic magmas. The magma must undergo high degrees of differentiation to sufficiently enrich incompatible W in the residual phase and (or) originate from partial melting of a W-rich protolith. Frequently, W orebodies are associated with other metals, including tin (Sn), molybdenum (Mo), rare metals (Li, Be, Ta, Nb), and to a lesser extent copper (Cu), gold (Au), and iron (Fe). This study investigates the association of W with different commodities and indicates that varying combinations result from distinct ore-forming processes.

Magmatic processes are recorded in the geochemistry of granitic rocks associated with W deposits. Therefore, we have compiled whole-rock geochemical data from granites associated to W deposits in 90 locations worldwide, forming a dataset with >2,600 analyses from 203 publications. Most W-related intrusions have high silica contents (>70 wt %), are peraluminous, and subalkaline. However, suites associated with Au display lower silica contents (66 ± 5 wt % SiO_2) and higher FeO total contents. Many W-Sn and W-rare metal associations share similar geochemical signatures including higher P_2O_5 and Na_2O contents and a calc-alkalic composition. W-related magmas are mostly magnesian, particularly those from W-Cu and W-Au association. Tungsten associated with Mo, Sn, Fe and rare metal displays a trend from magnesian to ferroan compositions with increasing silica contents. W-Cu, W-Au, and W-Fe associations are not highly fractionated (Rb/Sr <10) in contrast to most W-rare metal deposits. W-Sn and W-Mo display a broad range of Rb/Sr. Some Pan African and Variscan age tungsten-related intrusions hold high-fluorine concentrations (>2,500 ppm).

The World-Class Dolphin W Deposit, King Island, Tasmania – Exploration Tools and Infrared Spectroscopy

Angela I. Santos Costa, David R. Cooke, Lejun Zhang, Yamila Cajal
University of Tasmania - CODES, Hobart, TAS, Australia

Tasmania is well-endowed with tungsten (W), hosting ~26% of Australia's W resources distributed across several Devonian to Early Carboniferous granite-related Sn-W deposits. King Island, in the northwest Tasmania, hosts world-class W deposits near the town of Grassy. The premining resource for the Dolphin and Bold Head deposits has been defined as 23.8 Mt with 0.66% WO_3.

Given the encouraging exploration outcomes from recent drilling at the West targets on the western flank of the Sandblow Granodiorite and the return of mine operations at Dolphin in 2023, we have completed a systematic study to better constrain the geology, alteration, and mineralization of Dolphin, and to provide insights into the characteristics of West targets. The study was supported by systematic description, logging and sampling of diamond drill core from Dolphin and the West targets, and acquisition of HyLogger3 spectral data and short-wave ultraviolet (UV) images of selected samples. Geological domains for the skarn system and the correlation with the Grassy Group stratigraphy have been defined based on geological sections, 3D interpretations, infrared spectroscopy, and paragenetic investigations.

Following the development of biotite hornfels and spectacular early-stage barren reaction skarn at Dolphin, magmatic-hydrothermal fluids from the Early Carboniferous Sandblow Granodiorite pluton interacted with reactive carbonate-rich rocks from the Grassy Group, producing a prograde infiltration skarn assemblage containing andradite, hedenbergite-diopside, and the first stage of scheelite mineralization, which was Mo-rich (up to 8%). Retrograde skarn contains calcite with a strong pink UV fluorescence and a second mineralization stage defined by molybdenum-poor scheelite, along with epidote, chlorite and amphibole gangue. Late-stage hydrothermal processes produced calcite-ankerite and quartz veins, and minor sulfides (molybdenite, sphalerite, chalcopyrite, pyrite and pyrrhotite). A similar paragenesis has occurred at the West targets, but hornfels and reaction skarn are more common than infiltration and retrograde skarn, in contrast to Dolphin.

Searching for Tin and Tungsten in Cornwall, UK: Insights from Geochemistry and Geological Mapping

Gabriel Smith
Camborne School of Mines - University of Exeter, Penryn, United Kingdom

Tin and Tungsten are metals of increasing value due to their speciality use within developing industries including AI, battery technologies, defense, and space industries. The distribution of these resources means their supply has potential for disruption, deemed 'critical minerals' by many countries, including the US, UK, and the EU. Cornwall was once the largest global tin producer, with many of its shallow deposits having been extracted. Early miners did not have access to sophisticated modern day exploration techniques, thus deposits at depth or under thick cover may still be present. However, despite its prospective geologic setting, modern greenfield exploration has been limited, with drilling focused on brownfield sites.

Developing a camp-scale geologic model from geochemical, structural, and petrological data is critical for exploration of these deeper deposits. Cornwall has multiple granitic intrusions, which host polymetallic Sn, W, Cu, As, Ni, Pb, and Zn mineralisation. A recent field campaign conducted near a satellite pluton has confirmed the presence of multiple unmapped felsic intrusives, with vein-hosted base metal and Sn-W mineralisation observed at surface. A reassessment of the role of felsic intrusives within the SW England polymetallic ore field has subsequently been undertaken combining field and geochemical data.

During fieldwork, alteration zones, structural features, and lithological variations were mapped and in situ outcrop samples collected. The rock samples obtained were analysed to determine mineralogical and geochemical relationships. Integration of optical microscopy, ICP-MS, and pXRF have been used to determine the ore mineralogy, alteration signatures, and paragenesis. Granitic composition has been compared with data from proximal intrusions to determine the geochemical variations. Interpretation of the results alongside field data has produced a new camp-scale geologic model of the area. The results will be used to inform subsequent exploration within the area, targeting lithological and structural features delineated during the study.

The Importance of Pressure and Fractionation in Forming Compositionally-Diverse, Rare-Earth Rich Igneous Complexes: A Case Study of the Nechalacho Layered Suite, Canada

Caroline R. Soderman[1], Owen M. Weller[1], Charles D. Beard[3], Nicolas Riel[2]
1. University of Cambridge, Cambridge, United Kingdom, 2. Johannes Gutenberg University, Mainz, Germany, 3. Utrecht University, Utrecht, Netherlands

Alkaline igneous complexes can host important ore deposits (e.g., rare earth elements, REE) and contain a huge diversity of rock types, including silica-undersaturated (e.g., nepheline-syenite) and silica-oversaturated (e.g. quartz-syenite) compositions. Isotopic evidence frequently suggests that these rocks are derived from the same parental melt. Using recently developed thermodynamic phase equilibrium models we investigate the petrogenetic processes that generate these compositionally diverse complexes. Integration of trace-element partitioning models reveal and quantify mechanisms for REE enrichment during parental melt crystallisation.

We explore the petrogenesis of the Blatchford Lake Igneous Complex (BLIC), Northwest Territories, Canada, as a case study. The BLIC is a ~2.2 Ga intrusive complex that hosts the Nechalacho Layered Suite (NLS)—a world-class, syenite-hosted REE, Nb and Zr deposit—as well as cogenetic gabbro, granite, quartz- and nepheline-syenite. Using the most primitive BLIC gabbro as a model starting composition, we show that the observed diversity of rock types can be generated by fractional crystallisation occurring mostly at 4 to 5 kbar. This narrow pressure range represents a key 'tipping point' in the evolution of silicate melts, with shallower fractionation forming silica-oversaturated compositions, and deeper crystallisation driving residual melt to silica undersaturation. Importantly, this crystallisation pressure range coincides with a previous estimate of 4 kbar for the BLIC. Our modelling removes the need to invoke additional processes such as crustal assimilation or magma recharge, for which there is conflicting isotopic evidence. Finally, modelling of residual REE enrichment in the fractionating melts shows that pressure is also a key control in forming the heavy REE enrichment seen in the NLS. Variations in residual REE enrichment with pressure and extent of crystallisation are mostly driven by changes in abundance and composition of clinopyroxene, which has strongly composition-dependent REE partitioning. The results of this case study have global implications for the prospectivity of similar systems.

Fluorite Mineralization Associated with Albitite and Skarn: An Example of Jinmu-Mihara Fluorite Deposit, Sanyo Belt, SW Japan

Masahiro Sunada, Yasushi Watanabe, Takuya Echigo
Graduate School of International Resource Sciences, Akita University, Akita city, Japan

Fluorite deposits in Japan occur as vein-type or skarn-type, associated with the Cretaceous ilmenite-series Sanyo-Belt granitoids in the Inner Zone of southwest Japan. The Jinmu-Mihara deposit is one of the skarn-associated deposits in the belt. We conducted geological mapping, petrography and quantitative analysis of minerals using electron micro-probe analysis (EMPA) to clarify fluorite mineralization of the deposit and its relationship to the albititization and skarnization.

The host rocks of the Jinmu-Mihara deposit consist of limestone, chert, and amphibolite of the Jurassic Tamba Belt, present as a roof pendant of the Late Cretaceous biotite granite. Skarn is distributed between the host limestone and biotite granite, while albitite is present in the biotite granite. The albitite is disseminated with fluorite, with accessory gittinsite, thorite, thorianite, and Nb-bearing titanite. Albite, together with fluorite, replaces plagioclase in the granite. In skarn, fluorite occurs in bands of either of garnet-vesuvianite-fluorite, or clinopyroxene-hornblende-albite-fluorite-(vesuvianite). The core and/or mantle of some garnet is replaced by vesuvianite. Fluorite occurs between grains of vesuvianite, clinopyroxene, and hornblende. These banded structures indicate that the skarn minerals are replaced by fluorine-rich vesuvianite and fluorite. The fluorine contents of vesuvianite and hornblende are 1.8 to 2.4 wt % and 1.7 to 2.9 wt %, respectively.

Our results show that the fluorite mineralization of the deposit occurred as (1) fluorine-rich vesuvianite and fluorite formed by replacing skarn minerals (garnet, clinopyroxene) and (2) plagioclase-replacing fluorite in albitized biotite granite. These two processes occurred independently in different rock types, possibly from the same Na-F rich metasomatic fluids.

Tracing the Metal Sources of Iberian W-Sn Deposits Through Sr-Nd Radiogenic Isotopes

Iñigo Borrajo[1], Fernando Tornos[1, 2], John M. Hanchar[2], Jose M. Fuenlabrada[3]
1. Instituto de Geociencias (CSIC-UCM), Madrid, Spain, 2. Memorial University of Newfoundland, St John's, NL, Canada, 3. Facultad de Geologia - Universidad Complutense de Madrid, Madrid, Spain, 4. Consejo Superior de Investigaciones Cientificas, Madrid, Spain

Tungsten and Sn mineralization occurs worldwide and over much of geologic time. These mineralizations are spatially associated with felsic granitic intrusions and form a wide range of magmatic-hydrothermal deposits. However, the genetic link between specific suites of granitic rocks and the formation of hydrothermal wolframite- and cassiterite-bearing perigranitic quartz vein systems is still debated in most major Sn-W metallogenic provinces.

This study focuses on the isotopic geochemistry (Sr, Nd) of Sn-(Ta-Li) and W-(Sn) vein-type deposits within the Iberian Variscan Massif (Spain and Portugal) by doing novel isotopic analyses on pure mineral concentrates of cassiterite, wolframite, and scheelite.

The measured $^{87}Sr/^{86}Sr$ ratios and ε_{Ndi} values were recalculated to an initial common age of 300 Ma to correspond with the peak of Sn-W mineralization events in the region. The Sr isotopic results obtained underscore the complexity of fluid-rock interaction processes, and highlight the heterogeneity of the crustal sources involved in the genesis of these deposits. In contrast, the Nd isotope results more accurately trace the fluid and metal sources. Initial ε_{Ndi} values in samples from both the W-(Sn) and Sn-(Ta-Li) deposits in the northern area of the Iberian Variscan Massif range between -6.8 and -8 and support that Nd is primarily derived from the local dominant S_1 granites. The wolframite samples from a few W-(Sn) deposits, however, show initial ε_{Ndi} values higher than -4 suggesting a more juvenile Nd component likely derived from the local I-type granitoids. In the southern zone of the Iberian Massif, the W-(Sn) and Sn-(Ta-Li) deposits exhibit initial ε_{Ndi} signatures between -3.8 and -7.4, pointing to local S_2-type granites as the main source.

The Las Herrerias Deposit: Replacive IOCG Systems in the Ossa Morena Zone (SW Iberia)

Carmen Conde[1], David Rodriguez[2], Fernando Tornos[1], Dolores Garcia[2], Gonzalo Ares[1]
1. CSIC-IGEO, Madrid, Spain, 2. Atalaya Mining Plc., Zafra, Spain

The iron oxide-copper-gold (IOCG) mineral system encompasses a range of hydrothermal ore deposits characterized by abundant iron oxides, copper sulfides, and gold. These deposits can form through various geological processes, including the replacement of volcanic rocks and ironstones, brecciated fault zones, and the alteration of felsic igneous rocks.

The Ossa Morena Zone (SW Iberia) contains several ore deposits sharing features with the IOCG mineral system that are near enigmatic Fe-B-REE deposits. The Las Herrerías deposit is one of the best IOCG-like examples. The mineralization is hosted by an Early-Middle Cambrian rift-related sequence that includes limestone and bimodal volcanic rocks capped by shale and sandstone. These late siliciclastic rocks host a stratabound ironstone that is interpreted as being exhalative in an oxic third-order basin. During the Variscan deformation, the ironstone and host rocks were affected by major N-S shearing and formation of NW-SE extensional faults. These late faults controlled a second event of mineralization dated at 332 to 330 Ma that includes the widespread hydrothermal replacement of the ironstone by semi-massive chalcopyrite and bornite with magnetite-hematite, quartz, and carbonates. Host rocks have been affected by a pervasive phyllic alteration.

High $^{87}Sr/^{86}Sr_i$ and eNd_i ratios suggest that fluids forming the Cu-rich mineralization are significantly different from those related to the ironstone. They are of crustal derivation and probably were distal to coeval peraluminous granitoids. Numerical modelling indicates that the brittle ironstone is a preferential site for epigenetic copper-rich mineralization. Reaction of the hydrothermal fluids produces major changes in the fO_2 and pH, destabilizing the Cu aqueous complexes. However, the critical issue is the input of external reduced sulfur. In this case, it was derived from the abiogenic reduction of the barite in the ironstone.

A Geometallurgical Approach to Recovery of Critical Metals Tellurium, Bismuth, and Tungsten as By-products from a Swedish Gold Mine

Lara U. Tritton[1], Gawen R. Jenkin[1], Daniel J. Smith[1], Helena Moosberg-Bustnes[2]
1. University of Leicester, Leicester, United Kingdom, 2. Mandalay Resources, Skellefteå, Sweden

Rapid technological advances have led to the increasing importance of tellurium, bismuth, and tungsten. These critical metals are vital for key technologies, such as solar panels, but have vulnerable and limited supply chains. They are frequently enriched in gold deposits but are understudied and their geometallurgy is poorly understood. Mining and mineral processing are energy-intensive and have a large environmental footprint, and ESG standards are rising. This research focuses on assessing the possibility of recovering tellurium, tungsten, and bismuth from an existing mine. We aim to understand the occurrence of these critical metals, in the case study of Björkdal gold mine in Sweden, and their behaviour through the existing mineral processing, and ultimately aim for onsite by-product recovery. Geometallurgical characteristics of the feed and outputs of mineral processing (four concentrates plus tailings) are assessed through SEM-EDS, XRD, optical microscopy, and bulk geochemistry. Additionally, low-energy recovery using environmentally-benign deep eutectic solvents (DESs) is assessed through leaching experiments.

Target metals almost entirely deport to scheelite (tungsten – 99.8%) and bismuth tellurides (bismuth – 98% and tellurium – 97%). Concentrations and particle characteristics vary significantly between concentrates and tailings. Bismuth tellurides mostly mimic the behaviour of gold in the processing plant due to both mineralogical association with gold and the high specific gravity of these minerals. However, roughly 50% of bismuth tellurides and 70% of tungsten deport to tailings. One concentrate, a product of the gravity circuit, has been identified as a potential target for additional processing steps to recover these metals due to its relatively small volume and good concentrations of large, well-liberated particles of bismuth tellurides and scheelite. Leaching of this concentrate with DESs could allow the recovery of these critical metals on site, diversifying supply and providing extra revenue for the mine, whilst minimising the environmental impact.

A Process-Based Approach to Constraining Metal Enrichment in Alkaline-Silicate Magmatic Systems Using New Thermodynamic Models

Owen Weller[1], Tim Holland[1], Caroline Soderman[1], Eleanor Green[2], Roger Powell[2], Charles Beard[3], Nicolas Riel[4]
1. University of Cambridge, Cambridge, United Kingdom, 2. University of Melbourne, Melbourne, VIC, Australia, 3. Utrecht University, Utrecht, Netherlands, 4. Johannes Gutenberg University, Mainz, Germany

Alkaline-silicate magmatic systems are an important host for a variety of specialty metals, including rare earth elements (REEs). However, what controls whether a given system will become sufficiently enriched to form an ore deposit remains poorly understood. Phase diagram analysis provides a tool to investigate such controls, but currently it is not possible to apply this method to alkaline compositions due to of an absence of thermodynamic models for the constituent phases. Here we present new thermodynamic models that fill this knowledge gap and allow for the generation of phase diagrams for alkaline igneous compositions. The phase diagrams provide a novel tool to explore the role of variables such as pressure, fractionation, and crustal assimilation in the petrogenesis of these systems. Furthermore, by integrating trace-element partitioning, we can investigate the controls on residual enrichment of specialty metals such as REEs as precursors to magmatic ore deposits.

The thermodynamic models are calibrated using experimental phase relation data, and the model performance is benchmarked with independent experimental data sets, encompassing a range of pressure (0–22 kbar), temperature (680°–1350°C), and global compositional ranges observed in alkaline-silicate systems. The calculated phase diagrams successfully reproduce experimental crystallisation sequences and phase compositions, indicating that the models are well calibrated across this spectrum of conditions. Applying our models to alkaline-silicate bulk compositions (e.g. nepheline syenite) yields solid assemblages consistent with those reported for natural systems. Analysis of the liquid line of descent reveals that the final melt fraction and associated equilibrium mineralogy (e.g. pyroxene, feldspar) evolves to extremely alkali-rich compositions, consistent with alkali-rich 'fenitised' zones observed in and around some alkaline-silicate intrusions. Finally, examples will be presented that showcase which crystallisation conditions are conducive to the formation of REE-rich alkaline-igneous rocks. Importantly, this new methodology can be applied to any specialty metal of interest in magmatic systems.

Mineralogical Characterization and Location of Critical Metals in the VHMS: Reevaluating the Potential of the Tharsis Mine

Carmen Conde[1], Gonzalo Ares[1], Sergio Tenorio[2], Fernando Tornos[1]
1. CSIC-IGEO, Madrid, Spain, 2. THARSIS MINING SL, Huelva, Spain

The Iberian Pyrite Belt hosts some of the largest resources of base metals in Europe, hosted by what is considered the largest VHMS district worldwide. However, these deposits also have a significant anomaly in critical metals such as Co, Bi, Te, In, Ga, or Ge.

The Tharsis deposit is one of the most important VHMS deposits in the Iberian Pyrite Belt with original reserves of over 130 Mt. This shale-hosted deposit contains at least six massive sulfide orebodies and an underlying stockwork developed on chloritized dark shale. The ore exhibits various types of mineralization, mainly composed of pyrite with minor amounts of chalcopyrite, sphalerite, and galena. However, a facies enriched in siderite (carbonate ore) and the stockwork include a more complex mineral assemblage with significant contents of Co, Bi, Te, and Au.

This study aims to detail the mineral assemblage hosting these critical metals, their genesis, and to delineate the ore-rich zones within the stockwork. The stockwork consists of irregular veins composed of pyrite, minor quartz, Co–As–Fe–S minerals, and traces of Bi–Pb–Cu–(Sb) sulfosalts, tellurides, and native gold. LA-ICPMS shows that the pyrite is also variably enriched in these critical metals. Preliminary results show that there is a positive correlation between Bi-Au ($r = 0.94$) and Co-Au ($r = 0.74$). The data are consistent with the presence in the assemblage of cobalt-bearing minerals (e.g., cobaltite, alloclasite, and/or glaucodot) and Bi minerals (kobellite and/or bismuthinite) and strongly suggest that Co, Bi, and Au occur as solid solution or nanoinclusions. Spongiform and colloform pyrite is enriched in As and Au, likely formed in a high-temperature environment from a low fS_2– fO_2 fluid.

Heemskirk Tin Project, Western Tasmania, Australia: Characterization of Severn-Queen Hill Tin Deposits

Nelao N. Naimbale, David R. Cooke, Wei Hong, Robert Scott
University of Tasmania, Hobart, TAS, Australia

Western Tasmania hosts several Devonian granite-related Sn-W deposits, including (1) sulfide-bearing skarns and carbonate-replacement deposits (e.g., Renison Bell, Severn-Queen Hill, Mount Bischoff, Razorback), (2) silicate skarns (e.g., St. Dizier), (3) magnetite skarns (e.g., Mount Lindsay), (4) vein deposits (e.g., Pieman), and (5) greisen deposits (e.g., Federation). There is also one unusual skarn/hydrothermal Ni deposit (Avebury).

The Heemskirk Tin project is located near the town of Zeehan, western Tasmania, Australia. With a mineral resource estimate of 7.4 Mt at 1.04% Sn, Heemskirk Tin is the highest-grade undeveloped tin project in Australia and ranks as the third-highest globally. The project area consists of four Devonian granite-related sulfide skarn and vein deposits: Severn, Queen-Hill, Montana, and Oonah. Heemskirk Tin is primarily hosted within the Neoproterozoic to Cambrian sedimentary and volcaniclastic rocks of the Zeehan sub-basin within the Dundas Trough. Tin principally occur as cassiterite, with subordinate stannite. Fine-grained cassiterite grains are hosted in pyrite-pyrrhotite veins and disseminations, associated with pervasive siderite-sericite-chlorite-tourmaline ± fluorite ± topaz alteration. Tin mineralisation at Severn and Queen Hill is predominantly stratabound, occurring as stockwork/veins and replacement styles of pyrite-siderite ± pyrrhotite ± chalcopyrite ± galena ± sphalerite, closely associated with faults. Mineralisation is strongly zoned with late-stage galena-sphalerite-silver fissure veins located toward the periphery of the tin mineralisation zone.

Our ongoing study aims to aid near-mine exploration by understanding the controls on fluid flow and mineralisation at the Heemskirk Tin project through detailed geological, mineralogical, and geochemical characterisation. This will ultimately improve the understanding of the genesis of carbonate-hosted Sn mineralisation at Heemskirk and the deportment of critical and strategic metals at the project.

The 3D Distribution of Cobalt in VHMS: The Elvira Orebody, Iberian Pyrite Belt

Gonzalo Ares[1], Carmen Conde[1], Guillem Gisbert[2], Juan C. Videira[3], Fernando Tornos[1]
1. Instituto de Geociencias, Consejo Superior de Investigaciones Científicas, Madrid, Spain, 2. Departament de Mineralogia, Petrologia i Geologia Aplicada, Universitat de Barcelona, Barcelona, Spain, 3. Minas de Aguas Teñidas S.A (MATSA), Almonaster La Real, Spain

Cobalt plays a critical role in the energy transition, being a key component in lithium-ion batteries. The increasing demand for cobalt and ongoing instability in its global supply chain cause a pressing need to explore new sources of this element to satisfy the industry demands.

The Iberian Pyrite Belt (IPB) boasts the largest concentration of volcanogenic massive sulfide deposits worldwide. The first compilation data show that several orebodies of this area contain significant cobalt values (>300 ppm). But this information is scarce and lacks analytical contrast, and there is a lack of knowledge regarding distribution at deposit scale. For that, we conducted a comprehensive study and 3D modeling of the Elvira deposit based on 86 drill holes.

The Elvira deposit is a polymetallic (Cu-Zn-Pb) massive sulfide hosted by dark shale, with minor volcanic rock, located in the SE extension of the Sotiel-Migollas deposit. The main mineralization comprises a 400-m lens along a 60°E strike, dipping 40ºNW, with a thickness up to 50 m. The massive sulfide shows, on average, 250 ppm Co with some metre-thick intersections up to 1,200 ppm. Co tends to concentrate preferentially at the base of the massive sulfides and correlates positively with Cu. Above the massive sulfide, the bulk of the deposit includes a highly deformed 30-m-thick unit of interlayered semi-massive and massive sulfides and shale. On average, Co contents range between 50 and 100 ppm with metre-thick intersections of >500 ppm. Below the massive sulfides, there is an irregular stockwork that has similar average Co contents of ca. 100 ppm with metre-thick intersections of >500 ppm. Thrusted above the massive sulphides is an exotic stockwork with tectonic contacts. This unit, up to 100 m thick, contains high and variable Co concentrations ranging between 50 to 2,700 ppm, with rare cm-thick intersections up to 8,000 ppm.

The Ore Genesis of Okanjande Graphite Deposit, Central Zone, Damara Belt, Namibia

Nelao N. Naimbale[1], Shao-Yong Jiang[1], Fred Kamona[2], Daniel Machoko[3]
1. China University of Geosciences, Wuhan, China, 2. University of Namibia, Windhoek, Namibia, 3. Northern Graphite Corp, Otjiwarongo, Namibia

Okanjande is a flake-type graphite deposit hosted by amphibolite-grade banded, porphyroblastic, and migmatictic feldspathic quartzites of the Nosib Group, of the Pan-African Damara Belt, Namibia. The Central Zone is a high-temperature, low-pressure zone characteristic of sillimanite-cordierite metamorphic assemblages, numerous granitic plutons, and fold structures

Petrographic studies showed that quartzites are composed of quartz, K-feldspar, plagioclase, muscovite, biotite, chlorite, jarosite, sillimanite, rutile, pyrite, pyrrhotite, chalcopyrite, sphalerite, and molybdenite. Graphite forms isolated, well-crystallised, disseminated needle-like and platy flakes. Sillimanite is present as spherical to elongated and shear-like porphyroblasts.

δ^{13}Cgraphite from quartzites ranges from -13.36% to -21.58%, falling within the range of -10% to -40% for organic matter in the Proterozoic. On the other hand, δ^{13}Cgraphite from carbonate rocks yielded two different ranges of -4.54% to -7.01% and -11.8% & -22.68%, inferring values from devolatilization of carbonate minerals and carbonaceous materials as well as from the carbon isotopic exchange between graphite and calcite.

δ^{34}S compositions from pyrite, pyrrhotite, and chalcopyrite have values in the range of -4.4% to 3.7%, indicating a magmatic source of sulphur at Okanjande.

Graphite morphology studies from XRD indicated that graphite at Okanjande is well crystallised and matured (Lc (002) 233À - 544À; Graphitization Degree of 33À - 133À). The metamorphic temperatures calculated from δ^{13}Cgraphite (414°-684°C) and Graphitization Degree (459°-709°C) of graphite are in agreement with the range for Amphibolite facies (450°-700°C) as well as the previously estimated temperature of the Central Zone (432° to ±700°C). Our results support a syngenetic model for graphite formation at Okanjande.

Potential Sources of Rare Earth Elements (REEs) in Loch Borralan, Northwest Highlands, Scotland: Geochemical and Mineralogical Controls on REE Distribution in Carbonatite

Abdulkadir M. Abdulkadir
UCL Department of Earth Sciences, London, United Kingdom

Rare earth Eeements (REEs) are critical in low-carbon technologies, with a forecast of growing demand to switch to green technologies. Currently, REEs are mainly sourced from carbonatite for the light REEs (LREEs) and alkaline igneous complexes for the heavy REEs (HREEs). In this context, there has been a revived focus on the Loch Borralan Complex of Northwest Scotland for better understanding of REE enrichment processes in this complex. The complex is composed of mainly nepheline syenites, but small intrusions of carbonatite have been found and mapped in the south of the complex. Based on petrographic studies and microprobe analyses, our work has shown an enrichment in REEs in carbonatites and associated silicate rocks with whole-rock total REE oxide contents of up to 1.5 wt %. Based on textural evidence and microprobe results, this has resulted in the late magmatic crystallisation of LREE-bearing minerals, including bastnaesite, synchysite, ancylite, monazite, apatite, chevkinite, perrierite, zirconolite, titanite, REE-rich ilmenite, and allanite. Stable isotopic composition of carbon and oxygen measured in carbonate mineral separates from the different rock types in the complex suggest a primary magmatic origin. However, there is also evidence for more than one source of fluid components from the measured nitrogen and argon isotopic and elemental ratios of fluid inclusions by step crushing. Fluid inclusions and their stable isotope data show the interaction of carbonatite-derived fluids and externally derived fluids which may have exsolved or boiled, leading to fracturing of the country rock, permitting additional dilution of carbonatite-derived fluids by meteoric water. We suggest this as a means of precipitation of the REEs and other metals at Loch Borralan.

Magmatic-Hydrothermal Alteration and Deformation of Spodumene Pegmatites from Musha-Ntunga (Rwanda)

Jolan Acke[1, 2], Stijn Dewaele[2], Justin Uwiringiyimana[3], Renata Barros[4, 5], Anouk Borst[1, 6]
1. Department of Earth and Environmental Sciences, KU Leuven, Leuven, Belgium, 2. Mineralogy & Petrology Research Group, University of Ghent, Ghent, Belgium, 3. Trinity Metals Group, Musha Mines, Rwamagana, Rwanda, 4. Geological Survey of Belgium, Royal Belgian Institute of Natural Sciences, Brussels, Belgium, 5. Consortium for Battery Innovation, International Lead Association, Tervuren, Belgium, 6. Geodynamics and mineral resources, Royal Museum for Central Africa, Tervuren, Belgium

LCT-type pegmatites in the Mesoproterozoic Karagwe-Ankole Belt of Central Africa are intimately linked with granitic complexes that were emplaced approximately 1 billion years ago. This study focuses on drill core samples from spodumene-bearing pegmatites in the Musha-Ntunga region of eastern Rwanda, which are associated with the lake Muhazi granitic pluton. Our approach integrates petrographic, Raman spectroscopy, cathodoluminescence microscopy, elemental geochemical techniques (EPMA and LA-ICP-MS), and X-ray diffraction of spodumene grains to investigate the mineralization sequence, microtextural variations, and lithium distribution across key paragenetic stages. These stages encompass primary magmatic crystallization, magmatic-hydrothermal and hydrothermal alteration, and subsolidus deformation.

We have identified five textural types of spodumene. The first two are primary magmatic coarse-grained spodumene (Spd type 1) and spodumene with symplectitic spodumene-quartz intergrowth rims (SQI; Spd type 2). Spd types 3 and 4 are finer-grained intergrowths developed during consecutive magmatic-hydrothermal to hydrothermal alteration events. Additionally, the pegmatites underwent localized deformation, evident through various textures such as sigma-clast-shaped porphyroclasts ("spodumene fish") and boudinage structures of deformed Spd type 1 crystals, indicative of both brittle and ductile deformation processes. The strained Spd type 1 crystals experienced partial recrystallization to fine-grained elongated crystals (Spd type 5), occurring in bands with muscovite, quartz and apatite, and emphasizing the primary orientation of foliation.

Spodumene grains can be (partially) replaced by albite, microcline, muscovite, and quartz during magmatic-hydrothermal activity; and by cookeite, eucryptite, lithiophilite and kaolinite during late-stage alteration.

Montebrasite is identified both as a late primary magmatic phase coexisting with spodumene and as a secondary recrystallized phase formed during magmatic-hydrothermal alteration and subsequent deformation. Other ore minerals present, including cassiterite, columbite-tantalite, and monazite, further record the transition from a magmatic to a hydrothermal and deformational regime.

Mineralogy and Phosphate Paragenesis of the Ewoyaa Lithium Pegmatites, Southern Ghana

Salaam J. Adams[1], Marieke V. Lichtervelde[2], Prince O. Amponsah[1], Abdul-Razak S. Ballah[3], Iwan WIlliams[3], Lennard Kolff[3], Prosper M. Nude[1], Daniel K. Asiedu[1], Samuel B. Dampare[4]
1. University of Ghana, Accra, Ghana, 2. IRD/GET, TOULOUSE, France, 3. Atlantic Lithium, Mankessim, Ghana, 4. Ghana Atomic Energy Commssion, Kwabenya - Accra, Ghana

The presence of Fe-Mn phosphates and ferromagnesian silicates, such as biotite, garnet, and tourmaline, offers a chance to look at the connections between these two families of minerals. In this study, we determined the mineralogy of the Ewoyaa lithium-rich pegmatites. We also determined the paragenesis of the pegmatites using the phosphate minerals with emphasis on apatite.The pegmatite dykes (and quartz) came in as intrusions in the schists and are observed to be intercalating the schists. Three types of spodumene were identified; P1, P2, and P3. The P1 spodumene are coarse to very coarse grained and more abundant. The P2 spodumene are medium grained and less abundant as compared to the P2. The P3 are fine grained spodumene, which occurs as inclusions in the silicates. The most common of these ore minerals are apatite, columbite-tantalite, pyrite, uraninite, and zircon. The oxides are systematically associated with the silicates; albite, biotite, garnet, and spodumene. From the results, the type of apatite identified is fluorapatite. The manganese content ranges from 0.02 wt % to 8.46 wt % MnO, 0.02-16.43 wt % FeO, 0.01-0.31 wt % Na_2O, and 0.0-6.55 wt % Al_2O_3. Their mineral chemistry reveals low fractionation degrees, which is typically not of complex LCT pegmatites. Different phosphate species are crystallised in pegmatite melt. These species include primary phosphate and secondary phospahates. The Mankessim-Cape Coast pegmatite field and the Ewoyaa pegmatites could both be considered weakly to moderately evolved rare-element beryl pegmatites. The pegmatites are categorized as beryl and beryl-phosphate subtypes since the columbite-group minerals are insufficiently abundant in the Ewoyaa area while phosphate minerals are present in large amounts.

Petrography and Geochemistry of Pegmatite and Leucogranite of the Ntega-Marangara Area, Burundi, in Relation to Rare Metals Mineralisation

Quesnay de Jésus Akabahinga[1, 2], Anthony T. Bolarinwa[1], Seconde Ntiharirizwa[3]
1. University of Ibadan, Ibadan, Nigeria, 2. Pan African University, Life and Earth Sciences (Including Health and Agriculture), Ibadan, Nigeria, 3. University of Burundi, Bujumbura, Burundi

The Ntega-Marangara area, part of the Kanyaru supergroup, Western Domain (WD) of the Karagwe Ankole Belt (KAB), hosts numerous pegmatite veins related to the leucogranite. This investigation aims to characterise the granitoids, their alteration products, and rare metals mineralisation potentials. Quartz, plagioclase, microcline, albite, and muscovite are the essential minerals in both the pegmatite and leucogranite. The ΣREE values of the pegmatite and leucogranite are 147 and 102 ppm, respectively. The greisenised pegmatite with Rb, Cs, Ta, Li, Nb, and Sn of 5,940, 1015, >2500, 130, 1,595, and 671 ppm, respectively, are higher than the corresponding values of 636, 62, 32, 74, 58, and 110 ppm of the kaolinized pegmatites. This enrichment classifies them as lithium-caesium-tantalum (LCT) pegmatite. The mean K/Rb, K/Cs, and Nb/Ta in the leucogranites are 106.86, 2819.24, and 4.30; pegmatite is 35.33, 469.47, and 3.1; and greisen 14.05, 82.2, and 0.64, respectively, which suggest that greisenisation due to metasomatic reactions of late hydrothermal activity could have increased the potential for rare metals mineralisation of the altered pegmatite. Pegmatite of the Ntega-Marangara area is enriched in LREEs relative to HREEs, with a K/Rb ratio <100, indicating a high level of fractionation and mineralisation in Ta-Nb, Sn, and Li, in contrast to the leucogranite that is less fractionated and barren.

Petrography, Geochemistry, Tectonic Setting, and Rare Earth Elements Mineralisation Potentials of Selected Granitoids from Southwestern Nigeria

Shakirat Aminu[1, 2], Akinade S. Olatunji[2, 4], Maryam O. Abdus-salam[3], Victoria B. Omotunde[4], Oghenerume S. Emecho-Ted[4]

1. Kwara State University, Malete, Kwara, Nigeria, 2. Pan African University Life and Earth Sciences Institute, University of Ibadan (PAULESI), Ibadan, Nigeria, 3. The Polytechnic Ibadan, Ibadan, Nigeria, 4. University of Ibadan, Ibadan, Nigeria

The demand for rare earth elements (REEs) has necessitated the need to search for REEs from sources aside from carbonatites, leading to granitoids currently being assessed globally as possible sources of REEs. Granitoids from parts of southwestern Nigeria were studied based on field observations, textural features, and mineralogy; geochemical analysis of representative granitoid samples was done by LA-ICP-MS to determine major, trace, and REE concentrations, while mineral chemistry of rock units with abundant REE concentrations was analyzed with SEM-EDX to determine the REE-bearing minerals within them. The dominant rock units observed are granitic gneisses, granites, diorites, and monzonites with minor intrusions of pegmatites and quartz veins. Petrogenesis and tectonic setting classification done using geochemical data grouped the granitoids as calc-alkalic to alkalic, mostly metaluminous, and emplaced mainly as within plate tectonic rocks. The REE concentrations for the studied granitoids show higher concentrations of the light rare earth elements (LREEs) compared to the heavy rare earth elements (HREEs) with a negative europium anomaly indicating that these rocks have been formed from the crystallization of melted crustal materials with plagioclase fractionation. The ΣLREEs for the granites (ΣLREEs 13.10-1083.6 ppm, ΣHREEs 0.44-50.79 ppm) and monzonites (ΣLREEs 14.9-795 ppm, ΣHREEs 0.34-33.86 ppm), respectively, were found to compare favorably with the granites of China, (ΣLREEs 152-402ppm, ΣHREEs 16-53ppm), where REE exploitation is ongoing. REE-bearing minerals in granitic gneisses, granites, and monzonites include synchysites, allanites, monazites, apatites, and tritomites, from where economic concentrations of REEs can be extracted. These granitoids are therefore rock units of interest in understanding Nigeria's economic potential of REEs.

Nickel and Cobalt Enrichment Zone and Ore Potential Distribution in Sorowako Area and Surroundings, South Sulawesi, Indonesia

JJ A. Ayyala, Fadlin Fadlin
Jenderal Soedirman University, Banyumas Regency, Indonesia

The energy transition can be supported by optimizing the potential of metal elements in laterite deposit as mining materials. The downstream policy forces this sector to process low-grade nickel ore material (1-1.5%) and cobalt (>0.1%), which is known to have prospects for later production, so a detailed method is needed to recalculate the distribution of ore potential. The ordinary kriging geostatistical estimation method using Leapfrog Geo software is used to determine and visualize the distribution of enrichment zones and potential for nickel and cobalt ore. This research used collar and assay data for 218 drill points and also nickel and cobalt content values for 7,738 datums. Production cutoff grade (Ni 1% and Co 0.1%) is used as an indicator to simulate the minimum value of nickel and cobalt content in laterite deposit material. The optimal potential for laterite deposits is explored by inputting the geological control of the deposit domain and assay values (%Ni and %Co) through estimating the tendency of enrichment zone depth, as well as expressing the total volume distribution of nickel and cobalt content values to show a percentage of potential increase in nickel volume and potential cobalt volume. Based on the results of analysis in the research area located in Sorowako, South Sulawesi, it shows that the potential for nickel ore is 80% of the total volume of laterite deposits with low grade material (Ni 1-1.5%) of >3,500,000 m^3, high-grade (Ni >1.5%) of >1,800,000 m^3, and cobalt volume (Co >0.1%) of >1,600,000 m^3. Interestingly, this potential volume results in the prospect of increasing nickel production by around 200% and the utilization of cobalt as a new product. Laterite deposits thickness is around 30 meters with low-grade nickel (1-1.5%) and cobalt (>0.1%) depth at 5-25 meters, and high-grade nickel (>1.5%) depth at 20-30 meters.

Potential of Mineral Raw Materials for Battery Production in Hungary

Mohamed Badawi, Evane C. de Cuñha
Miskolc Egyetem, Miskolc, Hungary

Hungarian and international students from the Special College of Natural Resources Research and Utilization of the University of Miskolc, TEKH, undertook the data collection and preliminary evaluation of critical and strategic raw materials, as well as their potential occurrences of mineral raw materials in Hungary. Battery metals belong to the group of critical and strategic raw materials in the EU. In this grouping, they are still little known in Hungary, and there is a particular lack of knowledge regarding the occurrence of these materials in Hungarian geological environments. At the same time, the country has identified developing a world-class battery manufacturing industry as a priority national strategic goal. This requires increasing knowledge, research, and development of domestic sources of raw materials. The data collection and the evaluations that can be based on it can effectively help the assessment of the country's mineral resource resources with a new approach.

The overview is based on the data in the database under construction. For battery raw materials, lithium, cobalt, nickel, manganese, and graphite have been studied and searched on a nationwide level. The data are summarized in fact sheets by occurrences and accessible through the the TEKH website

Lithium has anomalous enrichment in geothermal brines under the Great Hungarian Plain. Mangenese with Co and REEs is found in two extensive manganese carbonate ore deposits: Úrkút and Eplény. Cobalt was also recorded in skarns of the Recsk base metal ore complex, as well as in Paleogene terrestrial sediments related to karst bauxite deposits. Finally, unexplored flake graphite occurrences are related to Paleozoic metamorphic rocks in the Szendrő Mountains.

Vanadium Enrichment in Silurian Black Shales of SE Sardinia (Italy): Comparison with Other Cases of Northern Gondwana

Maria Boni[1], Giuseppina Balassone[1], Nicola Mondillo[1], Giovanni Di Maio[2]
1. University of Naples Federico II, Naples, Italy, 2. GEOMED Geoarcheologia e Ambiente, Scafati, Italy

Vanadium is currently considered a critical metal (CM) for both the European Union and the United States. Black shale-hosted vanadium deposits account for about 80% of vanadium resources in the world, but only <2% V can be considered economic. The average Paleozoic black shales contain values of about 400 ppm of ∑Mo+Ni+Zn+Se+V, as well as 2–10% organic carbon.

In the Silurian of North Gondwana, significant black shale successions were deposited in a series of basins under strong anoxic conditions. These successions have a variable content of organic matter and metallic elements. Their vanadium amount ranges from subeconomic values (Northern Bavaria, Germany: 9,500 ppm V) to only scientifically significant concentrations (Turkey: up to 400 ppm V; Carnian Alps: 250-500 ppm V). Silurian black shales, still poorly investigated for V, are also present in several areas of Northern Africa, as in Lybia and Morocco.

Slightly metamorphosed Silurian black shales in SE Sardinia (Italy) have many paleontological, mineralogical, and geochemical similarities with other coeval successions of northern Gondwana. We detected a content in C/org ranging from 2.15 to 10.00%, as well as anomalous values of Fe, Ba, F, Pb, Zn, Cu, and Mo. Their amount in vanadium is fairly variable, ranging from 250 to 2,300 ppm V. Our preliminary data show that V is mainly contained in rutile and illite/muscovite. Both these minerals are associated with (sub-)greenschist facies metamorphism (SE Sardinia "External Nappes"); V may have been released from organic matter during metamorphism and incorporated in both phases. However, because mass balance calculations are insufficient to fully explain the V enrichment in rutile and fillosilicates, it is possible that organic matter has retained at least part of the V.

Mineralogical and geochemical research on V in Black Shales will be extended also to coeval shales in northern Africa.

LCT Pegmatite in the West African Craton: New Insights on the Metallogenic Context and Exploration Approach Based on the Issia District (Ivory Coast)

Christophe Bonnetti[2, 3, 4], Christophe Ballouard[1, 2], Hugo Watine[1], Arnaud Fontaine[2, 3], Anne-Sylvie André-Mayer[1, 2], Karl Akueson[4], Rémi Bosc[2, 3]
1. GeoRessources, Université de Lorraine, CNRS, Vandoeuvre-lès-Nancy, France, 2. LabCom THEIA, Nancy, France, 3. Arethuse Geology, Fuveau, France, 4. Switch Metals, Abidjan, Ivory Coast

The West African Craton has demonstrated to be highly prospective for lithium-cesium-tantalum (LCT) pegmatite, with recent discoveries of large lithium resources in Mali (Goulamina and Bougouni projects), Ghana (Ewoyaa project), and other exploration projects in Ivory Coast (Issia and Atex projects) and Senegal (Saraya project). New insights on the metallogenic context predominantly suggested an anatectic model for the petrogenesis of these LCT pegmatites either derived from metasedimentary or igneous protoliths and controlled by late deformation stages of the Eburnean orogeny (peak at c. 2050 Ma). At the Craton scale, mineral prospectivity analysis and predictivity modeling combining knowledge-driven and data-driven approaches identified a series of key criteria (density of mineral occurrences, age and typology of granite, late Eburnean structural pattern, and specific gravimetric and magnetic signatures) allowing the delineation of prospective areas. At the prospect scale, such as in the Issia area (Ivory Coast), and in contrast to well-exposed LCT pegmatite districts in Namibia and Zimbabwe, indirect geochemical exploration within undercover lateritic environment appeared to be a powerful tool to identify proxies towards LCT pegmatite corridors within prospective metallogenic context highlighted by predictivity modeling. The combination of stream-sediment, soil and rock-chip to trench sampling provided a strong support in targeting fractionated and rare metal-rich pegmatite source rocks, focusing on Li-Cs-Ta-Nb-Be-Rb-Sn-Mn-Ga geostatistical association and geochemical tracers such as K/Rb, Nb/Ta, Zr/Hf, and Rb/Sr ratios. In targeted anomalous areas, systematic mapping and sampling of outcropping pegmatites were preformed along with detailed petrographic description focusing on primary magmatic textures versus magmatic-hydrothermal alterations (albitization, greisenization) that could have positively or negatively impacted the economic potential of the lithium mineralization. This multimethod and multiscale approach has major implications for grassroots exploration of LCT pegmatite elsewhere in Paleoproterozoic terranes, especially where indicators of anatexis of fertile sources are still preserved.

Miocene Western Balkan Lithium-Borates Metallogenic Zone

Sibila Borojevic Sostaric, Tomislav Brenko
University of Zagreb - Faculty of Mining, Geology and Petroleum Engineering, Zagreb, Croatia

The western Balkan lithium-borate metallogenic zone extends over 1,500 km, parallel to the Sava-Vardar and Western Vardar zones, towards its Turkish counterpart, the Ismir-Ankara-Erzincan zone. In the past two decades, a series of new lithium-boron discoveries in the western Balkan countries have significantly enhanced our geological understanding, allowing us to summarise key geological features of this unique metallogenic zone.

1. Lithium-boron mineralization typically occurs in Miocene extensional lacustrine basins, often at depths exceeding 200 meters.

2. These lacustrine sequences comprise fine-grained siliciclastic sediments with intermittent evaporites and volcaniclastic deposits, suggesting proximity to volcanic activity.

3. The basins are typically controlled by normal faults, which likely serve as major fluid pathways.

4. Oligocene-Miocene granitoid rocks in the westernmost area host lithium anomalies and exhibit geochemical characteristics indicative of high-K calc-alkaline-shoshonite magma series.

5. Lithium and boron are likely associated with erosion of lithium and boron-enriched Oligocene-Miocene granitoid rocks in adjacent Miocene lacustrine basins.

6. Geological evidence suggests that adjected Miocene lacustrine basins were reheated during Middle Miocene exhumation, leading to the discharge of lithium and boron from the eroded material to the hydrothermal fluids and into the lake water during the lower and middle Miocene climatic maximum.

Monazite-Rich Carbonatite Dykes at Eureka, Namibia

Sam Broom-Fendley[1], Ed Loye[1], Pete Siegfried[1], Tolene Kruger[2], Chris Drysdale[2], Tim Smalley[2], Rainer Ellmies[3], Frances Wall[1]
1. University of Exeter, Penryn, United Kingdom, 2. E-Tech Resources, Halifax, NS, Canada, 3. Gecko Exploration, Windhoek, Namibia

Rare earth element (REE)-bearing dolomite carbonatites at Eureka Farm 99, Namibia, contain abundant low-Th monazite-(Ce) crystals, which can reach up to 20 cm in length. The locality was first documented in the 1960's, but a detailed understanding of the deposit has been hampered by Quaternary cover, compounded by the tendency of the carbonatites to occur in low, weathered depressions in the landscape. Diamond drilling by E-Tech Resources has shed new light on the genesis of this deposit and demonstrates that the carbonatite dykes consist of an assemblage of three rock types. These comprise a core unit of variably-mineralised dolomite carbonatite, with magnetite, pyrrhotite, monazite, and minor calcite; an outer zone termed "dalmatian carbonatite" consisting of calcite, olivine, serpentine, pyrrhotite, magnetite, and graphite; and a skarn rock, termed "GCARB," consisting of diopside, tremolite-actinolite, pyrrhotite, graphite, and large monazite crystals with locally developed allanite-apatite coronas. All of the carbonatites strike roughly parallel to the fabric developed within intensely deformed schists and quartzites of the Etusis Formation of the Damara Supergroup. Previously published $^{87}Sr/^{86}Sr$ isotope data support a mantle origin, with only minor crustal input, and U-Pb dates of 548 ± 4 Ma suggest a Pan-African related event was important in petrogenesis. It is possible that the monazite-(Ce) may have recrystallized at this time. Based on the contact relationships observed, and the published isotopic data, the deposit is interpreted as a metamorphosed carbonatite, with the surrounding monazite-bearing silicate rocks (GCARB) interpreted as a reaction product between the carbonatites and surrounding siliceous country rocks. This interpretation suggests other metamorphosed carbonatites may be present within Etusis age (and older) rocks of the Damara Supergroup, and further exploration of the Eureka dome, by E-Tech Resources Inc., has uncovered several new localities up to ~3.5 km from the main Eureka deposit composed of similar monazite-bearing calc-silicate rocks.

Rare Earth Element Mineralization in the Dongpao Deposit, Tamduong, North Vietnam

Tinh T. Bui[1, 2], Yasushi Watanabe[1]
1. Akita university, Akita, Japan, 2. Hanoi University of Mining and Geology, Hanoi, Vietnam

Dongpao is a giant rare earth element (REE) deposit in Vietnam, which is enriched by hydrothermal processes; it encompasses vein-type REE mineralization in syenite and carbonatite. The primary REE-bearing mineral in this deposit is bastnäsite, along with minor occurrences of parisite, monazite, and xenotime. REE mineralization took place during the later stages of hydrothermal activity, followed by a sulfide phase. Barite and fluorite are the principal gangue minerals associated with REEs. Ore samples from the carbonatite and vein-type ores exhibit high REE concentrations, ranging from 9,122 to 15,048 ppm and 39,700 to 131,188 ppm, respectively. The geological context and petrological attributes of the Dongpao deposit suggest that two primary factors contributed to REE enrichment: (1) fractures that facilitated the circulation of ore-forming fluids and provided space for REE precipitation and (2) the substantial presence of fluorite and barite in the ore, resulting in elevated concentrations of F^- and SO_4^{2-} ions in the ore-forming fluids, which promoted the transport and deposition of REEs.

Hard Rock-Hosted Lithium Deposits: A Process-Centric Review

Edward Bunker, Jean-Jacques Ferguson, Chris Lambert
CGG, Crawley, United Kingdom

Secure lithium supply is critical to meeting the energy storage demands of the energy transition. Lithium exploration has ramped up over the past decade in response to surging demand and has prompted exploration and development of lithium resources in various geological settings, including granites, pegmatites, clays, salars, and unconventional brine aquifers.

Despite the diversity of environments in which lithium accumulates, salars and hard-rock systems (such as pegmatites) remain the principal sources of commercial lithium concentrate. Hard-rock systems are likely to remain a key source of lithium, as they are widely distributed within permissive jurisdictions to mining. Hard-rock mining also utilises conventional mining and processing techniques, offering relatively low risk to explorers and investors.

Lithium exploration concepts have needed to evolve rapidly in recent years in response to increased interest from explorers. To date, a limited sample set of hard-rock lithium deposits have been studied in detail, many of which were originally investigated with an emphasis on other commodities such as tin. This historical research base, which has underpinned the understanding of hard-rock systems until recently, leads to a risk of bias within geological interpretations and exploration approaches.

Here, we review a range of hard-rock lithium deposits, including those currently being mined and others under development. We present a process-focused analysis of lithium systems that is driven by a mineral systems approach and review each system within this context. We consider three critical processes that lead to lithium concentration within the crust: partial melting, assimilation of crustal material, and fractional crystallisation. All deposits reviewed in this study involve at least one of these processes within their formation, but most represent a hybrid case. This work offers a holistic lens through which to view lithium mineral systems and introduces novel concepts with which to explore hard-rock lithium systems.

Strengthening Li-Bearing Pegmatite Exploration Through Applied Geophysics

Ana M. Carvalho[1, 2, 3], Ricardo Ribeiro[1, 2], Alexandre Lima[1, 2], Rui Moura[3, 4], Klaus Brauch[5], Axel Müller[6, 7]
1. Department of Geosciences, Environment and Spatial Planning, Faculty of Sciences of University of Porto, Porto, Portugal, 2. Institute of Earth Sciences, Faculty of Sciences of University of Porto, Porto, Portugal, 3. INESCTEC Centre for Robotics and Autonomous Systems, Superior Institute of Engineering of Porto, Porto, Portugal, 4. Department of Geosciences, University of Aveiro, University Campus of Santiago, Aveiro, Portugal, 5. Terratec Geophysical Services GmbH & Co. KG, Heitersheim, Germany, 6. Natural History Museum, University of Oslo, Oslo, Norway, 7. Natural History Museum, London, United Kingdom

Lithium is considered a critical raw material by the European Commission. Its prospection will be vital in the coming years. Lithium can be found in minerals such as spodumene or petalite. These minerals appear in hard rocks like pegmatites. These rocks can be found in the Barroso-Alvão pegmatite field (Portugal), which is one of the test sites for the Horizon2020 GREENPEG EU project.

In the frame of GREENPEG, a high-resolution helicopter-borne magnetic and radiometric survey was perfomed from Alijó to Bragado. Radiometric data revealed that across the investigated area, the Th signal shows higher values in the west and lower contents in the east. This observation may correlate with different country rocks; areas of higher Th are represented by siliceous rocks (quartz-phyllites) whereas areas with lower Th are underlain by less siliceous schists. At a smaller scale, negative Th anomalies are superimposing the general Th trend. These correlate with clusters of outcropping pegmatites. In addition, high total radiometric signals (U+Th+K) are observed along the area's two main rivers (Beça and Tâmega).

The interpretation of the airborne geophysical data is still in progress. Future work will include a spatial statistical analysis of the pegmatites, tectonic structures, lithological boundaries, and geochemical mapping results to relate this information to the detected geophysical gradients and anomalies. The final aim is to establish the 3D geological model of the area, considering geological, geochemical, and geophysical data. This 3D model will allow vectoring to areas with highest potential for Li-mineralized pegmatites.

Acknowledgments

This work is funded by the EU Horizon2020 GREENPEG project under the grant agreement no. 869274. This work is supported by National Funds through the FCT, projects UIDB/04683/2020 (doi:10.54499/UIDB/04683/2020) and UIDP/04683/2020 (doi:10.54499/UIDP/04683/2020). Ricardo Ribeiro and Ana Carvalho were financially supported within the compass of their Ph.D., SFRH/BD/140266/2018 and 2022.10315.BD, by national funds through FCT.

Metamorphic Upgrading of Cobalt in a Volcanogenic Massive Sulfide Deposit

Tarryn K. Cawood, Jan M. Peter, Duane C. Petts, Matthew J. Polivchuk
Geological Survey of Canada, Ottawa, ON, Canada

Intra-deposit remobilization of sulfides and their contained metals is well documented in many deformed and metamorphosed massive sulfide deposits. Significantly fewer examples are known where additional metals were added during tectonism. We use a combination of in situ microstructural mapping by electron backscatter diffraction (EBSD) and compositional mapping by laser ablation-inductively coupled plasma-mass spectrometry (LA-ICP-MS) to show that additional cobalt was added to the Windy Craggy Cu-Co-(Au-Ag) volcanogenic massive sulfide deposit, Canada, during deformation and greenschist-facies metamorphism.

At Windy Craggy, cobalt occurs in primary hydrothermal pyrite and pyrrhotite, as well as local metamorphic pyrite overgrowths and syn-deformational cobaltite. Cobalt substitutes directly for iron in pyrite and pyrrhotite, and is thus tightly bound in the crystal lattice. Extensive mechanical remobilization of hydrothermal pyrrhotite (by dislocation creep and dynamic recrystallization) did not affect its cobalt contents, whereas pyrite incorporated additional cobalt during local dynamic recrystallization. As such, no cobalt was lost from major cobalt-bearing phases within the deposit during tectonism, and the additional cobalt required to form the Co-rich metamorphic pyrite overgrowths and cobaltite must come from a source external to the deposit. We propose that the mafic (and possibly ultramafic) rocks that host the deposit provided the cobalt. During formation of the deposit, hot (~350-380°C) hydrothermal fluids leached cobalt from mafic ± ultramafic rocks in the underlying upflow zone, providing the cobalt that was incorporated into the original hydrothermal pyrite and pyrrhotite. During subsequent deformation and metamorphism, which occurred at similar temperatures (~370°C), metamorphic fluids extracted cobalt from a broader region of previously unaltered mafic ± ultramafic rocks. These fluids were channeled along localized faults, and where these structures intersected the Windy Craggy deposit, additional cobalt was added in the form of the metamorphic pyrite overgrowths and discrete cobaltite grains.

On the Physical Transport of Magmatic Sulfides: Insights from the Trace-Element and Isotope Signature of Sulfide Globule Halos

Maria Cherdantseva[1], Marco L. Fiorentini[1], Christopher M. Fisher[1], Antony I. Kemp[1], Laure A. Martin[2], Matvei Aleshin[2], Malcolm P. Roberts[2]
1. Centre of Exploration Targeting, School of Earth Sciences, The University of Western Australia, Perth, WA, Australia, 2. Centre for Microscopy, Characterisation and Analysis, The University of Western Australia, Perth, WA, Australia

The origin of volatile-rich minerals such as calcite, apatite, amphibole, and mica, as well as incompatible element-enriched minerals such as baddeleyite and zircon, which are commonly associated with magmatic sulfides in layered intrusions, mafic-ultramafic intrusions, and komatiites, has been debated for decades. Based on mineralogical and petrological studies, it was recently suggested that these minerals may form due to interaction between three immiscible liquids: silicate, carbonate, and sulfide. This hypothesis was confirmed by experimental data revealing the systematic envelopment of sulfide globules by carbonate melt, indicating their mutual affinity.

However, the source of the volatile-rich carbonate melt involved in this process remains unknown. To address this knowledge gap, we present data on isotopic signatures and trace element distributions of three minerals commonly found in spatial association with sulfides—calcite, apatite, and zircon—in selected samples from three mineralized intrusions. These are the Rudniy intrusion within the Torgalyk Complex surrounding the Tuva depression, Mongolia; the Valmaggia intrusion in the Ivrea Zone, Italy; and the Norilsk-Talnakh camp associated with the Siberian Traps Large Igneous Province, Russia. The rationale for the selection of these deposits is that these intrusions span in age, size, geodynamic setting, metal endowment, and emplacement depth, displaying a comprehensive series of characteristics that can be applied to deposits globally.

Our findings from the three selected localities indicate that both mantle and crustal sources play a role in the formation of these mineral phases, correlating with sulfur isotope systematics in magmatic sulfides. This relationship confirms the shared origin of sulfide, carbonate melts, and fluids during ore-forming processes. The trace element signatures underscore the complexity of apatite and zircon formation, reflecting a mixture of different melts over the intricate evolutionary history of magmatic sulfide systems in different settings and with diverse petrogenetic histories.

The Origin of Stratabound Cobalt Mineralization Along the Mines-R.A.T. Contact at Kinsevere, DRC, Central African Copperbelt

Robert S. Collar[1], Hervé Ilongo[2], Takalani Mandiwana[2], A. E. Williams-Jones[1]
1. McGill University, Montréal, QC, Canada, 2. MMG Ltd, Melbourne, VIC, Australia

The global energy transition requires a variety of metals in ever increasing quantities. One such metal is cobalt, demand for which is increasing rapidly because of its use in electric vehicle batteries. The Central African Copperbelt (CACB), particularly the Democratic Republic of the Congo, supplies more than 70% of the world's cobalt from sediment-hosted Cu-Co deposits. Despite the importance of the CACB to the global cobalt supply, the source of cobalt enrichment is unconstrained, as are the controls on the relative enrichment in copper and cobalt.

In the southern Congolese Copperbelt, one of the largest Cu-Co resources is Kinsevere, which consists of three écaille-hosted deposits. There, the Cu-Co ore is distributed across the Mines Subgroup. This mineralization is Cu rich and is found in veins and, to a lesser extent, disseminated. At the "Central" deposit, a distinct style of Co-rich mineralization occurs within the "Lower Orebody," along the contact between the Mines Subgroup and the underlying Roches Argileuses Talqueses (R.A.T.) Subgroup. This stratabound orebody consists dominantly of disseminated carrollite, with lesser Cu(-Fe) sulfides.

The Co-rich orebody occurs within 10 m of the Mines-R.A.T. contact and, on average, is centered on this feature. The highest Co grades, however, may be found well above and well below the contact and, in both cases, are similar, suggesting that lithology does not exert a strong control on the distribution of the mineralization. Furthermore, preliminary electron microprobe analyses of major and minor elements in carrollite suggest that there may be symmetrical zoning of the mineral across the orebody. These data suggest that the Co-rich stratabound orebody formed as a result of the precipitation of the ore minerals from a hydrothermal fluid flowing along, rather than across, the Mines-R.A.T. contact. The study emphasizes the contribution of late/postdiagenetic processes to the metal endowment of the CACB.

Advances in Exploring for Ni and V Ore Deposits Using Magnetite as an Indicator Mineral

Sarah Dare
Universiteé du Québec à Chicoutimi, Chicoutimi, QC, Canada

Mafic-ultramafic intrusions are important sources of Ni sulfide and V-rich magnetite deposits worldwide. Exploration in the northern hemisphere is challenging due to the thick cover of glacial till. However, indicator minerals, recovered as detrital grains from till, can help vector towards mineralisation. Trace element chemistry better fingerprints the source (deposit type, rock type) of indicator minerals. For magmatic ore deposits, magnetite is a useful indicator mineral because it is common in both mineralisation and host intrusion and easily recovered from till. However, magnetite is common in many other igneous rocks, including felsic rocks.

In this study, the trace element chemistry of magnetite, determined by laser ablation ICP-MS, is reviewed from a variety of Ni and V deposits and intrusions (mafic and felsic) from a range of tectonic settings. Through a better understanding of igneous processes that control trace elements in magnetite, a suite of discrimination diagrams is proposed to assist mineral exploration.

First, igneous magnetite is identified from hydrothermal magnetite using a Ti vs. Ni/Cr diagram. Second, igneous magnetite is divided into mafic and felsic origin using an Mg/V vs. V diagram; mafic magnetite has higher Mg-V than felsic magnetite. Third, the Mg/V vs. V diagram also distinguishes V mineralisation (oxide-rich) from unmineralized (oxide-poor) parts of mafic intrusion based on higher Mg/V. Fourth, mafic magnetite can be used to evaluate the fertility of mafic intrusions, to host Ni sulfide deposits at depth, using Ni depletion on the Ni-Cr diagram. Finally, the chemistry of magnetite from massive Ni sulfide ore is distinct from magnetite from mafic rocks. Therefore, this study advances the use of magnetite as an indicator mineral for the exploration of Ni and V deposits using detrital grains in till, not only in identifying a signature of the ore, but also to widen the footprint to prospective mafic intrusions.

Genesis of IOA and IOCG Deposits Along the Cordillera de la Costa, Peru: Insights from Iron (δ56Fe) and Triple Oxygen (δ18O, Δ17O) Isotopic Analysis of Magnetite

Jose C. Davila[1], Fredrik Sahlström[1], Fernando Tornos[2, 4], Valentin R. Troll[3], John M. Hanchar[4], Ilya Bindeman[5], Xin-Yuan Zheng[6]
1. Pontificia Universidad Católica del Perú, Lima, Peru, 2. Consejo Superior de Investigaciones Científicas, Madrid, Spain, 3. Uppsala University, Uppsala, Sweden, 4. Memorial University of Newfoundland, St. John's, NL, Canada, 5. University of Oregon, Eugene, OR, USA, 6. University of Minnesota – Twin Cities, Minneapolis, MN, USA

Iron oxide-apatite (IOA) and iron oxide-copper-gold (IOCG) deposits represent significant metal resources needed for the green energy transition. However, their genesis and possible relationships between these deposit types remain poorly understood and are a matter of intense debate. The Cordillera de la Costa of southern and central Peru hosts several IOA and IOCG deposits that have been comparatively little studied in a global context, and they may thus be able to provide new insights into the genesis of these enigmatic but economically important ores. We report new iron ($\delta^{56}Fe$) and triple oxygen ($\delta^{18}O$, $\Delta^{17}O$) isotope data for 23 magnetite samples collected from two IOA (Marcona, Hierro Acari) and two IOCG (Mina Justa, Raul-Condestable) deposits in Peru as well as from the nearby Pampa del Pongo iron skarn deposit. Petrographic characterization of magnetite-bearing samples from Mina Justa and Raul-Condestable was also done to provide a textural context for the isotope data. The measured magnetite isotopes ranges are $\delta^{18}O$ = 0.17 – 5.30 ‰ and $\delta^{56}Fe$ = -0.06 – 0.49 ‰. The values from all five deposits essentially overlap with the ranges determined for ortho-magmatic magnetite, thus indicating a predominantly magmatic origin. Nevertheless, the magnetite $\Delta^{17}O$ values (-0.097 – -0.016 ‰) are in part heavier and in part lighter than typical magmatic values. This suggests local alteration by low-temperature fluids and input from evaporites, respectively. Further analysis of our new isotope data will provide new insights into the genesis of and relationships between IOA, IOCG, and iron skarn deposits in Peru. Our insights will assist in the development of novel exploration strategies for these important metal resources in Peru and likely also on a global level.

Geochemical Characterization of K-Feldspar in Lithium-Rich Aplite-Pegmatites: Insights from the Barroso-Alvão Region, Northern Portugal

Filipa Dias[1, 3], Ricardo Ribeiro[1, 3], Filipe Gonçalves[1, 3], Alexandre Lima[1, 3], Encarnación Roda-Robles[2], Tânia Martins[4]
1. Universidade do Porto - Faculdade de Ciências, Porto, Portugal, 2. Dpto. Geología, Universidad del Pais Vasco/EHU, Bilbao, Spain, 3. Institute of Earth Sciences, Pole of University of Porto, Porto, Portugal, 4. Manitoba Geological Survey, Winnipeg, MB, Canada

Given the rising demand for lithium (Li) in the context of the energy transition towards electric vehicles, the identification and characterization of Li-rich aplite-pegmatites holds significant importance for sustainable resource development and the advancement of electric mobility. In the Barroso-Alvão region of northern Portugal, aplite-pegmatites enriched in Li constitute significant deposits of spodumene minerals, representing potential valuable economic resources. These igneous formations were emplaced during the Variscan Orogeny (Early Devonian to the Early Permian), and intruded Silurian metasediments. This study focused on the geochemical analysis of K-feldspar crystals collected from various aplite-pegmatites across the Barroso-Alvão aplite-pegmatite field. Using ICP-MS analysis, this research focused on the minor and trace elements of these K-feldspar crystals, offering new insights into the geological evolution of the region. The findings not only elucidate the spatial distribution of the aplite-pegmatites in relation to their Li mineralization but also provide crucial insights for differentiating between various types of aplite-pegmatites. Notably, this study establishes key chemical elements and their concentrations within K-feldspars, facilitating the characterization of the types of aplite-pegmatites in the Barroso-Alvão region: namely, Li-barren, spodumene-rich, and petalite-rich aplite-pegmatites. The implications of these results can be extended to mineral exploration, guiding future Li exploration campaigns.

This study was financially supported by FCT, I.P., in the framework of the ICT (doi: 10.54499/UIDB/04683/2020 and doi: 10.54499/UIDP/04683/2020). Filipa Dias and Ricardo Ribeiro were financially supported by their Ph.D. projects (doi: 10.54499/2020.05534.BD; SFRH/BD/140266/2018) and by national funds from MCTES, through FCT, co-financed by ESF through POCH and NORTE 2020.

Effects of Post-Mineralization Metamorphism and Deformation on Sulfides and their Remobilization in the Hotinvaara Ni Prospect, Pulju Greenstone Belt, Northern Finland

Juan C. Díaz Carreño[1], Petri Peltonen[1], Jaakko Puntalo[1], Christoph Beier[1], Kathleen A. Law[1], Jake Williams[2]
1. University of Helsinki, Helsinki, Finland, 2. Nordic Nickel Ltd., Perth, WA, Australia

The Pulju Greenstone Belt (PGB), located on the northwesternmost edge of the Central Lapland Greenstone Belt in Finland, bears significant potential for Ni sulfide deposits. The Paleoproterozoic komatiites of the Mertavaara Formation host the Hotinvaara Ni prospect, which is the most economically promising mineralization to date within the PGB. The Mineral Resource Estimate (MRE) completed for the Hotinvaara Ni-prospect by Nordic Nickel Ltd. indicates 418 Mt @ 0.21% Ni, 0.01% Co, and 53ppm Cu for 862,800 t of contained Ni, 40,000 t of contained Co, and 22,100 t of contained Cu (Nordic Nickel Ltd. 2024). The Hotinvaara mineralization occurs in MgO-rich olivine cumulates as moderately to weakly interconnected sulfide dissemination and as larger pentlandite-pyrrhotite blebs. In addition, Ni sulfides occur locally as massive, 10's-of-cm- to m-scale sulfide veins containing up to 9.61% Ni, 0.17% Cu, and 0.36% Co. Chalcopyrite, pyrite, mackinawite, violarite, and cobaltite-gersdorffite are present in subordinate amounts (<5%). All mineralization types have been subject to multistage deformation events (D_1-D_4) at lower-amphibolite facies metamorphic conditions, which led to the development of deformation textures, recrystallization, and annealing of sulfides as well as paragenetic relationships indicating remobilization of sulfides. On average, the sulfide phase contains 12.5 wt.% Ni and 0.57 wt.% Co, which is reflected in the sulfide mineralogy, as Co-bearing pentlandite is a ubiquitous sulfide in Hotinvaara. Nickel contents of the sulfide phases are extremely variable (1-36 wt.%), even between closely spaced samples, and this most likely reflects the effect of post-mineralization metamorphism that has resulted in the loss of S. Electron probe microanalyser (EPMA) data implies that Co is strongly partitioned into pentlandite (0.19-5.42 wt.%) and trace cobaltite-gersdorffite. LA-ICP-MS analyses will be conducted to examine the trace element (e.g., As and Sb) content in sulfides, in order to further constrain their compositional changes during post-mineralization thermal history.

SEG 2024 Conference: Sustainable Mineral Exploration and Development

Multi-Scale Characterisation of Cobaltite at Palokas, Rajapalot by X-Ray Computed Tomography and Electron Microscopy

Lily Dickson[1], Taija Torvela[1], Robert Chapman[1], Jukka-Pekka Ranta[2], Alice Macente[1]
1. University of Leeds, Leeds, United Kingdom, 2. University of Oulu, Oulu, Finland

Rajapalot is a Paleoproterozoic hydrothermal Au-Co deposit near Rovaniemi, northern Finland, with a cobalt mineralogy comprising cobaltite, linnaeite, cobalt pentlandite, and cobaltian pyrite. Cobaltite is the main cobalt mineral and is therefore the most important mineral for cobalt extraction. However, at present, the deposit model for the cobalt mineralisation at Rajapalot and details of the ore paragenesis remain unclear. In a geometallurgical study, the cobaltite at Rajapalot is described to have a consistent grain size of 23-28 μm, but initial results of this study reveal greater complexity. The objective of this study is detailed characterisation of cobaltite at Rajapalot using optical and scanning electron microscopy (SEM) combined with X-ray computed tomography (XCT). A combined 2D and 3D approach facilitates a rigorous characterisation of mineralisation; precise characterisation of the cobaltite will inform the paragenetic sequence for Rajapalot Au-Co and processes of remobilisation and deposition of cobalt.

Detailed petrography and analysis by SEM have revealed four styles of cobaltite mineralisation, with differing apparent grain sizes, morphologies, and mineral associations. Small (<5 mm) fragments of cobaltite rich ore were scanned by XCT. The high attenuation coefficient of cobaltite allows segmentation from pyrite and pyrrhotite, gangue minerals, and denser phases such as gold. The XCT scans show the three larger fractions of cobaltite and their 3D morphologies.

XCT analysis provides quantitative non-destructive volumetric data, while correlative 2D petrographic study by reflected light microscopy and SEM confirms the mineral species present in the sample and provides high resolution detail of mineral inclusions and microstructures. The data from this study highlights the added value and higher accuracy in ore characterisation through addition of 3D data analysis, especially for polyphase mineralisation, such as Rajapalot Au-Co.

SEG 2024 Conference: Sustainable Mineral Exploration and Development

Lithium and Graphite Mineralisation of the Kuldjuktau Mts., Uzbekistan

Alla Dolgopolova[1], Reimar Seltmann[1], Rustam Mirkamalov[2], Farid Divaev[2], Rustam Khalmatov[3], Rustam Uzokov[2], Yulia Mun[4], Nigora Isokova[5]
1. Natural History Museum, London, United Kingdom, 2. Institute of Mineral Resources, Tashkent, Uzbekistan, 3. Institute of Geology and Geophysics, Tashkent, Uzbekistan, 4. The Arctic University of Norway, Tromsø, Norway, 5. Centre for Advanced Technologies, Tashkent, Uzbekistan

Uzbekistan has vast mineral potential and is among the world leaders for production of Au, U, Cu, and base metals. Currently, critical raw materials are extracted incidentally as by-products or by small-scale mining. The territory of Uzbekistan is located at the junction of three continental blocks (Kazakhstan, Afghan-Karakum, and Ustyurt), separated by a fold-thrust belt formed at the closure site between Turkestan and Zeravshan oceans. Here we assess metallogeny of the Kuldjuktau Mts., which are the westernmost spur of the Zeravshan Ridge.

Dominant mineralisation types include gold, rare metals (Li), graphite. Rare metal (Li) mineralisation is linked to pegmatite and aplite veins. The Aktosty lithium occurrence is associated with lepidolite-spodumene pegmatites cutting through the gabbroids of the Shaidaraz intrusion. The Aktosty-Shaidaraz pegmatite field stretches SE-NW for 6 km. There are two mineralogical types of pegmatite bodies: oligoclase-microcline-tourmaline (predominant) and quartz-feldspar-lepidolite-spodumene. Analyses of lepidolite samples showed following concentrations (%): Li_2O – 1.95; Rb_2O – 1.0; Cs_2O – 0.094; Ta_2O_5 – 0.006; Nb_2O_5 – 0.014. The Taskazgan graphite deposit is represented by gabbro, norites, harzburgites, and peridotites. Along the contact of gabbroids with host limestones, a 10- to 15-m-thick graphite-bearing melange zone develops. Drilling has established that the gabbro massif can be traced to a 500 m depth. Graphitisation is superimposed due to carbon-containing (CO_2 and CH_4) mantle fluids penetrating along deep fault zones and depositing in favourable structures. The carbon isotopic composition of carbonized gabbro (δC^{13} = -16.39‰) indicates mantle origin.

Acknowledgments

This publication has been produced within the framework of the grant “Resourcing low-carbon technologies for green economy of Uzbekistan” (REP-24112021/70), funded under the MUNIS Project, supported by the World Bank and the government of the Republic of Uzbekistan. The statements do not necessarily reflect the official position of the World Bank and the government of the Republic of Uzbekistan.

Punta Corna Co-rich Vein System (Italy): Late-Alpine Hydrothermal Circulation and Metallogenesis

Giulia Domenighini
università di Torino, Torino, Italy

The Punta Corna Fe-Co-Ni-Ag-Bi deposit is located in the Lanzo Valleys, Piedmont (Italy). It consists of five-element mineralization exploited for Fe and Co until the 19th century and then abandoned. The global concern for strategic raw material supply, such as Co, has raised a renewed interest in old mining sites where economic potential was not fully investigated.

Hydrothermal activity started with silicification and sericitization of the host rock, followed by the precipitation of Fe-rich carbonates (siderite and ankerite), calcite, quartz, and baryte. This first stage concludes with the deposition of base metal sulfides (pyrite, chalcopyrite, galena). The Fe-Co-Ni arsenides comprise nickeline, overgrown by safflorite, then euhedral skutterudite, and final rammelsbergite in siderite and quartz gangue. Native Bi and As may be recognized in the core of skutterudite and safflorite (Moroni et al., 2019). The final hydrothermal stage is characterized by sulfides and sulfosalts such as tetrahedrite, chalcopyrite, sphalerite, galena, pyrite, and bournonite. This stage, at the very end, also includes Bi-Sb sulfosalts such as the emplectite-chalcostibite solid solution, horobetsuite and wittichenite in siderite, quartz, ankerite, and baryte.

Fluid inclusion studies employing microthermometry, cathodoluminescence, and micro-Raman spectroscopy revealed the presence of different fluid types, suggesting a mixing process as the cause of ore precipitation. These fluids exhibit distinct temperatures, salinities, and composition, indicating diverse provenance sources. The evolution of fluids associated with Stage I mineralization and those linked to the arsenides align with existing models, stating that the transition from base metal sulfides to Fe-Co-Ni arsenides is triggered by a process of rapid reduction caused by the oxidation of a reducing agent (Markl, 2016). The presence of Bi-Sb sulfosalts, appearing in the late stages of the mineralization, may represent a connection to the hydrothermal Au-Bi-Sb veins in the nearby Gran Paradiso Massif (Cevales et al., 1961).

Constraints on the Genesis of the Orbicular Rocks and Sulphide Mineralization in the Koperberg Suite, Namaqualand Metamorphic Complex, South Africa

Senamile S. Dumisa[1], Grant M. Bybee[1], Paul A. Nex[1], Ayesha B. Jogee[1], Ayesha B. Jogee[1]
1. University of the Witwatersrand, Johannesburg, South Africa, 2. University of Johannesbug, Johannesburg, South Africa

The 1020-1060 Ma Koperberg Suite (KS) rocks contain zones with orbicular textures, thought to form by magmatic, metasomatic, or metamorphic processes. Some orbicular zones are also associated with Cu mineralization, suggesting a possible link between orbicule formation and metallogenesis. This study documents a selection of orbicular localities from the KS to understand their genesis using petrography, in-situ mineral chemistry, and isotopic data. Four different orbicule localities are described including Orbicule Koppie (OK), Henderson North (HN), Henderson South (HS), and Hoogskraal Lease (HL).

Orbicules occur in a spectrum of lithologies of the KS, ranging from granite to diorite. The orbicules are characterized by coarse-grained (2-6 mm) felsic cores. Sharp contacts mark the transition from cores to the fine-grained (0.2-1 mm) shells that exhibit polygonal textures. Sharp boundaries mark the transition from core to shells and coarse-grained matrices. Some of the orbicule localities contain deformed orbicules trending parallel to steep structures. Plagioclase in the HN (An_{47-65}) and HS (An_{52-62}) orbicules are more calcic than plagioclase at HL (An_{35-52}) and OK (An_{41-57}). Biotite in the HN (x = Mg#77) and the OK (x = Mg#75) orbicules are more magnesian than those at HS (x = Mg#62) and HL (x= Mg#68) orbicules. Plagioclase in the HN orbicules are characterized by elevated initial $^{87}Sr/^{86}Sr$ ratios (I_{Sr}) (0.709675-0.723517) relative to other orbicular localities (0.705608-0.724905).

A variety of textural and geochemical constraints rules out a metasomatic origin. These include chemical similarity between orbicules and the magmatic matrices, as well as a variety of synmagmatic textures. This, along with isotopic data, indicate that distinct pulses of orbicular-forming magmas make up the KS, where variable amounts of magma mixing, mingling, and assimilation occurred. These differentiation processes possibly took place during magma ascent and emplacement, possibly triggering the ideal conditions for orbicular rock and sulphide formation.

Vanadium Distribution in Soil of Okahandja Map Sheet (2117 DA – DD) in Otjozondjupa Region, Namibia

Wilka Embula
Geological Survey of Namibia, Ministry of Mines and Energy, Windhoek, Namibia

The 2116-Okahandja Map Sheet (1:250,000) was sampled in 2003-2004 as part of the Regional Geochemical Sampling Project. A total of 2,700 soil and stream sediment samples were collected on 32 small-scaled map sheets at 1:50,000.This study was undertaken on four of these map sheets, namely 2117 DA-DD, comprising 192 soil samples. Sampling was done at a density of 1 sample/10 km^2, with a composite sample obtained from a minimum of five sample pits at a depth of 25 cm and screened through for coarse fraction (<2 mm) and fine fraction (<180 µm) samples. Wooden and plastic digging tools were utilized to avoid contamination during the survey. Concentrations of elements including vanadium were determined using the handheld Niton energy dispersive X-ray fluorescence spectrometer for 150 seconds at the Geological Survey of Namibia laboratory. A homogenized portion of each 180-µm fine fraction sample was milled to <64-µm size before analysis, and QAQC duplicates and standards were routinely inserted into batches every 20th interval.

Spatial distribution patterns of vanadium were generated using ESRI ArcGIS software and interpreted in conjunction with geological maps and sample site description data recorded. Natural and anthropogenic activities such as rock weathering, soil erosion, and agrochemical usage contributes to vanadium content in soil. Vanadium in this area ranges from 57.67 to 243.49 ppm, where the presence of amphiboles, gneiss, granites, pegmatites, marble, magnetite, quartzites, and schists played a role. High V concentrations were observed in association with structures such as basins and fold hinges, whereas low V values are concentrated within the domes. The mean (115.07 ppm) and median (111.12 ppm) were found to be higher than the world soil average (90 ppm) and crustal abundance (97 ppm).

Characterization of Fluids Related with Alkalic Porphyry Deposits in British Columbia, Canada

María Paula Estrella, Pilar Lecumberri-Sanchez, Xinyue Xu
University of Alberta, Edmonton, AB, Canada

Alkalic porphyries are Cu-Au high-grade deposits also known to host important PGE resources in localized metallogenic provinces such as those in British Columbia (BC) and eastern Australia. Alkalic deposits share some similarities with their calc-alkalic counterparts but also clear differences in terms of metal budget, timing with respect to subduction, vein composition, and alteration mineralogy.

This project aims to (1) identify the likely source of the mineralizing fluids and (2) characterize the physicochemical properties of fluids as a function of their distance to likely sources for some of the most alkalic porphyry deposits in British Columbia .

Mount Polley and Galore Creek are two large Late Triassic-Early Jurassic alkalic porphyries in BC, in which mineralization occurs as Cu-Au(-Ag) ore associated with hydrothermal-breccia bodies and coeval volcanic rocks. In the Mount Polley deposit, unaltered diopside phenocrysts in the alkalic volcanic units host crystallized melt inclusions consisting of sulfate, carbonate, silicate, and salt (?) phases. Apatite crystals within extensively altered and mineralized zones host aqueous fluid inclusions that are dominantly vapor-rich or liquid-rich. From preliminary observations in Galore Creek, fluid inclusion types similar to those at Mount Polley are expected, but hosted in diopside, anhydrite, and apatite crystals within the mineralized and altered regions of the deposit

Sulphide Mineralization and Associated Platinum Group Elements in Mafic-Ultramafic Units of the Tantalite Valley Complex, Southern Namibia

Daniel Ferreira, Bjorn von der Heyden, Martin Klausen
Stellenbosch University, Stellenbosch, South Africa

Magmatic sulphide deposits are an important source of base-, semi-, and precious metals such as platinum group elements (PGEs), nickel, copper, and cobalt. These elements are crucial in facilitating the green energy transition, and finding more of these deposits is to be considered of critical importance. The study investigates the Tantalite Valley Complex in southern Namibia and evaluates the mineralization potential of this complex and by extension the Kum Kum Suite. This complex is geologically located within the Pofadder Shear Zone in the Richtersveld Subprovince in the Namaqua sector of the Namaqua Natal Metamorphic Province.

Research to date has determined that sulphur within the complex is derived from the mantle, characterized the sulphide assemblages and their trace element makeup, and determined an R-factor of roughly 3,000, suggestive of moderate degrees of sulfide/silicate melt interaction. With the present data set, it is speculated that sulphide immiscibility was brought about due to high degrees of crystal fractionation, with sulphide melt preferentially concentrating within an orthopyroxenite layer in the southeastern sector of the complex. This study reports for the first time the occurrence of PGEs and platinum group minerals within the sulphide assemblages belonging to the Tantalite Valley Complex, highlighting the prospectivity of the region for this deposit type and the need for further research and exploration in a relatively underexplored mafic magmatic suite.

The Role of Carbonate Melts in the Transport and Concentration of Metal-Rich Magmatic Sulfides

Marco Fiorentini[1], Maria Cherdantseva[1], Isra Ezad[2], Michael Anenburg[3], John Mavrogenes[3], Stephen Foley[2]
1. University of Western Australia, Perth, WA, Australia, 2. Macquarie University, Sydney, NSW, Australia, 3. Australian National University, Canberra, ACT, Australia

Sulfur and metal mobilization during mantle metasomatism and magmatic transport of dense, metal-rich sulfide liquid have been intractable problems for decades. A better understanding of how these processes operate at multiple scales may provide insights into the genesis of critical resources such as nickel, copper, and platinum group elements. We investigated experimentally the role of incipient carbonate melts in mobilizing sulfur and metals at mantle conditions. Results show that carbonate melts can sequester sulfur in its oxidized form as sulfate, as well as base and precious metals from mantle lithologies as sulfide. Accordingly, incipient carbonate sulfur-rich melts may play a first-order role in the metallogenic enhancement of localized lithospheric domains.

We subsequently addressed the problem of sulfide transport in magmas derived from partial melting of variably metasomatized mantle domains. Whereas recent studies emphasize adherence of volatiles to sulfide liquids to lower their densities and favor transport across extensive magmatic plumbing networks, we show that the presence of volatiles alone cannot generate the mineral assemblages documented in natural rocks. Experimental results indicate that carbonate melt wrapping around sulfide globules is an alternative process to the previously proposed gas bubble to aid dense sulfide liquid on its journey across the lithosphere.

We conclude that carbonate melts act as effective agents to dissolve, redistribute, and concentrate metals within discrete domains of the mantle and into shallower regions of the Earth, where dynamic physicochemical processes can lead to ore genesis at various crustal depths. The recognition of a greater role for carbonate in enhancing physical processes in the transport and concentration of metal-rich sulfide liquid may lead to a paradigm change in exploration for magmatic sulfide deposits, as a series of targeting proxies related for example to crustal contamination that we currently utilize may not be essential in ore genesis.

Determining Uranium Mineral Compositions by Electron Probe Microanalysis

Guy G. Freemantle
University of the Witwatersrand, Randburg, South Africa

Uraninite is one of the most abundant uranium ore minerals. It occurs in west central Namibia, in the world's largest mineable granite-hosted deposits. It is the primary ore mineral in late- to post-orogenic sheeted leucogranites (SLGs) that intruded in multiple generations into structurally prepared granulite-facies Neoproterozoic metasediments. Contemporary interpretations for the ore genesis favour models of oxidised, low-%, partial melt of weakly uraniferous source material intruding and reacting with reducing units of the metasedimentary sequence. The primary driver for uraninite crystallisation within the SLGs is redox buffering of dissolved U^{6+}-complexes by reductants in the reactive Damaran metasediments. Resultant insoluble U^{4+} has combined with available impurities and U^{6+} to bond with O^{2-} and crystallise impure uraninite. The uranium repositories in the source rocks likely included common accessory minerals such as zircon, monazite, biotite, and apatite that retain or release U, Th, lanthanides, and other incompatible elements into the silicate melt and magmatic fluids. Variable proportions of the uranium-bearing refractory minerals betafite and brannerite reflect diverse local conditions within deposits during and subsequent to ore genesis.

Future Lithium Supply: Towards a Robust Quantitative Assessment

Laura Buarque Andrade[1], Max Frenzel[1], Carolin Kresse[2], Britta Bookhagen[2], Michael Schmidt[2], Dirk Sandmann[3]
1. Helmholtz Institute Freiberg for Resource Technology, Freiberg, Germany, 2. Deutsche Rohstoffagentur (DERA) in der Bundesanstalt für Geowissenschaften und Rohstoffe (BGR), Berlin, Germany, 3. ERZLABOR Advanced Solutions GmbH, Freiberg, Germany

There has recently been considerable controversy over whether primary lithium (Li) sources will be able to supply the rapidly growing needs of the electric mobility transition. Unfortunately, existing assessments of future supply are hampered by a lack of consideration for the decision-making processes occurring at the level of individual mining projects and how these respond dynamically to changing market conditions. In the present contribution, we demonstrate how these processes, as well as associated uncertainties, can be incorporated into an assessment of the likely future evolution of global primary Li supply, including probable price trends.

Our method uses Monte-Carlo simulations to achieve this goal. A global database of existing Li mining projects (all development stages, including case histories) is used to build models that can estimate the likelihood of each project proceeding to the next development stage in any given year, depending on specific project characteristics such as location, deposit type, and ore grade, as well as market conditions. A series of stochastic simulations (1,000) are then run, in which projects are moved through the development pipeline according to these estimated likelihoods, year-by-year, up to 2050. New discoveries are also included to achieve realistic results over the relatively long timespan covered. A few simple demand scenarios (low, intermediate, high) are used as fixed inputs to these simulations to allow for easier interpretation.

Results show increasing divergence between demand and supply for demand scenarios incorporating more rapid growth. Some of this divergence may be offset by increasing recycling contributions to overall supply or through demand destruction as Li prices increase due to the divergence.

In Situ Geochemistry of Mica in Exomorphic Spodumene Pegmatite Halos: Implications for Mineral Exploration

Elena J. Geiger[1, 2], Julian F. Menuge[1, 2], Teimoor Nazari-Dehkordi[1], Maurice Brodbeck[1]
1. University College Dublin (UCD), Dublin, Ireland, 2. SFI Research Centre in Applied Geosciences (iCRAG), Dublin, Ireland

Spodumene pegmatites intrude peraluminous granite of the Ordovician-Devonian Leinster Batholith and its immediate mica schist country rocks within the East Carlow Deformation Zone in South Leinster, Ireland. Pegmatites are often surrounded by exomorphic metasomatic halos formed by magmatic hydrothermal fluids expelled from crystallizing pegmatite. Some halos are detectable by whole-rock analysis 10s to 100s of metres from pegmatite contacts, making them useful exploration vectors, but others are much more spatially restricted. Micas are ideal for monitoring hydrothermal processes related to granitic pegmatite systems, as they can incorporate a wide range of halo elements. This work investigates the viability of using mica geochemistry in restricted halos surrounding spodumene pegmatites in South Leinster to spatially extend halo detectability and to better understand pegmatite crystallization.

White mica is the main metasomatic phase identified in the Leinster halos. It was characterized by quantitative SEM analysis and LA-ICP-MS chemical mapping and spot analysis. The compositional series ranges from muscovite in the unaltered granite to lithian ferroan muscovite in the halo. Intermediate compositions are found within structures interpreted as closed fractures several meters away from the pegmatite. Although white mica appears homogeneous, LA-ICP-MS mapping reveals two distinct types with complex zoning patterns. The first is proposed to represent altered original muscovite with a pristine core zone; the second newly formed mica with core zones enriched in Cs, Li, and Rb. Whereas Cs and Li concentrations decrease with distance, Ta and Nb are concentrated in distant halo mica. These observations have implications for exploration purposes. Methods like LIBS, able to measure element concentrations or ratios in individual crystals, can use the abundances of Li, Cs, Ta, and Nb to vector towards nearby pegmatite. However, the compositional heterogeneity of the mica could cause a problem if depleted core zones are analysed and mistakenly classified as background.

Trace Elements in Alteration Minerals: A Case Study from the Kandilka Prospect, Eastern Rhodopes, SE Bulgaria

Yana Y. Georgieva[1], Rossitsa Vasileva[2], Istvan Marton[1], Stefan Metodiev[1]
1. Dundee Precious Metals, Sofia, Bulgaria, 2. Geological Institute, Bulgarian Academy of Sciences, Sofia, Bulgaria

The Kandilka occurrence is situated 5 km westwards from the Ada Tepe open-pit gold mine in Eastern Rhodopes, SE Bulgaria. The mineralization style is alike the adjacent Surnak deposit, which is closely related to major N-S-striking sub-vertical structures and shallow-dipping to the NNE listric faults. The sulphide mineralization is hosted by the pre-Alpine basement rocks of the Rhodope Massif, presented in this part of low-grade metamorphic rocks (paragneiss, amphibolite, and marble) and in the overlying Paleocene-Eocene sedimentary rocks (polygeneous breccia, conglomerate, and sandstone).

The current study is focused on the quartz-pyrite ore-bearing veins, which crosscut the foliation of the metamorphic rocks and are spread among the sedimentary rocks. Field mapping and sampling of drill cores identified mineralized veins with elevated Au (0÷0.7 ppm), Ag (0÷10.5 ppm), As (0÷3020 ppm), and Sb (0÷147 ppm). Detailed petrography, SEM, and LA-ICP-MS analyses of the quartz-pyrite veins revealed two generations of pyrite (Py1, Py2). On surface, Qz-Py1 veins are presented by collomorph-banded texture, milky white quartz, and euhedral pyrite, while Qz-Py2 veins are mainly brecciated, with grey opaque quartz and fine-grained pyrite-marcasite aggregates. In Py1 have been detected higher concentrations of Co and Ni, up to 2%, and Cu ~0.8%, while in the carbonate-silicate matrix (Ca-Fe-Mg carbonates, K-feldspar, plagioclase) of the sediments, Cu is ~200 ppm and Co and Ni enhance to ~200 and ~600 ppm, respectively. Py2 is considered as arsenian pyrite; it has approximately 4% of As and Sb, ~300 ppm of Ag, and traces of Au.

The presence of collomorph microcrystalline quartz in the veins suggests episodes of high degrees of supersaturation of silica. The mineral association of marcasite and arsenian pyrite relates to precipitating conditions of acidic solution. Based on textural peculiarities, mineralogical paragenesis, and trace element composition, we consider the Kandilka occurrence as a low-sulphidation epithermal prospect.

Mineralogy and Deportment of Nickel, Cobalt, and Scandium in the Laterite Deposits of Sebuku Island (SE Kalimantan, Indonesia)

Michele Giorno[1], Alkiviadis Kontonikas-Charos[2, 1], Doreen Ebert[1], Joachim Krause[1], Robert Möckel[1], Dirk Sandmann[3], Ernowo Ernowo[4], Agata V. Kindangen[5], Yoseph Swamidharma[6], Yudhi Krisnanto[6], Max Frenzel[1]
1. Helmholtz-Zentrum Dresden-Rossendorf, Helmholtz Institute Freiberg for Resource Technology, Freiberg, Germany, 2. School of Earth and Atmospheric Sciences, Queensland University of Technology, Brisbane, QLD, Australia, 3. ERZLABOR Advanced Solutions GmbH, Freiberg, Germany, 4. National Research and Innovation Agency, Bandung, Indonesia, 5. Geological Agency of the Ministry of Energy and Mineral Resources, Bandung, Indonesia, 6. Sebuku Iron Lateritic Ore, Sebuku, Indonesia

As nickel (Ni), cobalt (Co), and scandium (Sc) play crucial roles in driving essential technologies for sustainable energy and transportation, Indonesian laterites have emerged as prominent sources or potential sources for these metals. In the present case study, new geological and mineralogical data on the laterites from Sebuku Island (SE Kalimantan, Indonesia) are presented.

The Sebuku ore deposits extend over an area of more than 30 km^2 and contain a JORC-compliant resource of ~390 Mt at 42.5 wt.% Fe, 0.9 wt.% Ni, and 0.15 wt.% Co. Orebodies are mostly limonitic and oxide-dominant. They were formed by the weathering of Jurassic-Cretaceous ophiolites. Although mining has been underway since 2006 (primarily for iron ore), limited mineralogical and geochemical data exist, hindering the optimization of beneficiation processes and the efficient recovery of Ni, Co, and Sc.

Typical laterite profiles at Sebuku include a weathered bedrock composed of serpentinized dunites and harzburgites, overlain by a saprolite zone ranging from 0.2 to 7 meters thick, a yellow limonite zone spanning 2 to 8.5 meters, and a red-limonite zone measuring 1 to 3.5 meters thick. These soil horizons exhibit complex relationships amongst each other, characterized by irregular boundaries, lenticular interbeds, and lateral variations.

X-ray diffraction (XRD) and mineral liberation analysis (MLA) data reveal a complex ore mineralogy, encompassing oxides and oxi-hydroxides such as goethite, gibbsite/bayerite, chromite, various iron oxides (maghemite, hematite, magnetite), manganese minerals (asbolane and lithiophorite), and silicates comprising serpentine/talc, chlorite, pyroxene, garnet, quartz, and smectites. Combined MLA, electron probe microanalysis (EPMA), and X-ray fluorence (XRF) allowed us to determine the deportment of Ni, Co, and Sc in those minerals. Based on these new findings, potential beneficiation strategies are being developed, with the ultimate goal of enhancing the efficiency of the existing processing flowchart and recovering strategic metal by-products (i.e., Sc).

Advancing Greener Economy: Insights from Studies on Bauxite Deposits in Bosnia-Herzegovina (Jajce and Posušje) and Porphyry Copper System in Bulgaria (Assarel)

Marko Holma[1, 2, 3], Šime Bilić[4], Ivica Pavičić[4], Franjo Šumanovac[4], Gordana Deljak[5], Ivan Grubišić[6], Irena Peytcheva[7], Stoyan Georgiev[7], Elitsa Stefanova[7], Dimitrina Dimitrova[7], Atanas Hikov[7], Desislav Ivanov[8], Ivaylo Vasilev[8], Pasi Kuusiniemi[14, 3], Timo Enqvist[14, 3], Gergely Surányi[15], László Bálazs[15], Sandór Demők[15], Jarmo Korteniemi[16, 3], Giulio Casini[9], Eduard Saura[9, 11], David Marti[9, 12], Helena Seivane[13], Martin Schimmel[13], Markku Pirttijärvi[10]

1. Kerttu Saalasti Institute, University of Oulu, Oulu, Finland, 2. Muon Solutions Oy Ltd, Rovaniemi, Finland, 3. International Virtual Muography Institute, Tokyo, Japan, 4. University of Zagreb, Faculty of Mining, Geology and Petroleum Engineering, Zagreb, Croatia, 5. Rudnici boksita Jajce d.d., Jajce, Bosnia and Herzegovina, 6. Rudnici boksita d.o.o. Posušje, Posušje, Bosnia and Herzegovina, 7. Geological Institute, Bulgarian Academy of Sciences, Sofia, Bulgaria, 8. Assarel-Medet JSC, Panagyurishte, Bulgaria, 9. Lithica SCCL, Girona, Spain, 10. Radai Oy Ltd, Oulu, Finland, 11. Universitat Autonoma, Barcelona, Spain, 12. IGME, CSIC, Madrid, Spain, 13. Geo3BCN, CSIC, Barcelona, Spain, 14. Muon Solutions Oy Ltd, Pyhäjärvi, Finland, 15. Muon Solutions Oy Ltd, Budapest, Hungary, 16. Muon Solutions Oy Ltd, Turku, Finland

A transition towards sustainable energy sources requires the exploration and development of minerals critical to electronics and energy storage and delivery systems. Funded under Horizon Europe (Grant Agreement N°101058178), the AGEMERA project is a three-year initiative addressing critical raw materials (CRMs), including their exploration and extraction. The conventional methods applied in the project encompass structural geological fieldwork and modeling, mineralogical studies, trace element geochemistry, and geochronology. The innovative exploration methods include drone geophysics, passive seismic surveys, and cosmic-ray muography, occasionally combined with other geophysical techniques. These methods are applied in varying combinations across different areas, depending on technical capabilities and specific needs.

The deposit types studied for their CRM content or potential include orogenic gold with atypical metal associations, sediment-hosted stratiform copper, karst bauxite, porphyry copper, epithermal gold, polymetallic veins, volcanic-hosted massive sulfide, and iron oxide-copper-gold. The countries involved in the geological studies are Bosnia-Herzegovina, Bulgaria, Finland, Germany, Poland, Spain, and Zambia.

This abstract focuses on the bauxite deposits in the regions of Jajce and Posušje, Bosnia-Herzegovina, and the porphyry copper system in Assarel, Bulgaria. The Late Cretaceous Jajce bauxites are predominantly boehmitic in composition, with rare diaspore in some localities. Hematite is the dominant Fe mineral. These deposits locally exhibit up to 1,353 ppm of REEs. The Early Eocene Posušje bauxites chronologically align with the Paleocene–Eocene Thermal Maximum (PETM) around 56 million years ago. Boehmite is their primary Al-bearing mineral, jointly with gibbsite at some localities. Hematite remains the dominant Fe phase, and the total REEs vary between 357 and 1,112 ppm. Our preliminary studies on the CRM potential of the Assarel porphyry deposit reveal high concentrations of several CRMs in some ore and alteration minerals. For example, enriched Te, Co, Ag, Ge, and Bi contents were defined in a high-sulfidation mineral association.

Temporal Evolution of the Majuba Hill Cu-(Mo)-(Sn) Deposit, Pershing County, Nevada

Alexander P. Holmwood[1], Matthieu Harlaux[2], Philipp Ruprecht[3]
1. Freeport-McMoRan, Phoenix, AZ, USA, 2. Bureau de Recherches Géologiques et Minières, Orléans, France, 3. University of Nevada, Reno, Reno, NV, USA

The Majuba Hill intrusive complex of northwestern Nevada exhibits two superimposed hydrothermal mineral deposits. Quartz vein-hosted Cu-(Mo) mineralization occurring in Jurassic (~160 Ma) granodiorite was later overprinted by Oligocene (~25–28 Ma) subvolcanic rhyolite magmatism and related Cu-(Mo)-(Sn) mineralization.

The Jurassic hydrothermal system is characterized by early barren to molybdenite-bearing granular quartz veins and barren comb-textured quartz veins reopened by later tourmaline, calcite, and chlorite. Fractures in early veins provided pathways for tourmaline, calcite, chlorite, and later sulfide deposition. Hydrothermal Cu occurs as anhedral chalcopyrite overprinting earlier veins and gangue minerals.

Oligocene hydrothermal deposition commenced with granular quartz + (molybdenite) veins and barren comb-textured quartz veins crosscutting subvolcanic rhyolites. Euhedral cassiterite + tourmaline + quartz mineralization (~25 Ma) followed and accompanied significant tourmaline + quartz + sericite + fluorite deposition and alteration of porphyritic rhyolites. Anhedral chalcopyrite deposited as disseminated crystals overprinting altered K-feldspar phenocrysts and by infilling portions of open space-filling veins. Chalcopyrite also crosscuts euhedral quartz-arsenopyrite veins. Subequal amounts of pyrrhotite and minor sphalerite are spatially associated with chalcopyrite but occurred earlier in the paragenetic sequence.

Shallow portions of the Majuba Hill deposit were affected by a supergene fluid that deposited digenite + (covellite) along the margins of hypogene chalcopyrite crystals. Minor cuprite and native Cu were also observed in drill hole MHB9 but appear less common than digenite and covellite.

Paragenetic sequences from both the Jurassic and Oligocene hydrothermal systems suggest the sulfidation state of parent hydrothermal fluid(s) increased over the duration of each system. The genetic association between cassiterite, hydrothermal tourmaline + quartz, and emplacement of subvolcanic rhyolites suggests Sn endowment may be derived from a peraluminous and fractionated parent intrusion.

A Fresh Look at the World-Class “Five-Element” Vein Deposits, Cobalt Embayment, ON (Canada): Fluid and Metal Sources and Systematics, and Mineralization Ages

Joshua Jackman[1], Jacob Hanley[1], Daniel Kontak[2], Robert Creaser[3], Richard Stern[3], Jonathon Toma[4]
1. Department of Geology, Saint Mary's University, Halifax, NS, Canada, 2. Harquail School of Earth Sciences, Laurentian University, Sudbury, ON, Canada, 3. Department of Earth and Atmospheric Sciences, University of Alberta, Edmonton, AB, Canada, 4. Department of Earth and Planetary Sciences, Yale University, New Haven, CT, USA

The world-class “five-element” (Ag-Co-Ni-As±Bi) Paleoproterozoic vein deposits of the Cobalt Embayment (Ontario, Canada) have not been rigorously studied for decades. Here we present the results of a fully integrated approach that employs Re-Os dating, mineral-chemical constraints (EDS, XRF, LA-ICP-MS), and fluid inclusion (FI) measurements (microthermometry, laser Raman, cathodoluminescence (CL), and in-situ $\delta^{18}O_{quartz}$ SIMS), which has important implications for fluid and metal sources, process systematics, and ore system initiation that collectively impact the conventional genetic model for this high-tenor ore setting.

Early- (pre-Co-Ni-Ag-As-Bi ore) and late- (syn- to post-ore) stage vein quartz preserves two dominant fluid types as homogeneously and heterogeneously trapped FI assemblages (FIAs). The FI types are single-phase carbonic FIs ($CH_4 \pm N_2$) and aqueous FIs (liquid±vapour±halite) of variable salinity (0.2 to 50 wt. $\%_{NaCl\ eq.}$). The secondary and pseudosecondary timing of FIAs record numerous episodes of interaction between methane-dominant fluids with aqueous fluids.

CL imaging of early- and late-stage quartz shows pronounced primary growth zones and basal-terminated fracture plumes, indicating pervasive dissolution-recrystallization during and after vein formation. In-situ $\delta^{18}O_{quartz}$ data reflects a steady increase of values between primary growth zones and secondary features hosting FIs with <100°C of cooling during early vein mineralization. This, in tandem with fluid mixing, was a major control on mineralization.

Elemental mapping of accessory Fe sulphides reveals complex enrichment/depletion patterns with discordant zones overprinting primary growth features. Metal distribution indicates variable enrichment, decoupling of Ag from Ni-Co-As, and oscillations among Ni-Co-As, Ag, and base metal (Cu-Zn-Pb) mineralization stages. These observations are inconsistent with the traditionally documented paragenesis for five-element vein mineralization.

Re-Os radiometric dating constrains Ni-Co-As arsenide/sulpharsenide, and Cu sulphide mineralization to ~2232Mya, pre-intrusion age for the modelled heat-source (Nipissing Diabase intrusions) by ~13Ma, thus precluding the latter as a source of localized heat and/or metal, and necessitating a full-scale reevaluation of the deposit model.

Lithium-Caesium-Tantalum-Mineralised Granitic Pegmatite Systems in the Palaeoproterozoic Bedrock of Sweden

Erik Jonsson[1, 2]
1. Geological Survey of Sweden, Uppsala, Sweden, 2. Department of Earth Sciences, Uppsala University, Uppsala, Sweden

The Palaeoproterozoic bedrock of Sweden is a core component of the Fennoscandian Shield in northernmost Europe. It is well known for its granitic pegmatites after several centuries of research and discovery of both new metals (e.g., lithium in 1818 and many of the rare earth elements, starting with "yttria" in 1794) and minerals (e.g., gadolinite, spodumene, petalite, tantalite minerals) as well as in the development of scientific characterization and understanding of these fascinating and potentially economically relevant rocks. Of the presently known different rare element-enriched pegmatites and pegmatite fields, primary interest today is focused on those of a lithium-caesium-tantalum-enriched type (LCT-pegmatites).

The most important provinces hosting LCT-pegmatites are the eastern part of the Skellefte district in northern Sweden (including the classic, previously mined Varuträsk pegmatite), the Bothnian Basin pegmatite fields together with the Räggen and Mårdsjön areas in west central Sweden, and the Stockholm region, including its archipelago, in southeast central Sweden. The recent discovery of an LCT-pegmatite near Bergby in east central Sweden led to the recognition of this region as a new prospective area, and exploration there with a focus on lithium (± tantalum) is on-going, with significant new discoveries made during the past year. All major LCT-pegmatite fields in Sweden are located within low- to medium-grade metamorphosed metasupracrustal successions, and typically associated with late-orogenic, c. 1.8 Ga S-type granites such as the Stockholm, Härnö, and Skellefte granite suites. In addition to these areas, less-evolved LCT-type pegmatites occur sparsely in northernmost Sweden (e.g., at Suorravaara and Kluntarna), as well as in southwest Sweden (e.g., at Skuleboda-Essljung and Skantorp). Presently, all major provinces in Sweden known to have significant evolved granitic pegmatite systems of LCT type are seeing exploration work on various scales, mainly for lithium, but to some extent also for tantalum and other commodities.

The Structural Setting and Relative Timing of the Bergby LCT-Pegmatite Field, Central Sweden

Karin Högdahl[1], Erik Jonsson[2, 1], Magnus Leijd[3], Anders Zetterqvist[3], Pablo Petri[1]
1. Uppsala University, Uppsala, Sweden, 2. Geological Survey of Sweden, Uppsala, Sweden, 3. Bergby Lithium AB, Bollnäs, Sweden

The Bergby LCT-pegmatite field is the latest discovered LCT-pegmatite field in Sweden and the Fennoscandian Shield and the most promising Li project in the country. It covers an area of c. 50 km^2 and is located in the west-central part of the Paleoproterozoic Svecokarelian orogen. Here we suggest a temporal link between the emplacement of the LCT-pegmatites and late-orogenic shear activity.

The host rock to the LCT-pegmatites is a refolded metasupracrustal syncline with a thrust contact to the surrounding c. 1.85 Ga granitic gneisses. The metasupracrustal sequence comprises mica schists succeeded by 1.89 Ga metavolcanic rocks with compositions ranging from rhyolitic to basaltic that are discordantly overlain by quartzites originating from <1.85 Ga clastic sediments. Regional metamorphism affecting the metasupracrustal rocks reached low-P amphibolite facies with an increase in temperature from the andalusite stability field in the east to sillimanite in the western part of the area, where barren granitic pegmatites are ubiquitous. Peak metamorphism at 1.83 Ga was accompanied by ductile deformation characterised by westward-verging overturned F1-folds and thrusts. Changes in convergence direction refolded the older structures to a large-scale, easterly-dipping, reclined F2-synform. Strain partitioning resulted in kilometre-wide, steep 1.81 Ga shear zones arrested at the limbs of this fold.

The known LCT-pegmatites are found in the most competent rocks in the area, represented by amphibolites, the quartzite, and the high-strain zones. They strike in the same direction as the main foliation, but in the F2 hinge zone they dip to the west; i.e., in the opposite direction to the foliation. In the shear zones, the LCT-pegmatites are steep and variably affected by deformation such as boudinage and shear folding. Occasionally they display an internal ductile strain. The suggested temporal link between the LCT-pegmatites and the shear activity implies that they were formed and emplaced at 1.81 Ga.

High Magma Differentiation Controls Rare Metal Mineralization: Traces by Tourmaline Major and Trace Elements, and Boron Isotopes

Haiquan Li[1], Youye Zheng[2], Xin Chen[2]
1. Institute of Geological Survey, China University of Geosciences, Wuhan, China, 2. School of Earth Resources, State Key Laboratory of Geological Processes and Mineral Resources, China University of Geosciences, Wuhan, China

Tourmaline is a common and crucial mineral in highly evolutional rocks due to being rich in boron. We research the structures, major and trace elements, and boron isotopic compositions of tourmalines in tourmaline-muscovite leucogranite, tourmaline leucogranite, barren pegmatite, beryl pegmatite, and spodumene pegmatite from the Rongxia-Qongjiagang area, greater Himalayan orogenic belt. All samples exhibit schorl characters and are divided into four groups by different host rock types. The tourmalines from the tourmaline-muscovite leucogranite exhibit obvious difference by the core-rim textures in a sample. The cores are characterized by highest $\delta^{11}B$ ratios (-10.04~-7.74‰), yet the rims exhibit high $\delta^{11}B$ ratios (-11.79~-10.49‰) close to another sample ($\delta^{11}B$ ratios are between -12.50~-12.06‰). All tourmaline-muscovite leucogranites exhibit higher MgO contents than other. The tourmaline leucogranite (-14.54~-13.88‰) and barren pegmatites (-14.02~-12.93‰) exhibit visible low $\delta^{11}B$ ratios. The beryl pegmatite is composed of lowest $\delta^{11}B$ ratios (-15.69~-14.28‰) with high B_2O_5 contents and slightly enriched Pb, Nb, Ta, and light rare earth elements. The spodumene pegmatite is composed of varied $\delta^{11}B$ ratios (-14.59~-10.25‰) with high Al_2O_3, B_2O_5, Pb, Nb, Ta, and light rare earth element contents, yet low SiO_2 contents. The structures, major elements, and B isotopes display the magmatic genesis of all tourmalines and record the magmatic differentiation process from early tourmaline-muscovite leucogranite to later mineralized pegmatite. Especially, the tourmaline core of tourmaline-muscovite leucogranite maybe generated in low-evolution magma conditions; the tourmaline of spodumene pegmatite may be affected by hydrothermal fluids, and the tourmaline formation records the independent enrichment of different rare metals. Thus, the tourmalines are useful tools to reconstruct the rare metal mineralization process controlled by high magma differentiation, not only in the Himalayan orogenic belt but also at other crucial metal mineralization belts.

Copper and Molybdenum Isotope Variation Along a Vertical Transect Through the Epithermal-Porphyry Valeriano Cu-Au System, El Indio-Maricunga Belt, Northern Chile

Stephanie Lohmeier[1], Bernd Lehmann[2], Jan Pašava[3], Lukas Ackerman[3], Ondrej Šebek[3], Alexandre Andronikov[3], Andrew Hodgkin[4], Francisco Bravo[4]
1. Clausthal University of Technology, Mineral Resources/ Mining Engineering, Clausthal-Zellerfeld, Germany, 2. Clausthal University of Technology, Mineral Resources, Clausthal-Zellerfeld, Germany, 3. Czech Geological Survey, Prag, Czech Republic, 4. ATEX Resources/ SBX Consultores, Santiago de Chile, Chile

The Valeriano epithermal-porphyry Cu-Au system in the Miocene El Indio-Maricunga belt, Vicuña segment, has been intermittently explored for about 35 years, initially for its El Indio-style gold potential. The deep porphyry-Cu(-Au) system at a depth of ~1,000 to >1,800 m, identified in 2012, is overlain by a shallow silicic lithocap, with low-grade epithermal Au-Ag(-Cu) mineralization. A deep drill hole (VALD-D13-14) from the surface down to the porphyry system allows the systematic evaluation of Cu and Mo isotope profiles over the full vertical drill length of ~1,900 m. There is no systematic relationship between δ^{65}Cu (-6.46‰ to +2.67‰) and δ^{98}Mo (-0.32‰ to +0.27‰) isotope composition and lithology, metal (Cu, Au, Mo) content, and Cu mineral assemblages, but there are relationships between δ^{65}Cu-δ^{98}Mo and alteration mineral assemblages.

From top to bottom, copper isotopes show an increasing trend in the epithermal system (-6.48‰ to -0.54‰) characterized by dominantly advanced argillic alteration, followed by a narrow enrichment zone with Cu contents up to 2 wt % and δ^{65}Cu of -2.84 to +2.67‰, with dominantly phyllic alteration. The transition to the deep porphyry system is characterized by systematic increase in δ^{65}Cu values (-2.12‰ to +2.67‰) in phyllic altered rock. The deeper porphyry system has in general a narrow δ^{65}Cu range between -0.10‰ and 0.86‰ in mostly potassic altered rock. Propylitic alteration domains in the epithermal system and in the transitional section have heavier Cu isotope compositions, while domains with advanced argillic alteration have lighter Cu isotopes. The epithermal system is characterized by the highest δ^{98}Mo values (+0.27‰), whereas δ^{98}Mo decreases systematically in the transitional section downwards towards the deep porphyry, a behaviour which can be attributed to Rayleigh fractionation combined with boiling. Advanced argillic alteration displays heavier Mo isotope values, while propylitic alteration has the opposite effect.

Different Sn-W Vein Textures: The San Finx Deposit, NW Spain

Ivan Losada[1], Iñigo Borrajo[2], Fernando Tornos[2]
1. University of Salamanca, Salamanca, Spain, 2. IGEO Instituto de Geociencias, Madrid, Spain

Tin and tungsten have been included by several countries in the list of critical and strategic minerals for the energy transition. These minerals primarily occur as granitic-related deposits including porphyries, greisens, skarns, pegmatites, stratabound deposits, and perigranitic vein systems.

The San Finx Sn-W vein-type deposit is located in northwest Spain, within the Sn-W metallogenic district of the Iberian Massif. This province includes notable mines such as Panasqueira (Portugal) and La Parrilla and Barruecopardo (Spain).

The mineralized, dominantly exogranitic, quartz veins at San Finx extend 3,500 m in length and are approximately 50 m wide on average, striking N30-50°E, N10°W with a steep dip. The cassiterite and wolframite (hubnerite) mineralization is coarse grained and high grade. U-Pb dating on apatite and ^{40}Ar-^{39}Ar dating on muscovite have yielded ages of 281.7 ± 1.03 Ma and 287.75 ± 1.03 Ma, respectively.

Wolframite-deplete, cassiterite-rich, brecciated sections of up to 80 m length occur included within the dominant quartz veins hosting the main cassiterite-wolframite mineralization. These segments include (1) an ultragreisen-type assemblage dominated by muscovite and likely developed under conditions of retrograde quartz solubility; and (2) a feldspars assemblage, primarily composed of potassic feldspars and depleted in quartz. The transition between the feldspar sectors with the dominant quartz assemblage within the veins is always sharp. Moreover, minor feldspar veinlets are also commonly present in the brecciated surrounding wall rock. This supports that the feldspar sectors originate from boiling of an original magmatic-hydrothermal fluid, where phase separation accelerates cooling and alkalinization of the resulting high-density aqueous phase, promoting the precipitation of feldspar instead of quartz.

Magmatic Evolution and Sulfide Saturation Mechanisms Forming the Ultramafic-Hosted Stormi Ni-Cu-(Co) Deposit of the Vammala Belt, SW Finland

Jakob Machleidt[1], Karsten Haase[1], Manuel Keith[1], Reiner Klemd[1], Christoph Beier[2], Patrick Hehn[1]
1. Friedrich-Alexander University Erlangen-Nuremberg, Erlangen, Germany, 2. University of Helsinki, Helsinki, Finland

The Proterozoic (~1.88 Ga) Stormi Ni-Cu-(Co) deposit is hosted in the ultramafic Vammala Ni belt in SW Finland. Interpreted as arc magmas along a paleosubduction zone related to the Svecofennian Orogeny (~2.0-1.8 Ga), the orthomagmatic sulfide deposit can be compared to more recent Phanerozoic analogues, such as the Aguablanca Ni-Cu-(PGE) deposit in Spain. Assimilation of black schists is often considered as an important mechanism driving sulfide saturation in silicate melts to form large orthomagmatic ore deposits, which has also been suggested for Aguablanca. A similar hypothesis has been proposed for the Vammala Ni belt. However, few available $\delta^{34}S$ data of magmatic sulfides that range from -0.79 to +12.5‰ rather exclude that black schist assimilation induced the sulfide saturation. Here, we test whether Proterozoic arc magmas may have been more reducing than those from the Phanerozoic, causing early sulfide saturation during fractional crystallization in the absence of black schist assimilation.

Drill cores of magmatic sulfide-bearing olivine cumulates with minor clinopyroxene and chromite from the Stormi Ni-Cu-(Co) deposit were sampled at the Geological Survey of Finland. The main sulfide phases are pentlandite, pyrrhotite, and chalcopyrite with accessory Co-Ni arsenides, cubanite, and Bi-Te-Ni phases. The ultramafic cumulates were serpentinized and affected by amphibolite facies metamorphism. However, remnants of magmatic minerals are preserved. Olivine forsterite contents of ~82% suggest that the cumulates do not refer to a primitive magma source. In combination with the Ni concentration, the potential for olivine as an exploration tool for orthomagmatic sulfide deposits in arc crusts will be investigated. Additionally, trace element LA-ICP-MS data, along with new S and O isotope data, help to further constrain the magma evolution and timing of sulfide segregation.

Geochemical Fingerprinting of Lithium Pegmatites: A Whole-Rock Perspective

Ryan J. Manton[1], Rebecca M. Montsion[1], Anusuriya Devaraju[1], Jessica M. Walsh[2]
1. CSIRO, Perth, WA, Australia, 2. Geoscience Australia, Canberra, ACT, Australia

A global compilation of whole-rock geochemistry from Li pegmatites is underdeveloped relative to other more abundant, fine-grained rock types, such as arc-related granites and oceanic basalts. A globally representative whole-rock geochemical database comprising mineralized spodumene pegmatites, with associated analytical and geological metadata, has been compiled for later use in machine learning and statistical exploratory analyses. Analytical results of multi-element geochemistry from a range of laboratory workflows as well as tectonic setting at time of formation, host-rock metamorphic grade, and mineralogical characteristics of zones intra-pegmatite have been compiled from international journal articles as well as government and private company databases from Australia. This global database will provide the foundation to (1) test commonly applied classification and economic assessment workflows; (2) evaluate existing lithium pegmatite paragenesis models; and (3) compare between differing intrusive phases, melt sources, and geological settings of critical mineral prospective pegmatites globally.

Overview of the Potential of Strategic Minerals in Bahia State - Brazil

Hevelyn E. Martins, Maria J. Mesquita
Universidade Estadual de Campinas, Campinas, SP, Brazil

The energy transition to a low-CO_2 economy is occurring now, and the global demand for strategic metals and minerals is exponentially increasing because of their use in solar energy, electric vehicles, and batteries. Investing in the product chain from mineral exploration to production will be necessary. The strategic metals nickel, cobalt, and copper are highly required. Brazil accounts for 8% of the Ni production in the world. Bahia state has numerous resources of Ni, Co, Cr, Cu, and V, mainly associated with the layered mafic intrusions (LMI), including the Santa Rita mine. The Santa Rita (Ni-Cu-platinum group element [PGE]) deposit is situated within the Archean-Paleoproterozoic high-grade metamorphic sequences in the São Francisco craton and is an example of a high-Ni sulfide tonnage deposit (88 Mt @0.61% NiS) developed in LMIs. Its age of 2.06 Ga is similar to the age of several large world-class deposits such as Bushveld and Sakati. In the center of Bahia state, mafic-ultramafic complexes occur associated with Cr mineralization as Jacurici Valley, Campo Formoso, and the Riacho do Mocambo. The western region of the state, close to Barreiras city, has been highlighted due to the numerous occurrences of Mn-Co with anomalous values of cobalt (200–5,000 ppm). The genesis of cobalt has not yet been discussed, pointing to the dramatic need for investments in mineral exploration projects. The LMIs of Bahia State present a valuable geologic and economic resource, with Ni, Cu, Cr, and Co deposits playing a pivotal role in the industries and energy transition. More than 40 deposits were described, but few studies about metalogenic process. We intend to present an overview of Bahia state's layered mafic complexes and prospective maps about critical minerals mineralization and its potential for Bahia state, with the intention of encouraging investments in new mineral projects.

Carbon Isotopic Studies of Calcite Veins Associated with U Mineralization in the Jahaz Deposit, North Delhi Fold Belt, Northwestern India

Priyanka Mishra[1], Rajagopal Krishnamurthi[1], Baskaran Saravanan[2]
1. Indian Institute of Technology Roorkee, Roorkee, UK, India, 2. Atomic Minerals Directorate for Exploration and Research, Hyderabad, TS, India

Jahaz is one of the promising U deposits along the "albitite line" in the North Delhi Fold Belt, India, and is hosted within the Meso-Proterozoic metamorphic rocks of the Delhi Supergroup. The deposit is medium tonnage, low grade, metasomatic type, and structurally controlled. Quartzite, quartz biotite schist, and amphibolite are found to be deformed and albitized. The altered rocks have been classified into two groups based on petrographic studies: (1) less to moderately altered (LTMA) and (2) intensely altered (albitized). The major metamorphic mineral assemblages include hornblende, biotite, bytownite, quartz, muscovite, and almandine with a minor amount of sillimanite, andalusite, and ilmenite. Two types of calcite (calcite-1 and calcite-2) are distinguished based on the mode of occurrence and texture. Calcite-1 formed due to the replacement of parent mineral associated with the LTMA rocks. Calcite-2 occurs as vugs/veinlets in the albitized rocks. Calcite-2 is found to be in close association with uraninite and sulfide minerals (pyrrhotite, pyrite, chalcopyrite). Hence, we attempted the carbon isotopic analyses of representative calcite-2 veinlets using the isotope ratio mass spectrometers (IRMS) technique to infer the nature of hydrothermal fluid responsible for metasomatism and related U mineralization. The ranges of variation in $\delta^{13}C_{VPDB}$ and $\delta^{18}O_{VPDB}$ are relatively narrow, ranging from −6.18 to −6.51‰ and −19.93 to −21.03‰, respectively. The isotopic data matches with the meteoric-derived hydrothermal fluid, and such fluid was responsible for metasomatism and U mineralization in the study area. The close association of uraninite and sulfides (pyrite, pyrrhotite, chalcopyrite, molybdenite) within albitized zones can be used as an effective exploration guide to delineate rich zones of uranium in similar geological settings.

Estimation of Lithium Content in Brines of La Union Salt Flat, Located in Ahumada, Chihuahua, Mexico

Héctor Montes Carreón, Perla Rodríguez Contreras
Universidad Autónoma de Ciudad Juárez, Ciudad Juárez, Chihuahua, México, CHH, Mexico

Lithium was defined as a critical element for the energy transition, and despite being present in the upper continental crust at a concentration of 30.5 ± 3.6 ppm, only certain geological contexts contain deposits that could be profitably exploited. Lithium brine deposits represent more than half of the global lithium source. In Mexico, several government agencies and mining companies have explored the presence of lithium in the country. The Mineral Resources Council detected 283 ppm of lithium in water samples from the evaporation ponds of La Unión Salt Flat, but no subsequent studies were conducted. The site of interest is in El Barreal Playa Lake, a closed basin with characteristics associated with lithium brine deposits such as arid climate, tectonic subsidence, suitable lithium source rocks, and an aquifer with a high degree of salinity. The magmatism of the area is related to the migration of a Cretaceous-Cenozoic magmatic arc, derived from the subduction of the Farallon plate beneath the North American plate. The rhyolitic tuff outcrops in the study area correspond to the Oligocene. The objective of this project is to estimate the amount of lithium in brine. The methodology includes the identification of the aquifer horizons with possible brine accumulations through the interpretation of previous geohydrological studies and multiparameter well surveys in the study area. Additionally, it involves the determination of the concentration of lithium in brine through sampling of wells and evaporation ponds of the La Unión Salt Flat, for subsequent ICP-MS analysis. This contribution to knowledge of the region is considered relevant, since it provides additional information about the site, which would be useful for subsequent studies.

Application of Apatite Chemistry in LCT Pegmatite Exploration in the Late Archean Yellowknife Pegmatite Province, NWT Canada: A Preliminary Analysis

Gilles N. Ngoran[1], David R. Lentz[1], Gideon Lambiv Dzemua[2]
1. Department of Earth Sciences, University of New Brunswick, Fredericton, NB, Canada, 2. Government of Northwest Territories Geological Survey, Yellowknife, NW, Canada

Primary apatite has been proven to record the evolution of granitic magmas and can thus be used in lithium pegmatite exploration. The Slave Craton in the Northwest Territories of Canada is host to numerous Archean LCT pegmatites, which occur as dykes, commonly in clusters, in granitic and metasedimentary host rocks. Variably mineralized pegmatites occur together in the same clusters, thereby necessitating individual assessment.

Detailed petrography aided by microXRF-EDS elemental mapping of polished sections followed by compositional analysis *in situ* of magmatic apatite by LA-ICP-MS was performed to assess the mineralization potential of five pegmatite dykes (Best Bet, Hidden Lake, Moose II, Pancho, and Riber) from the Yellowknife Pegmatite Province in the southern Slave Craton. All the dykes, except Riber, have reported spodumene mineralization.

A total of 121 spots on 117 apatite grains in polished sections of the pegmatites were analyzed by LA-ICP-MS. Median Y values increase from 0.4, 15, 52, 116 and 995 in Pancho, Moose II, Best Bet, Hidden Lake, and Riber, respectively. All samples reported positive Eu anomalies, except Riber. Median ΣREE, Th, U, Fe, and Mn concentrations in Best Bet (123 ppm, 5.2 ppm, 20 ppm, 254 ppm, 469 ppm), Hidden Lake (243 ppm, 22 ppm, 485 ppm, 2174 ppm, 6855 ppm), Moose II (24 ppm, 1.2 ppm, 27 ppm, 417 ppm, 1316 ppm), Pancho (3.5 ppm, undetected, 32 ppm, 31,768 ppm, 38,546 ppm), and Riber (888 ppm, 5.9 ppm, 174 ppm, 3317 ppm, 13,871 ppm), respectively. Results reveal low Fe/Mn and Th/U values, generally below 1 and 0.05, respectively, suggesting these pegmatites are very highly fractionated. This is supported by K-Rb-Cs-Li-Sn-Ta systematics in muscovite-K-feldspar, which revealed Best Bet as the most fractionated and Riber as the least.

These findings confirm the suitability of using apatite chemistry to evaluate LCT pegmatite fractionation, Li-Ta-Sn-Cs prospectivity, and U-Pb geochronology.

SEG 2024 Conference: Sustainable Mineral Exploration and Development

Raw Materials from Geothermal Fluids

Simona Regenspurg[1], Lydia Olaka[2, 3], Franziska Wilke[1], Katrin Kieling[1], CRM-geothermal Team
1. GeoForschungsZentrum-Potsdam, Potsdam, Germany, 2. University of Nairobi, Nairobi, Kenya, 3. Technical University of Kenya, Nairobi, Kenya

The energy and digital transitions require a large amount of mineral and raw materials which are considered 'critical' worldwide. Interestingly, there is an untapped resource in geothermal fluids, some of which contain significant amounts of CRMs.

The EU-funded CRM-geothermal project therefore proposes to combine the extraction of raw materials and geothermal heat, a renewable energy resource from the ground which is available 24 hours per day. The combined extraction of heat and raw materials from geothermal reservoirs offers a series of advantages like minimising the environmental impact, maximising returns on investment, avoiding additional land use, and enabling domestic supplies of CRMs.

Although CRMs are known to occur in geothermal fluids, there are still many uncertainties concerning their occurrence in different geological settings and the sustainability of their extraction. The actual extraction process is still a challenge requiring technology development. The project therefore aims to:

- Establish an overview of the potential for raw materials in geothermal fluids for a large range of CRMs.
- Determine the source of selected CRMs, their mobility, and potential for sustained extraction from geothermal brines.
- Develop and optimise innovative extraction technologies for selected CRMs from geothermal brines that can form a business case.
- Assess the environmental-social-economic viability, create transparent and traceable value chains, and foster ethical sourcing of CRMs.
- Demonstrate at a pilot site the extraction technology and system's sustainability for at least one CRM at the scale of a mini-plant.

The combined extraction technology will help developing a more resilient and ethical CRM supply chain. The proposed solution could bridge the gap between societal resistance to domestic raw materials extraction and the increasing demand for raw materials. Finally, the combined extraction of minerals and heat will also increase the number of viable geothermal projects, fostering the green transition and diversifying the worldwide energy portfolio.

Survey for Critical Raw Materials in Tanzanian East African Rift Geothermal Areas

Franziska Wilke[1], Simona Regenspurg[1], Anna Jentsch[1], Timmu Kreitsmann[2], Martin Zimmer[1], Michael Bau[3], Samuel Niedermann[1], Lydia Olaka[4, 6], Albano Mahecho[5]
1. GeoForschungsZentrum-Potsdam, Potsdam, Germany, 2. University of Tartu, Tartu, Estonia, 3. Constructor University, Bremen, Germany, 4. University of Nairobi, Nairobi, Kenya, 5. Tanzanian Geothermal Development Company–TDGC, Dar es Salaam, Tanzania, 6. Technical University of Kenya, Nairobi, Kenya

The energy and digital transitions require ample critical raw materials (CRM), which can be found in geothermal fluids. In the EU-funded project "CRM-geothermal," we aim to environmentally co-produce CRMs while harnessing geothermal energy. In the East African Rift System (EARS), significant levels of rare earth elements (REEs), Sr, F, etc. are anticipated in water and solids, in areas with carbonates. Kenya benefits most from the eastern EARS branch, where hot rift water migrates along permeable faults, providing around 80% of its energy demand. A 2022 survey in Tanzania along the EARS sought to determine CRM content in geothermal water, gas, and surrounding rocks.

Water samples were collected from hot springs (26-74°C) and nearby alkaline lakes, with lake waters ranging from pH 7.6 to 10.0 and spring waters from pH 6.9 to 9.7. These samples exhibit high carbonate alkalinity, elevated chloride, sulphate, and fluoride content, with sodium concentrations up to 9,000 mg/kg. The highest total rare earth elements and yttrium (∑REY) concentration was found in northern Lake Eyasi at 12 µg/L, around 500 times higher than in Lake Malawi, south Tanzania. Springs display variable rare earth element patterns compared to lakes, with alkaline springs showing more variability. Noble gas analysis indicates a mixed crustal-mantle contribution, with helium concentrations reaching up to 7 vol % and $^{3}He/^{4}He$ ratios up to 3.4 Ra, mainly in the southern part. Two geological settings are dominant in Tanzania: basement with gneisses ± granulites and alkaline volcanics associated with rifting. Geothermal sites are found in both settings, with craton-based samples dominated by Na-K±Ca silicates rich in Fe- and Ti-oxides. Minerals and rocks hosting elevated REEs, alkali-rich sulphates, and Sr-carbonates were sampled at and near Lake Eyasi in the north and in the south along the Mbaka fault. Thorium-rich minerals occur predominantly in central Tanzania.

Insight from Field and Petrographic Studies of Saltpond and Winneba Pegmatites and Other Associated Rocks: Implications to Ghana's Lithium Occurrence

Theophilus Owusu Agyapong, Mary N. Selby, Samuel Nunoo, Prince O. Amponsah
University of Ghana, Accra, Ghana

The study of pegmatites along the coastal belt of the Kibi-Winneba belt is becoming an emerging geological expedition. This interest is driven by the recent discovery of lithium-bearing spodumene that has commercial value in the green technology. The study is part of an ongoing larger research focused on understanding the mechanisms of fluid migration and interaction of potential Li-bearing pegmatite and tourmaline-rich gabbro. Rock samples collected near/in Saltpond and Winneba provide clues about the geologic history and processes that contributed to the formation of the Li-bearing pegmatites. The Saltpond pegmatites exhibit a distinctive graphic texture, characterized by an intricate intergrowth of quartz and alkali feldspar, and greisen-like muscovite. Additionally, the pegmatite contains acicular to needle-like muscovite laths intergrowth with plagioclase. The specific mechanism behind the formation of this needle-like muscovite is still being investigated, but the kink pattern suggests mineral realignment from postformation deformation. Winneba pegmatite is composed of interlocking and near equigranular grains of muscovite, feldspars, quartz, and spodumene, and dark phases suspected to be Columbite Group minerals. In contrast, the tourmaline-rich gabbro shows ophitic texture, and is composed of crystals of tourmaline, pyroxene, feldspar, and secondary muscovite. While the observed textures suggest the rocks were likely formed from an igneous process, the secondary minerals hints at a later hydrothermal and metasomatic alteration event, a role in concentrating lithium-bearing fluids. Such alteration likely destabilized feldspars, converting them into secondary mica. The schist host rock acted as the potential country rock interacted by fluid from late-stage granitization and magmatism. Graphic texture suggests the presence of water and other volatiles, which lowered the interfacial energy between quartz and feldspar, facilitating their intergrowth. The graphic texture has been recognised as indicator for the potential presence of valuable minerals like lithium, beryllium, or tantalum.

A Conceptual Mineral Systems Model for the LCT Pegmatites

Shanta Panja, Alok Porwal, Majid Keykhay-Hosseinpoor
Indian Institute of Technology Bombay, Mumbai, MH, India

The LCT (lithium-caesium-tantalum) pegmatites are of uttermost importance for hosting minerals having the highest concentrations of lithium, a critical metal. Consequently, there is an urgent need to understand the genesis, emplacement, and preservation of these deposits within the framework of mineral system analysis for effective regional-scale exploration targeting. Through an exhaustive review of literature encompassing pegmatite occurrences worldwide, this study conceptualizes the targeting criteria for the formation of lithium-bearing pegmatites: (i) volcanic-arc granites, post-collisional granites, and the granites derived from the anatexis of the metapelites, derived from sediments that have undergone extreme chemical weathering and residual enrichment, making them the fertile sources for Li;(ii) compressional zones and post-collisional extension zones as the prospective geodynamic settings that initiate the mineralization process; (iii) crustal-scale faults in the aforementioned tectonic settings that facilitate the focusing and transportation of these melts; and (iv) the post-emplacement tectonic regime that preserves the deposits. Additionally, this research identifies spatial proxies based on the aforementioned criteria that are accessible from publicly available geoscientific data sets, including geological and topographic maps, airborne geophysical data, geochemical data, and remote sensing data sets. These spatial proxies can be integrated using methods like fuzzy inference model for the targeted modelling of lithium-bearing LCT pegmatites, thereby highlighting the most prospective areas for further detailed exploration.

Lithium Clays in Volcano-Sedimentary Deposits

Francesco Putzolu[1], Robin Armstrong[1], Tom Benson[2], Chris Emproto[3], Rebecca Ball[4], Isabel Abad[5], Adrian Boyce[6], Laura Hepburn[6], Nicolas Bompard[6], Richard Herrington[1]
1. Natural History Museum, London, United Kingdom, 2. Lithium Americas (Argentina) Corp, Vancouver, BC, Canada, 3. University of Michigan, Ann Arbor, MI, USA, 4. Jindalee Resources Limited, Perth, WA, Australia, 5. Universidad de Jaén, Jaén, Spain, 6. Scottish Universities Environmental Research Centre, Glasgow, United Kingdom

The increasing lithium demand for the manufacture of Li-ion batteries is driving the interest of the mining industry toward the exploration of new types of resources, in addition to more typical salar and spodumene-rich pegmatite deposits. Volcano-sedimentary Li deposits represent an unconventional and still untapped Li resource, where Li is bound in mineralogically complex assemblages largely dominated by clay minerals, and more rarely by other exotic phases such as borosilicates and phosphates. Lithium has been known about in volcano-sedimentary systems for several decades, although the nature of clay-type Li mineralization (e.g., small particle size and historical processing difficulties) has hampered the development of this ore type, driving the industry investments towards more amenable spodumene and brine deposits. However, the ongoing increase of the Li demand coupled with the development of low-temperature acid leaching technologies has led to a resurgence in the interest towards volcano-sedimentary Li ores. Therefore, in this contribution we describe the nature of Li-rich clay minerals from selected volcano-sedimentary deposits including Jadar (Serbia), the McDermitt deposits (US) (i.e., Jindalee Ltd. and Lithium Americas tenures), and other globally significant systems. We summarize the mineralogical and geochemical diversity to be expected in clay-type mineralization, and the consequent implications for genetic processes and Li grade continuity.

A Mineral Systems Approach for Li Volcano-Sedimentary Deposits: Commonalities with Other Li Deposit Models

Robin N. Armstrong[1], Francesco Putzolu[1], Tom Benson[2], Richard Herrington[1]
1. The Natural History Museum, London, United Kingdom, 2. Lithium Americas (Argentina) Corp, Vancouver, BC, Canada

The formation of lithium volcano-sedimentary (VS) deposits requires the juxtaposition of a series of critical processes to develop grade and to be preserved in the geological record. The mineral systems approach provides an ideal framework to examine the interplay of these processes and their relative weightings across the diversity of Li-VS deposits. Through the examination of key examples—Jadar (Serbia); the deposits of the McDermitt Caldera (USA); Rhyolite Ridge (USA); and Shavazsay (Uzbekistan)—we describe these critical and constituent processes. Key amongst these are favourable melt/magma types, regional and local structural settings, and environments favourable for formation of clay and other Li phases. This analysis allows us to establish the specificity and sensitivity of the various mappable criteria, which may be used by exploration teams seeking to discover new VS deposits. An emergent theme of this work is the commonalities in some of the critical processes between Li-VS and other Li deposit types such as rare metal granites and pegmatite systems.

Unveiling the Geological Potential of Bauxite Deposits in the Democratic Republic of Congo (DRC) for Sustainable Aluminium Production and the Green Transition

Derick U. Ramazani, Dominique W. Osomba
University of Kinshasa, Kinshasa, Democratic Republic of the Congo (Kinshasa)

Among the global shift towards a greener economy, the demand for sustainable raw materials has surged, with bauxite emerging as one of the cornerstones due to its integral role in renewable energy infrastructure and green technologies. The Democratic Republic of Congo (DRC) holds one of the world-class bauxite provinces (100 to 150 million tonnes with 42.6% of Al associated with gibbsite mineral), but it is not exploited, with deposits such as Sumbi, Mayedo, Kinzoki, and Kimvela strategically located near vital facilities like the largest Inga hydropower and the Atlantic Ocean. This study aims to delineate and characterize the geological features of these deposits through comprehensive petrographic analyses, X-ray diffraction (XRD), and geochemical investigations. Petrographic and mineralogical examinations unveil a progression of supergene alterations within dolerite and basaltic rocks, culminating in bauxite formation. These formations exhibit an evolving mineralogical assemblage made up of kaolinite, gibbsite, quartz, anatase, and goethite. The main Al oxide mineral, gibbsite, peaks at 47.06%, observed at the Mayedo deposit. The geochemistry across deposits shows wide variability in different profiles with SiO_2 (3.51-48.37%), Al_2O_3 (8.68-44.60%), Fe_2O_3 (17.17-57.39%), and TiO_2 (1.96-12.48%). Comparative analysis of the major oxides with other worldwide bauxite deposits reveals the predominance of the classification typical of the kaolinic and bauxitic zones, with significant variations in the alteration indices (33.32-100%). Geochemical analyses further underscore the unique characteristics of these deposits, emphasizing the extensive supergene fluid flow and complex leaching processes that have transformed the protolith into gibbsite-rich minerals. These findings not only underline the vast untapped potential of the DRC's bauxite resources but also highlight their pivotal role in catalyzing sustainable aluminum production and fostering the global transition towards a greener future.

Palladium Enrichment and Mobilization through Alteration and Replacement of Base Metal Sulfides in the Parisien Lake Zone, East Bull Lake Intrusion, Ontario, Canada

Marc Rochette[1], Daniel Layton-Matthews[1], Dave Peck[2], Reid Keays[3]
1. Queen's University, Kingston, ON, Canada, 2. Grid Metals Corp., Toronto, ON, Canada, 3. Monash University, Melbourne, VIC, Australia

The East Bull Lake intrusion occurs approximately 80 km west of Sudbury, Ontario, and hosts a widespread, disseminated sulfide mineralization that has been classified as contact-type PGE-Cu-Ni mineralization. Contact-type mineralizations occur at the base and margins of most of the co-eval mafic-ultramafic intrusions in the greater Sudbury region, though none of these occurrences have proven to be currently economically viable. The lowermost stratigraphic units of the East Bull Lake intrusion hosts Pd grades in the range of 1-10 g/t. The Pd mineralization is associated with an altered, chalcopyrite-dominated disseminated sulfide mineralization. Units containing sulfides associated with secondary hydrous silicates are strongly correlated to high Pd whole-rock grades. Core samples were taken from seven DDH with >1.0 g/t Pd at the Central Parisien Lake Zone, which is >1.0 km^2. DDH were quartered, with half used for whole-rock geochemistry, and the remaining material was billeted for 54 thin sections. Optical mineralogy and automated mineralogy scanning electron microscopy (MLA-SEM) were used to analyze mineralogical and textural characteristics of the samples for in-situ trace element analyses. Laser-ablation inductively coupled plasma-mass spectrometry (LA-ICP-MS) was used to define trace element chemistry of sulfides and platinum group minerals (PGMs). The aim of these analyses is to identify PGE mineral hosts and to compare sulfide chemistry changes during alteration to better understand its effects on the Pd mineralization. Mineral chemistry results indicate that Pd occurs mostly as a stoichiometric replacement in sulfides such as (in order of highest Pd concentration to lowest) pentlandite, chalcopyrite, bornite, pyrrhotite, and pyrite. Alteration enriched Pd in sulfides, before being remobilized on the millimeter scale to form PGMs such as kotulskite (Pd(Te,Bi)). By further characterizing the alteration effects within the Parisien Lake Zone of the East Bull Lake intrusion, we hope to develop more robust techniques for future exploration.

Overview of Uranium Activities in Namibia and the African Continent

Gabi Schneider
Namibian Uranium Institute, Swakopmund, Namibia

After years of low uranium prices, the strongly improved acceptance of nuclear power together with stockpile depletion and the current geopolitical situation resulted in a substantial increase in the commodity's value. This has put the Namibian uranium sector into a phase of resurgence and sparked renewed interest in projects throughout Africa. The Rössing uranium mine has been in operation for 48 years and is the largest and longest-operating uranium open cast mine in the world. Investment in pit and TSF extension, plant refurbishment, and infrastructure upgrades are ensuring the current LoM to 2036, and possibly beyond. Swakop Uranium's Husab mine has an annual mining capacity of more than 100 million tonnes, enabling an output of 6,000 t U_3O_8. The combined production of the two mines was 8,238 t of uranium oxide in 2023. The Langer Heinrich mine completed its restart project, and first production was achieved in the first quarter of 2024. Bannerman Energy and Reptile Mineral Resources both have DFSs in place and were granted mining licenses for their projects in December 2023. Investment decisions are expected during the course of 2024. Other advanced exploration projects include Zhonghe and Valencia, with grassroots exploration having been carried out by Elevate Uranium and Uranium One for a number of years. In addition, there are a number of new exploration companies who have taken over nuclear fuel EPLs in Namibia. The Namibian Uranium Association supports these activities by ensuring best practice environmental, social, and governance performance and hence the sustainability and public acceptance of the sector. Besides the established mining activities in Niger and South Africa, the African continent has exploration activities in Algeria, Botswana, Burundi, Cameroon, CAR, Chad, Djibuti, Egypt, Equatorial Guinea, Ethiopia, Guinea, Kenya, Libya, Madagaskar, Mali, Mauretania, Morocco, Nigeria, Senegal, Sudan, Tanzania, Tunisia, Zambia, and Zimbabwe.

Exploring Lithium Metallogenesis Within the Groundwaters of the Arid-Hyper Arid Environments of the Northern Cape, South Africa

Mary N. Sithole, Khuliso Masindi
University of the Witwatersrand, Johannesburg, South Africa

Ongoing demand for critical metals such as lithium and rare earth elements for the green energy transition necessitates exploration of other lithium deposits in various parts of the world. Due to the fact that, with the growing momentum of the transition, supply has not kept apace with the demand, known lithium deposits are being exhausted and facing environmental as well as social concerns. On a global scale, lithium production is from brines and hard rocks; the former is generally low-grade and the latter high-grade ore. In southern Africa, lithium mining is limited to hard rocks because of the high-grade ore and limited extend of the mineralized rock. The low-grade lithium ore in brines in South Africa is underexplored. With the increased demand for lithium brines and development in mining technology, brines could be a valuable source for lithium in the future. This research looks at the potential of extracting lithium from groundwater in Namaqualand, South Africa. The area is known for its lithium-caesium-tantalum pegmatites, a potential lithium source, as well as its arid climate making it favourable for brine formation. This research reports on the Li content in groundwater from the Northern Cape in relation to the brines. It also looks at the potential resources and further discusses the possible social and environmental effects extraction would have on the communities living in water-sparse areas.

Mineralogical and Geochemical Characteristics of the Carbonatite Laterite at Sokli, Finland

Martin P. Smith[1], Martin Y. Li[1], Sam Broom-Fendley[2], Tuomas Gråsten[3], Frances Wall[2]
1. University of Brighton, Brighton, United Kingdom, 2. University of Exeter, Exeter, United Kingdom, 3. Sokli Oy, Helsinki, Finland

The Devonian Sokli carbonatite complex, Finland, is composed of an early ultramafic intrusive unit, a phoscorite-carbonatite unit, and a late carbonatite unit. Up to 100 m of laterite occurs on top of the bedrock, comprising a major P, Nb, and REE resource. The laterite formed during the Miocene under a warmer climate and is overlain by up to 5 m of glacial deposits. The laterite is composed of residual primary minerals, dominated by apatite, magnetite and pyrochlore, variably cemented by goethite, hematite, crandallite-group minerals, francolite, and britholite, together with pyrochlore-group and residual niobate minerals. Lenses of francolite ± britholite + vermiculite occur at varying depths. These are locally brecciated and cemented by goethite and Mn oxides. Coarse-grained phlogopite-rich intervals occur at varying depths, locally weathered to vermiculite. Assay data from Sokli Oy indicates REE concentrations are highest (up to over 20,000 ppm) in the shallow part of the laterite and either greatly decrease down the profile or fluctuate, with zones of 10-100x enrichment relative to bedrock, owing to mobilisation during weathering and lithological variation. There are only minor differences in the bulk REE pattern between laterite and bedrock, except for Fe oxide-rich laterite, which is enriched in the light REEs. REE concentrations positively correlate with Fe and Nb. Phosphate mineralogy controls enrichment trends with P, with secondary francolite zones having very low REE concentrations, whilst crandallite-group minerals correlate with strong REE enrichment, indicating at least two stages of weathering and phosphate mineralisation. The detailed paragenesis will the subject of further investigation, alongside in-situ analysis to help determine the REE budget in the laterite.

Compositional Features of Lithium-Bearing Pegmatites of the Sinsikourou and Kola Areas Around the Bougouni Region, Southwestern Mali

Elhadji Mory Traore[1, 2], Akinade S. Olatunji[1, 2], Mory Sidibe[3], Seko Sanogo[4]
1. Pan African University of Life and Earth Science Institute (including Health and Agriculture), Ibadan, Nigeria, 2. Department of Geology, University of Ibadan, Ibadan, Nigeria, 3. University of Segou, Segou, Mali, 4. Laboratoire de Minéralogie et de Pétrologie, Faculté des Sciences et Techniques, Université des Sciences, des Techniques et des Technologies de Bamako, Bamako, Mali

The resurgence of global demand for rare metals—notably lithium, which is essential in energy transition—has stimulated renewed interest in exploring economically viable deposits. This study focuses on the Paleoproterozoic pegmatites located in the Sinsikourou and Kola areas of the Bougouni region, southwestern Mali. Selected pegmatite samples from the study areas were prepared for petrographic investigation; whole-rock pegmatite analysis using the inductively coupled plasma-optical spectrometer (ICP-OES) and inductively coupled plasma-mass spectrometer (ICP-MS) was also undertaken. Petrographic studies reveal quartz, plagioclase, microcline, muscovite, and spodumene as major mineral constituents, while garnet, apatite, and tourmalines occur as accessory minerals. Geochemical analysis reveals a high silica value (SiO_2 = 72.3-80.91 wt.%) in the whole-rock pegmatite, while low compositions were observed for the other oxides. Lithium shows significant enrichment values ranging from 3,400 ppm to >10,000 ppm and low to moderate enrichment ranging from 19 ppm to 871 ppm, revealing both lithium-enriched and barren pegmatites in the investigated areas. Additionally, rare earth elements such as Cs (11-203 ppm); Ta (2-149 ppm); Nb (9.4-90.8 ppm); Sn(50,8-561 ppm); and Be (52-668 ppm) are fairly enriched. Depletion patterns in rare metals are evident, as illustrated by Ta vs. Nb and Ta vs. K/Cs indices. Overall, these findings suggest the viability of the Paleoproterozoic pegmatites in the Sinsikourou and Kola areas as sources of rare metals, particularly lithium. Further exploration and analysis may be warranted to assess the economic feasibility of exploiting these deposits.

Assessing Proximity to ISCG and IOCG Mineralisation Using REE-Bearing Phosphate Chemistry

Travis Batch[1, 2], Caroline Tiddy[1, 2], Adrienne Brotodewo[1, 2], David Giles[1, 2], Michael Taylor[3], Vaclav Metelka[3]
1. Future Industries Institute, University of South Australia, Mawson Lakes, SA, Australia, 2. Mineral Exploration Cooperative Research Centre, University of South Australia, Mawson Lakes, SA, Australia, 3. AIC Mines Ltd., Subiaco, WA, Australia

Copper is a critical metal required for the green transition, and can be sourced from iron sulfide-copper-gold (ISCG) and iron oxide-copper-gold (IOCG) deposits. However, finding copper deposits is becoming increasingly difficult, especially in areas where sedimentary cover dominates the landscape. Geochemical tools that use the chemistry of resistate minerals to assess proximity to mineralisation can aid exploration in covered terrains.

This study presents geochemical discrimination schemes for ISCG and IOCG deposits based on the chemistry of REE-bearing phosphates (monazite and rhabdophane) from mineralised and barren rocks. ISCG and IOCG deposits in the Cloncurry District of the Mount Isa Inlier, north-west Queensland, Australia, are used as case studies.

The ISCG discrimination scheme uses samples from Jericho and Kulthor ISCG deposits, which preserve REE-bearing phosphates with higher S/Th and Ca/Th molar ratios compared to their unmineralised counterparts and consistent total REE concentrations across deposits. Jericho is the only deposit in which rhabdophane is identified. The discrimination scheme for IOCG deposits shows that samples from mineralised and barren areas within the Osborne and E1 IOCGs can be distinguished based on their light REE (La, Ce, Nd) and Th content, which supports previous investigations in using monazite chemistry as an indication of IOCG mineralisation (Tiddy et al., 2021). Phosphates from mineralised and barren samples from the SWAN IOCG deposit can be discriminated using both schemes.

These results indicate elevated S and Ca, and depleted Th concentrations characterise REE-bearing phosphate chemistry from ISCG mineralisation, while elevated La + Ce and depleted Nd and Th are typical of REE-bearing phosphate chemistry associated with IOCG mineralisation. These criteria offer the advantage of using the resistate mineral phase, monazite, which can be integrated into cover sequence materials, thereby enhancing the geochemical signal of buried mineralisation. It is noted that the weathering potential of rhabdophane is not yet realised.

Is the Tennant Creek Mineral Field an IOCG Province? Insights from Chlorite Chemistry

Damian Braize[1], Jeffrey A. Steadman[1], Sebastien Meffre[1], Ana Liza Cuison[2]
1. CODES, University of Tasmania, Hobart, TAS, Australia, 2. Emmerson Resources Ltd, West Leederville, WA, Australia

Recent studies in mineral geochemistry have shown that chlorite can delineate distal and proximal zones in porphyry Cu deposits. However, it is not known whether chlorite from other mineral deposit types exhibits the same characteristics. The Proterozoic Tennant Creek mineral field (TCMF), Northern Territory, Australia, has produced over 200 t gold and 350,000 t copper since 1932. Although the deposits are classified as iron oxide-copper-gold (IOCG) type, the Tennant Creek deposits differ from typical IOCG systems in that they have spatially restricted chlorite-rich hydrothermal alteration haloes and are enriched in Bi. Laser ablation-inductively coupled plasma-mass spectrometry (LA-ICP-MS) analyses have been carried out on chlorite associated with several deposits from the TCMF and compiled with chlorite data from different ore deposit types (such as porphyry, IOCG, and volcanic massive sulfide deposits).

Results from univariate and multivariate analyses show distinct mineral chemistry domains of IOCG, porphyry, and VMS chlorite. The Tennant Creek population overlaps with other Australian IOCG deposits, highlighting the strong potential of chlorite geochemistry for explorers and researchers to help recognise ore deposit types. Compositional variations in chlorite chemistry proximal to known deposits in Tennant Creek permit it to be used as a geochemical vector towards mineralisation. Preliminary results show different trace element content between chlorite from barren and mineralized ironstones, enabling a distinction that is not possible with traditional exploration methods. Moreover, strong Fe and Co content in some chlorite could allow workers to explore the potential for critical metal mineralisation in the TCMF. These advances confirm that chlorite geochemistry is a powerful tool that could improve exploration models and lead to future world-class discoveries in Tennant Creek and other IOCG deposits.

Geochemistry and Origin of the Magnetite-Vonsenite Mineralization of the Monchi Mine (SW Iberia)

Alberto Campanero, Carmen Conde, Gonzalo Ares, Fernando Tornos
Geosciences Institute (IGEO-CSIC), Madrid, Spain

The Monchi Mine (SW Iberia) is a high-grade iron deposit with an unusual mineral assemblage that is currently interpreted as a sub-type of magnetite-(apatite) systems. The mineral assemblage includes dominant massive magnetite and vonsenite along with smaller amounts of cobaltite, löllingite-safflorite, allanite, uraninite, cassiterite, and monazite; the ore contains only sparse amounts of pyrite, chalcopyrite, and gold. The mineralization is located at the western margin of the Burguillos del Cerro Plutonic Complex, a Variscan, concentrically zoned pluton from gabbro to granodiorite. This part of the complex is adjacent to a significant N-S–trending dextral strike-slip shear zone. Within this zone, Ediacaran to early Cambrian metapelitic and calc-silicate hornfels and marble crop up. The magnetite-vonsenite mineralization is adjacent to the metasedimentary units forming sizeable lens-shaped bodies with sharp contacts with the intrusive rocks and is locally mylonitized. Subordinately, a U-REE–bearing Fe-pyroxenite is intimately associated, which is locally brecciated and cemented by a pegmatite composed of albite, K-feldspar, quartz, clinoamphibole, biotite, and axinite. Within the metasediment layer, there is a post-tectonic exoskarn, also with magnetite mineralization. Allanite U-Pb TIMS dating from the U-REE–rich pyroxenite yields 337.13 ± 0.99 Ma, within the age range of the Burguillos Plutonic Complex (335-340 Ma). Whole-rock analysis on the pyroxenite shows an enrichment in REEs (La, Ce, Pr, Nd) and U (up to 47 ppm). Ore geochemistry analysis reveals high values of Co (6 %), Ni (1 %), Sn (501 ppm), U (518 ppm), and Ta (127 ppm). The origin of the mineralization remains unknown, although could be explained by I) crystallization of a B-bearing iron melt, similar to the magnetite-(apatite) deposits but with boron as the fluxing agent, and being immiscible with the pyroxenite melt; or II) the result of a syn-magmatic interaction of Ca-rich rocks with deep-sourced fluids focused along thrusts.

Vectoring in Metamorphosed VMS Deposit: Insights from Chlorite, White Mica, and Garnet

Cendi Dana[1], Steven Hollis[1], Darryl Podmore[2], Megan James[2], Ian Butler[1], Florian Fusseis[1], Adrian Boyce[3], Riquan Azri[2]
1. School of GeoSciences, The University of Edinburgh, Edinburgh, United Kingdom, 2. Black Raven Mining, Perth, WA, Australia, 3. Scottish Universities Environmental Research Centre (SUERC), Glasgow, United Kingdom

Most ancient volcanogenic massive sulphide (VMS) deposits have been subjected to metamorphism particularly those Archean in age. Despite countless advances in recent years, exploration for (VMS) remains challenging. This is particularly the case in the Yilgarn Craton of Western Australia, where there is a paucity of outcrop, and weathering is deep and extensive. Recent significant breakthroughs in the accessibility of low-cost hyperspectral data to industry offer a great opportunity as an exploration tool in geologically complex terranes. In this study, we investigate the King Zn deposit using detailed mineralogical and hyperspectral analysis of both the orebody and host rocks. Mineralogical and geochemical data suggested that the footwall stratigraphy consists of several basaltic volcanic rocks that have been metamorphosed into garnet amphibolite and various types of felsic-intermediate schists. The immediate hanging-wall stratigraphy consists of metaexhalite and argillaceous metasedimentary rocks.

The chemistry of chlorite and white mica is strongly controlled by lithology. Chlorite is dominated by mixed Fe-Mg compositions throughout the footwall stratigraphy, with localised zones of Fe- and Mg-rich compositions related to distinct lithologies and mineralization. Muscovitic white mica compositions predominantly occur in footwall basaltic rocks with more paragonitic compositions in dacitic rocks directly underlying massive sulfides, whereas the hanging-wall white mica is phengitic. Fluorine and manganese concentrations in chlorite increase systematically towards the ore horizon before dropping sharply in the hanging-wall strata. Similarly, fluorine concentrations in white mica increase systematically through the footwall stratigraphy. Garnet composition is dominated by almandine end member, with increasing manganese content towards the ore proximity. Integrating mineral chemistry and hyperspectral data offers a rapid and cost-effective method for identifying potential ore horizons during the initial exploration phase. These findings have significant implications for enhancing exploration strategies in geologically complex regions.

Constraints on the Pre-Eruptive Metal and Volatile Content of Magmas Associated with Archean Volcanogenic Massive Sulfide Deposits from Zircon-Hosted Melt Inclusions

Priyal Daya[1], Jacob J. Hanley[1], Kevin Neyedley[1], Sarah Speight[1, 2], Stephen J. Piercey[2], Alexandra Tsay[3], Zoltan Zajacz[3]

1. Department of Geology, Saint Mary's University, Halifax, NS, Canada, 2. Department of Earth Sciences, Memorial University of Newfoundland, Saint John's, NL, Canada, 3. Department of Earth Sciences, University of Geneva, Geneva, Switzerland

Archean VMS systems in the Abitibi Subprovince of Canada differ greatly in metal tenor, particularly in Au endowment. In this first study of silicate melt inclusions (SMI) in such environments, primary melt inclusions in zircon hosted in pre-, syn-, and post-VMS ore felsic volcanic lithologies will provide compositional constraints on the initial metal and volatile chemistry of the magma before eruption, allowing a comparison of the precursor metal budgets of magmas that actively degassed and/or were passively leached to supply metals to the deposits. Primary methodologies conducted in this study include (i) SMI petrography (inclusion and host zircon origin, preservation, accidentally trapped phases, etc.); (ii) SEM-EDS for determination of bulk inclusion compositions on exposed SMIs; and (iii) LA-ICP-MS to quantify major/trace element composition of SMIs and host zircon chemistry.

Magmatic zircons from volcanic rocks hosting the LaRonde-Penna VMS deposit in the Abitibi Subprovince contain accidentally trapped mineral inclusions (e.g., apatite, quartz, biotite, oxides; 50% of all inclusions), SMIs (40%), and sulfide (melt and/or mineral) inclusions (~10%). SMIs (up to 35 μm diameter) occur as microcrystalline (recrystallized) and monomineralic to polymineralic. They are composed of quartz, alkali-feldspars, albite, epidote, and apatite daughter phases in varying proportions. Sulfide inclusions (1.5-15 μm) are composed of galena and/or chalcopyrite, and in some lithologies, zircon-hosted inclusions contain sulfide melt co-entrapped with silicate melt, indicating sulfide saturation.

LA-ICP-MS data of SMIs show a wide range of compositions ranging from sub-alkaline rhyolite to alkaline intermediate melts (trachyte to trachyandesite); notably, this is prevalent within single lithologies indicating that zircons preserve melt aliquots along a variety of possible evolutionary pathways involving fractionation, contamination, and/or magma mixing. Metal contents of SMIs generally increase from Lower Bousquet units to Upper Bousquet units, and low metal values in the Footwall unit are possibly due to magmatic degassing prior to ore formation.

Combined Micron-Scale EBSD Study and LA-ICP-MS Analysis Reveals the Evolution of Pyrite and Cu-Fe-Sulphides in the Olympic Dam Cu-U-Au-Ag Deposit, South Australia

Samuel A. King[2], Kathy Ehrig[1, 2], Nigel J. Cook[2], Cristiana L. Ciobanu[2]
1. BHP Copper SA - Olympic Dam, Adelaide, SA, Australia, 2. School of Chemical Engineering, The University of Adelaide, Adelaide, SA, Australia

The breccia-hosted Olympic Dam IOCG-type deposit, South Australia, displays a systematic inwards and upwards zonation of Cu-Fe-sulphides from pyrite-chalcopyrite, bornite-chalcopyrite, to bornite-chalcocite assemblages, interpreted as a preserved primary feature. This includes lower-grade pyrite-chalcopyrite disseminations (outer shell) enclosing the resource. Superimposed tectonothermal events have modified deposit mineralogy and ore textures since mineralisation at ~1.6 Ga but remain imperfectly constrained. To better understand both syn- and post-formational deformation, we combine electron backscatter diffraction (EBSD) with laser ablation-inductively coupled-plasma mass spectrometry (LA-ICP-MS) element mapping of pyrite and chalcopyrite to reveal microstructures and associated trace element (re)distribution.

Pyrite grains from representative outer shell assemblages either preserve marginal dislocation glide or dislocations formed by dislocation creep, both indicative of ductile deformation >260°C. Two stages of ductile deformation are inferred from such microfabrics, separated by intermittent fracturing. Dislocations display characteristic enrichment in Ag-Au-Pb-Bi-Te and are shown to disrupt an earlier As-Co-Ni compositional zoning.

Chalcopyrite from pyrite-chalcopyrite and chalcopyrite-bornite assemblages is revealed as fine aggregates despite its homogenous appearance. Silver, Bi, Pb, and Sb are present along the aggregate grain boundaries and, to a lesser extent, along twin planes. Chalcopyrite also preserves (220) and (001) crystallographic preferred orientations, {110} and {112} twinning, grain boundary migration recrystallization, and foam textures, which can be correlated to P-T conditions.

Coupled EBSD-LA-ICP-MS analysis provides valuable new insights into ore genesis and superimposed events. Our findings represent the first evidence for ductile deformation in pyrite from a brecciated IOCG deposit. Novel application of EBSD to chalcopyrite shows its suitability for unravelling low-medium–temperature stages of ore evolution, complementing data from coexisting pyrite, a mineral routinely used to depict higher-temperature deformation. Grain-scale elemental distributions confirm the key role played by microstructures in Cu-Fe-sulphides as traps for critical/precious metals (Ag, Bi, Sb, Au, Te), and deleterious Pb.

Funding: Australian Research Council LP200100156.

Geology and Mineralogy of Pb-Ag-Cu Veins of the Central Plateau Deposit of the Jbel Addana District, (Western Anti-Atlas, Morocco)

Zoubair El Ouad[1, 2], Mustapha Souhassou[1], Moha Ikenne[2], Hibti Mohamed[3], Ilmen Said[4], Abia El Hassan[2], Younes Moundi[5], Atmane Madi[5], Abdelhak Karfal[5]
1. L2GBEI, Research team EGERNE, Polydisciplinary Faculty of Taroudant, Ibn Zohr University, Taroudant, Morocco, 2. LAGAGE, Faculty of Sciences, Ibn Zohr University, Aagadir, Morocco, 3. L2GCE, Faculty of Sciences and Technologies, Cadi Ayyad University, Marrakech, Morocco, 4. CAG2M, Polydisciplinary Faculty of Ouarzazate, Ibn Zohr University, Ouarzazate, Morocco, 5. MANAGEM Group, Twin Center, Casablanca, Morocco

Jbel Addana is a branch-parallel anticline of the Ordovician belt of Jbel Bani. It located about 35 km south-east of the Precambrian inlier of Tagragra d'Akka in the western Anti-Atlas of Morocco, where it dominates the Drâa plain for some 40 km following a NE-SW trend. Geologically, this anticline consists of alternating poorly metamorphosed layers of sandstone and clay, which represent the Ktaoua, Rouid-Aissa, and Second Bani Formations of Upper Ordovician ages (Caradoc and Ashgill). In metallogenic terms, this anticline corresponds to a Pb-Ag-Cu ore district comprising several deposits, where the orebodies are represented by a system of en échelon veins with a vertical dip. The Central Plateau deposit is located southwest of the anticline and considered the widest in the district. Mineralized bodies consist of parallel veins extending in a WNW-ESE–trending direction. Macroscopic paragenesis is mainly composed by massive galena associated with a quartz-iron gangue, with an altered assemblage including malachite, iron oxides-hydroxides. Mineralogical analysis under microscope has revealed a paragenetic succession comprising three stages: i) an early hydrothermal stage with quartz, siderite, calcite, galena, pyrite, and chalcopyrite; ii) a late hydrothermal stage composed of quartz II, calcite II, and galena II; and iii) a typical lead-copper to iron supergene alteration stage characterized by concentric-textured cerusite and anglesite replacement around galena, with collomorphic-textured malachite, fibrous coronadite, and an iron assemblage with goethite and iron oxides.

Mineralogy and Trace Element Composition of Sulfides from Porphyry-Stage and Minor Late Polymetallic Mineralization in the Gaby District, Northern Chile

Alexandre Cugerone[1], Kalin Kouzmanov[1], Lorenzo Tavazzani[2], Lluís Fontboté[1], Sergio P. Henríquez[3], Juan C. Toro[3], Rubén Pardo[3], Carolina Rodriguez[3]
1. University of Geneva, Geneva, Switzerland, 2. Institute of Geochemistry and Petrology, ETH Zürich, Zürich, Switzerland, 3. CODELCO, Santiago, Chile

Metal precipitation in porphyry systems involves multiple hydrothermal events forming a continuum of porphyry-style alteration and mineralization commonly overprinted by a late polymetallic stage. Textural variations and trace element content of major sulfides remain largely unexplored in porphyry systems, despite their significant potential for understanding ore-forming processes. Chalcopyrite, bornite, pyrite, and sphalerite composition from both porphyry mineralization and a minor polymetallic stage, as well as supergene chalcocite, are investigated in the Gaby district (northern Chile) using optical microscopy, automated mineralogy (QEMSCAN), electron microprobe, and laser ablation-inductively coupled plasma-mass spectrometry (LA-ICP-MS). Copper mineralization is developed in a suite of Eocene porphyries intruding a Carboniferous-Permian volcano-sedimentary sequence and a Permo-Triassic plutonic complex. An early stage of copper introduction, coeval with potassic alteration, occurs in veinlets and disseminations with chalcopyrite, bornite, molybdenite, and minor pyrite intergrown with K-feldspar and secondary biotite. A second stage of copper precipitation is paragenetically associated with white mica alteration. Chalcopyrite, molybdenite, and locally pyrite intergrown with white mica and chlorite, form a dominantly disseminated mineral association, commonly overprinting the potassic alteration. Sulfides occur also in early halo veinlets that partly reopen previous structures. Traces of luzonite formed at the expense of high-temperature Cu-Fe sulfides are also observed. Polymetallic intermediate-sulfidation mineral associations are sporadically observed at Gaby, consisting of Fe-poor (<0.4 wt% Fe) sphalerite, galena, chalcopyrite, tennantite-tetrahedrite, pyrite, locally wittichenite, and native gold, intergrown with early Ca-Mg-Fe carbonates, minor rhodochrosite, late siderite and Mn-poor calcite, and traces of kaolinite. Secondary chalcocite is formed during supergene processes and can be detected down to >500 m from the present-day surface. LA-ICP-MS results indicate higher In content (5-20 ppm) in porphyry-stage chalcopyrite compared to the polymetallic veins and high Co content (up to 2 wt. %) in pyrite from polymetallic veins in volcanic rocks.

White Mica Alteration in Porphyry Systems: Insights from the Gaby Porphyry Cu Deposit, Northern Chile

Alexandre Cugerone[1], Kalin Kouzmanov[1], Lluís Fontboté[1], Luis Krampert[1], Alexandra Tsay[1], Zoltan Zajacz[1], Sergio P. Henríquez[2], Juan C. Toro[2], Rubén Pardo[2], Carolina Rodriguez[2]
1. University of Geneva, Geneva, Switzerland, 2. CODELCO, Santiago, Chile

In porphyry systems, spatial distribution of white mica alteration frequently correlates with economic Cu-Mo(-Au) mineralization. White mica occurs either i) as coarse-grained infill of open-space veins with quartz, found in apical to root zones of the system; ii) as pervasive alteration or alteration halos commonly in association with chlorite and chalcopyrite, or iii) as a sericitic alteration halo together with quartz and pyrite along porphyry D-veins.

In the Gaby porphyry copper deposit (northern Chile), white mica alteration is well developed in Eocene porphyries, intruding a Carboniferous-Permian volcano-sedimentary sequence and a Permo-Triassic plutonic complex, correlating generally with relatively high Cu grades. Using a combination of chemical in-situ and mapping techniques and integrating both textural and chemical characteristics, we evidence two stages of white mica alteration. Early stage is associated with high Cu grades, shows low Mg/Fe ratios, and is divided in four substages: E1 and E2 developed as coarse-grained white mica in the apical zone of the porphyry near USTs, as pervasive alteration, with high Fe content, and infill of mica-quartz veins with high W and Sn, respectively; E3 is intergrown with chlorite and chalcopyrite as disseminations or early halo veins, with high Li, or enriched in Ti, Sc, V, and Cr when developed on volcanic rocks. Late white mica alteration has lower Cu grades, high Mg/Fe ratios, and is divided into two substages: L1 forming thin veinlets with chlorite is rich in Li, Fe, Mg; L2 is related to D-veins with pyrite and has low Mg and Fe content. Chlorite is present in both early, Cu-rich and late, Cu-poor white mica alterations, but in different proportions. Within the alteration halo, the chemical composition of white mica combined with quantitative mineralogy provides evidence of extensive vein reopening, highlighting the significance of telescoping in hydrothermal alteration styles in porphyry systems.

Alteration Assemblages and Quantitative Mineralogy of the Boundary Zone Zn-Pb ± Ag Deposit, Yukon, Canada

Haruna M. Grema[1, 2], Joseph M. Magnall[1], Sarah A. Gleeson[1, 2], Jack E. Milton[3], Anja M. Schleicher[1], Alicja Wudarska[1, 4], Hans-Martin Schulz[1]
1. GFZ German Research Centre for Geosciences, Potsdam, Germany, 2. Institute of Geological Sciences, Freie Universität Berlin, Berlin, Germany, 3. Fireweed Metals Corp., British Columbia, BC, Canada, 4. Institute of Geological Sciences, Polish Academy of Sciences, Warsaw, Poland

The Boundary Zone Zn-Pb ± Ag deposit, Selwyn Basin, Canada, is a recently discovered clastic-dominated (CD-type) occurrence with two distinct ore mineralization stages within the Late Ordovician-Early Silurian Duo Lake and Middle-Late Devonian Portrait Lake Formations. Understanding the intricacies of the alteration assemblages allows for developing vectors toward the base metals that are essential for meeting net-zero targets and renewable energy generation. Seventy-three drill core samples from nine drill holes were analyzed using quantitative X-ray diffraction (QXRD), electron probe microanalysis (EPMA), and U-Pb secondary ion mass spectrometry (SIMS) to examine alteration assemblages linked to the sulfide mineralization.

Sulfide formation at the Boundary Zone occurs with quartz, pyrite, barian-mica, fluorapatite, and siderite, with specific assemblages associated with different host rocks (biosiliceous mudstones, conglomerates, and volcaniclastics). Stratabound ore-stage I mineralization in the Duo Lake and Portrait Lake Formations involves the replacement of radiolarian-rich mudstone beds. This mineralization is coeval with the transformation of Opal-A to chalcedony and microquartz in and around the mineralized intervals. Additionally, in the Portrait Lake Formation, authigenic barite replacement by pyrite and barian-mica is observed. The dissolution-reprecipitation of quartz, likely preserved porosity and coupled with barite dissolution, generated permeability facilitating metal-bearing fluid flow.

Ore-stage II sulfides formed via hydrothermal fluid-induced brecciation and veining of the host rocks with stylolitic margins and are accompanied by fluorapatite, megaquartz, and pyrobitumen. Mg-siderite formed together with early vein sphalerite and as late cavity fillings contemporaneous with bladed megaquartz in the Portrait Lake Formation. Fluorapatite in one vein that crosscut ore-stage I records a Middle Jurassic U-Pb age; however, timing related to ore-stage II is undetermined. Collectively, the multiple mineralizing events at the Boundary Zone indicate a protracted history of ore formation and host rock alterations, likely spanning from diagenetic stages to periods of Cordilleran-related deformation within the Selwyn Basin.

Factors for Classification of the Precious and Base-Metal Mineralization in Epithermal Deposits: Evidence from Apatite Chemistry in Ore-Related Igneous Activity

Masaya Koseki[1], Shota Satori[2], Yasushi Watanabe[1], Takuya Echigo[1], Shogo Aoki[1]
1. Akita University, Graduate School of International Resource Sciences, Akita, Japan, 2. National Institute of Advanced Industrial Science and Technology (AIST), Geological Survey of Japan, Tsukuba, Japan

Metal assemblages concentrated in epithermal deposits are related to Cl and S contents in hydrothermal fluids. The differences of Cl and S contents in hydrothermal fluids may reflect the difference of ore-related magma compositions. We estimated Cl and S contents exsolved from magmas by using apatite chemistry in the felsic rocks associated with the Miocene Kitano-o Au-Ag and Inaushi Cu-Pb-Zn deposits in northeastern Hokkaido, Japan.

The EPMA analysis indicated that Cl content in the apatite is 0.56–0.97 wt% in the Inaushi dacite, while it is 0.37–0.45 wt% in the Kitano-o rhyolite. Based on the partition coefficients between apatite and melt, the calculated maximum Cl contents in the Inaushi and Kitano-o melts are 1.48 wt% and 0.96 wt%, respectively. Both apatites contain ~100 ppm S, indicating <10 ppm of S in the Inaushi and Kitano-o melts.

We also traced the Cl in the melt during magmatic fractionation of the Inaushi dacite by using apatites of different crystallizing stages, which are determined by paragenetic sequence of apatite-hosting minerals. Maximum Cl content is 0.78 wt% during the early plagioclase crystallization and 0.97 wt% during the amphibole crystallization, and then it decreases to 0.87 wt% and 0.83 wt% during titanomagnetite and groundmass crystallization, respectively.

The significant difference of Cl content and similar small S content between the Inaushi dacitic and Kitano-o rhyolitic melts estimated by apatite chemistry suggest that the Cl content in the epithermal fluids for the Inaushi base-metal and Kitano-o precious metal mineralization was inherited from the those in the magmas associated with these deposits. At Inaushi, Cl content was estimated to have exsolved from melt during amphibole-titanomagnetite crystallization stages.

Fluid History of Zambian Copper Deposits Based on Metal Zonation in Pyrite

Ross Large
University of Tasmania, Hobart, TAS, Australia

This study was aimed at determining the textural and chemical zonation in pyrite from the black shale Copper Orebody Member at the Nkana deposit and the footwall Mindola Clastics in the Konkola deposit. Pyrite in fifteen black shale samples from Nkana and eighteen footwall sandstones from Konkola were mapped in detail by LA-ICPMS to determine the timing of the various metals in the ore system. The resulting pyrite maps demonstrate that the original sedimentary pyrites appear to be low in Co, Ni, Cu, and As. The first fluid phase appears to be rich in Co and As, shown by the penetration of Co and As into the structure of early pyrite. Copper mineralization follows the Co-As phase, but only penetrates pyrite or diffuses into pyrite around the pyrite rims. The Cu fluid phase is accompanied by minor Pb, Zn, Ag, and Bi, which penetrate the rims, internal fractures, and grain boundaries of the pyrite. Significant conclusions are 1) the sequence of fluid events was a Co-As fluid first, followed by a Cu-rich fluid with traces of Pb, Zn, Ag, and Bi to follow; 2) the Cu-silfides did not replace pre-existing pyrite, indicating that sulfur for the copper minerals was most likely sourced from anhydrite beds and nodules in the sedimentary rocks; 3) Cu and Ag contents of pyrite are not a reliable indicator of proximity to ore; 4) cobalt-alteration of pyrite appears to increase in intensity when approaching the copper ore zones; and 5) poorly developed Pb and Bi rims and grain boundary alteration up to 100 ppm is present on all pyrites investigated whether proximal or distal to ore. Thanks to Mawson Croaker and Nicky Polington, who originally collected the legacy samples used in this study.

Origin of the Magnetite-(Apatite) and IOCG Deposits of the Coastal Cordillera of the Andes

Ignacio Ledesma, Fernando Tornos
IGEO CSIC-UCM, Madrid, Spain

The Andean Coastal Range between Santiago (Chile) and Lima (Peru) hosts dozens of magnetite-(apatite) (MtAp) and iron oxide-copper-gold (IOCG) deposits, forming one of the major clusters of these styles of mineralization. Both of them are related to an Upper Jurassic-Lower Cretaceous magmatism. The MtAp mineralization is always located in extensional zones within arc-parallel strike-slip faults and is characterized by the presence of massive magnetite-actinolite/diopside-apatite mineralization, many times capped by a pegmatite. IOCG deposits replace mafic volcanoclastic rocks or the previous MtAp mineralization; the assemblages are somewhat similar but have no apatite and are sometimes enriched in sulfides, including chalcopyrite, pyrite, and bornite. Vertically, they evolve into quartz-calcite-sulfide–rich veins.

In southern Peru, Marcona hosts significant magnetite mineralization crosscutting the basement, with massive tabular orebodies showing magmatic relationships with coeval dacitic dykes; vertically, it evolves into a massive diopside-apatite pegmatite. Above the Precambrian-Mesozoic unconformity, the giant orebodies of Marcona Cobre and Mina Justa host dominantly stratabound IOCG mineralization. The nearby MtAp Hiero Acarí deposit is located along subvertical tensional veins hosted by quartz-diorite; here magnetite-rich parts alternate with actinolite-magnetite-apatite pegmatite zones. In both cases, mineralization is hosted by a wide aureole of Ca-Fe-K hydrothermal alteration. Locally, stratabound IOCG-like mineralization alternates with large calcic and magnesial skarns. .

Radiogenic isotope geochemistry indicates significant crustal contamination in both types of deposits. MtAp rocks seem to be derived from contaminated magmatic systems but inherit a significant mantellic Nd signature; IOCG deposits track a more crustal derivation. Preliminary geochronology confirms that major MtAp mineralization took place around 165-154 Ma, with IOCG mineralization ca. 20 Ma later. However, there's a discrepancy in ^{40}Ar-^{39}Ar dating, suggesting a complex hydrothermal evolution for the mineral systems.

Understanding Overprinting Mineralizing Systems of the Ruby Hill Deposit, Eureka, NV: Implications for Future Exploration

Alyssa Lindsey, Simon Jowitt
University of Nevada, Reno, Reno, NV, USA

The Ruby Hill deposit in Eureka, NV, hosts gold, silver, base metal, and critical mineral mineralization. The mineralization within the Eureka district includes Zn-Pb skarn, pollymetallic carbonate replacement deposit (CRD), distal disseminated Ag-Au, and Carlin-type Au. Although historical mining focused on the CRD-type mineralization from the 1880s to the 1950s and Carlin-type gold from the 1990s onward, recent exploration has outlined additional CRD mineralization beneath cover that was previously unrecognized within the greater Ruby Hill system. Understanding the temporal, spatial, and genetic relationships between the different styles of mineralization at Ruby Hill is crucial for future exploration in the region.

The initial research undertaken during this project focuses on a total of 166 diamond core and reverse circulation drill holes previous logged by Ruby Hill geologists within the north end of the district. These data provide a guide for spatial and mineralogical relationships between skarn, CRD, and distal disseminated mineralization at Ruby Hill. Preliminary geochemical analysis indicates that the Ruby Hill skarn and CRD systems are geochemically zoned with distinct overlaps between these different systems. Initial interpretation of whole-rock geochemical data for the intrusions in the Ruby Hill area has also identified multiple phases of granodiorite intrusion, although further research is needed to determine the specific phase(s) of these intrusions that are responsible for the different styles of mineralization at Ruby Hill. Further research in the area will include detailed paragenetic and geochronological studies to determine the relative and absolute timing of the formation of the different mineral systems at Ruby Hill as well as any potential genetic links between these systems.

Facies-Constrained Lithogeochemistry in the Irish Midlands: A Comparison Between the Collinstown Prospect and the Stonepark Zn Deposit

Joseph Magnall[1], Sarah Gleeson[1], Philip Rieger[4], Mark Holdstock[2], Kacie Wellington[3], David Furlong[2], Murray Hitzman[4]
1. GFZ Potsdam, Potsdam, Germany, 2. Group Eleven Mining and Exploration Ltd., Dublin, Ireland, 3. Teck Ireland Ltd, Mullingar, Ireland, 4. iCRAG, Dublin, Ireland

Lower Carboniferous marine sedimentary rocks are host to some of the most significant Zn-Pb deposits in the world. In the Irish Midlands, more than 25 economic and sub-economic base metal deposits (>20 Mt Zn metal) are hosted by a succession of Lower Carboniferous carbonate rocks host. Lithogeochemical data is now routinely generated on mineralized rocks during mineral exploration and project development, resulting in large quantities of multivariate data. However, there is a marked drop-off in resolution and quality of data immediately beyond ore zones. Facies-constrained sampling approaches that extend from ore zones to unmineralized rocks could help to define more sensitive vectors towards ore, but remain time consuming, costly, and subjective. In this study, legacy data has been compiled from two sites in the Irish Midlands to evaluate i) the geochemical footprint of mineralization in the Limerick Basin, where the Stonepark deposit is located; and ii) the potential for undiscovered mineralization at depth in the Dublin Basin, which is peripheral to the world-class Navan deposit. A workflow is presented for extracting maximum value from these data sets, which involves the imputation of missing data and evaluation of data structure using statistical methods (e.g., PCA). The legacy data has then been augmented with more discrete sampling of individual sedimentary facies in stratigraphic intervals that extend beyond ore zones and prospective intervals, to constrain whether specific beds might preserve cryptic alteration. Importantly, facies containing pyrite are shown to be enriched in trace elements with a known hydrothermal affinity (e.g., As).

Mineral Prospectivity Mapping for Porphyry Copper-Gold Mineralization Style Using Fuzzy Logic Method in the Arthurs Seat Project, Jamaica

William Miguel Ortiz
UNIVERSIDAD NACIONAL DE INGENIERIA, LIMA, Peru

The Arthurs Seat project lies in one of the “Central Inliers” of Cretaceous rocks in Jamaica, approximately 45 km northwest of Kingston. The project area is overlain by Lower Cretaceous conglomerates, volcanic breccias, intercalated ash, and crystal tuffs and andesitic to basaltic andesite lava flows rocks of the Arthurs Seat Formation. The Upper Cretaceous comprises a sequence of limestone, limestone-siltstone, siltstone, andesite-tuff, conglomerate, and limestone rocks of Peters Hill, Black River, Dawburn Content, Main Ridge, Slippery Rock, and Guinea Corn Formation. The Cretaceous rocks are covered by Eocene sandstone, sandstone-limestone, limestone of the Lichfield, Chapelton, and Troy Formations.

The Fuzzy Logic method was successfully used to map areas of Cu-Au porphyry mineralization potential in the Arthurs Seat project. Proximity to geological features is transformed into fuzzy membership functions based upon qualitative and quantitative knowledge of spatial associations between known Cu-Au porphyry occurrences and geological features in the area. Through this approach, favourable lithologies, geophysical, structural, and geochemical criteria were integrated into a prospectivity map. The most favorable lithologies are the intrusive rocks hosted by Arthurs Seat formation. Total Magnetic Intensity Reduced to Pole was used to map circular features, and First Derivative of Total Magnetic Intensity Reduced to Pole combined with factual faults was used to map lineaments. Soil data was leveled by formation to apply Principal Component Analysis (PCA). Five blind porphyry Cu-Au targets were identified in the Arthurs Seat project and were validated by mineral occurrences in rock chip samples greater than 1% in Cu.

The Characteristics of Chromite in the Lateritic Nickel Deposit from East Sulawesi Ophiolite, Indonesia

Ibnu Munzir, Marek Michalik
Institute of Geological Science, Jagiellonian University, Krakow, Poland

The chromite present in the peridotite mantle in ophiolites offers valuable insights into the evolution and formation of ophiolites. Podiform chromites have been identified in the lateritic nickel deposit within the Phanerozoic East Sulawesi Ophiolite (ESO) located at Sulawesi Southeast Arm. As a result of metamorphism and significant deformation, the majority of chromite grains in the ESO peridotites are present in various types, including chromite and aluminum-rich chromite. Most of the Al-rich chromites are found in the primary chromite, while Cr with positive correlations with Fe are found in secondary chromite. Chemically, Fe substitutes the Al in the primary chromite during the weathering process causing the increasing value of Fe with the lowering of Al. The values of major and trace elements largely vary from both types, indicating that the chromites have experienced a strong alteration and element mobilization. Major, minor, and trace elements that occurred in the chromites are calculated to determine the tectonic origins and the peridotites mantle evolution in the ESO. The chromite characteristics from the Phanerozoic ESO peridotites mantle have higher Cr# values and lower Mg# values, indicating a higher degree of Mg mobility. In-situ major, minor, and trace element characteristics of podiform chromites in the ESO show a strong signature of peridotite geological evolution.

The Essential Role of High-Volume Base Metals: Catalysts of Modern Industry

Chawanangwa S. Mwansa
University of Zambia, Lusaka, Zambia

The demand for vital high-volume base metals, including copper, aluminum, nickel, zinc, and lead, continues to shape industries worldwide due to their indispensable roles in various sectors. This abstract provides a comprehensive overview of these metals, highlighting their significance, applications, and evolving market dynamics. Copper, known for its conductivity and malleability, is crucial in electrical wiring, plumbing, and electronics. Its role in renewable energy systems, such as solar panels and wind turbines, underscores its importance in the transition to sustainable energy. Aluminum, prized for its light weight and corrosion resistance, dominates industries like aerospace, automotive, and construction. Its recyclability further enhances its appeal in eco-conscious markets. Nickel, essential in stainless steel production, is experiencing increased demand due to its vital role in electric vehicle batteries and emerging technologies like hydrogen fuel cells. Zinc, valued for its anti-corrosive properties, finds widespread use in galvanizing steel, alloy production, and the pharmaceutical sector. Lead, despite declining usage in some applications due to environmental concerns, remains indispensable in batteries for automotive and industrial purposes. Factors such as technological advancements, geopolitical dynamics, and sustainability initiatives are reshaping the landscape of these base metals. The transition to electric vehicles, infrastructure development, and renewable energy expansion are driving demand for these metals, with implications for supply chains, pricing, and environmental regulations. The presented data provides valuable insights into the multifaceted nature of vital high-volume base metals, highlighting their pivotal roles in modern economies and their evolving trajectories in a rapidly changing world.

Allanite U-Th-Pb Geochronology of IOCG and Zn-Ag-Pb Deposits in the Mount Isa Inlier, Australia

Luís Portela[1], Coralie Siégel[2], Charlotte M. Allen[3], Jim R. Austin[4]
1. University of Aveiro, Aveiro, Portugal, 2. Discovery Program, Mineral Resources, Commonwealth Scientific and Industrial Research Organisation (CSIRO), Australian Resources Research Centre, Perth, WA, Australia, 3. School of Earth and Atmospheric Sciences and Central Analytical Research Facility, Queensland University of Technology, Brisbane, QLD, Australia, 4. Potential Fields Geophysics, Mineral Resources, Commonwealth Scientific and Industrial Research Organisation (CSIRO), Lindfield, NSW, Australia

The Mount Isa Inlier in Northwest Queensland (Australia) is renowned for containing multiple world-class ore deposits. The Inlier is composed of Early-Middle Proterozoic sedimentary, volcanic, and intrusive rocks, heterogeneously affected by polyphase metamorphism, deformation, and metasomatism spanning the Barramundi, Wonga, and Isan orogenies (1900-1500 Ma). Allanite is common in these deposits and, given its high closure temperature for the U-Pb system, presents a valuable opportunity to provide insights into the timing of ore-forming processes.

This study reports new LA-ICP-MS U-Th-Pb ages and compositions of allanites from several deposits from the Cloncurry District in the Mount Isa Eastern Succession, including IOCG (Brumby, Canteen and SWAN) and Broken Hill-type (BHT) Pb-Zn-Ag deposits (Altia and Cannington), and new ages for the Cameron River prospect (IOCG) in the Mary Kathleen Domain. Allanites show distinct composition in the studied deposits: (a) very low La/Sm and Th/U and positive Eu anomalies in Cannington; (b) low and homogeneous La/Sm (~10) and wide-ranging Th/U (up to 440) at Altia; (c) high La/Sm (>80) and low Th/U in SWAN and Canteen; (d) intermediate compositions at Cameron River and Brumby.

Allanites from the Cameron River prospect yield a well-defined age cluster at ca. 1.74 Ga, consistent with the emplacement of the nearby Wonga Suite. Allanites from deposits of the Eastern Succession revealed consistently younger results. Brumby, Canteen, SWAN, and Cannington allanite U-Pb ages converge to ~1.5 Ga, overlapping in time with the emplacement of the Williams Batholith (1516-1493 Ma). Altia allanites define a lower intercept age of ~1.35 Ga. The ages obtained for the BHT deposits are younger than expected age of 1.7-1.67 Ga and are interpreted to record the timing of a late skarn overprint and/or retrograde hydrous alteration. The ~1355 Ma ages yielded by Altia allanites are consistent with those reported for Mo-REE mineralisation in the district.

Copper Bio-Enrichment: Results from Isotopic Analysis at the Las Cruces Deposit, Spain

J. Javier Rey-Samper[1], Ryan Mathur[2], Fernando Tornos[1]
1. Spanish National Research Council (CSIC), Madrid, Spain, 2. Juniata College, Huntingdon, PA, USA

The Las Cruces is a volcanogenic massive sulfide deposit located in the easternmost part of the Iberian Pyrite Belt. The primary stratabound massive sulfides of late Devonian- Early Carboniferous age have been affected by a pervasive supergene alteration, giving rise to a unique secondary mineralization. The secondary rocks include the widespread reduction of the former gossan and the formation of a cementation zone dominated by massive chalcocite and covellite. The formation of the supergene mineralization is probably due to the reaction, from ca. 7 My to Present, of different types of fluids that circulated through fault-controlled and unconformity-related confined aquifers below a thick (>200 m) sequence of Messinian marl. The key feature of the deposit is the existence of a complex microbial ecosystem that dramatically modified the pre-Alpine assemblages.

The ore grade at the cementation zone has made Las Cruces one of the richest copper deposits worldwide. For this study, rock samples from fresh outcrops and water samples from the aquifers in and outside the mine have been analysed in an Isoprobe multicollector ICP-MS at the University of Arizona. $\delta^{65}Cu$ values for ore samples range from 9.2 to +4.3‰, while water values range between -0.01 ‰ to +4.5‰.

The extremely low $\delta^{65}Cu$ values and their distribution are consistent with a microbial control on the precipitation of the secondary copper sulfides. Our best interpretation of the $\delta^{65}Cu$ values are that the Las Cruces had a small cementation zone below a gossan cap and previous to the burial at 7 My; later on, sulfate and copper reducers coexisting with other microbes in the deep anoxic environment were able to trap copper transported as Cu^{2+} by the water, lowering the originally high $\delta^{65}Cu$ values. The likely source of copper is the dismantling, oxidation, and dissolution of massive sulfides located upstream.

Lithostratigraphic Controls on the Mineralizing System at the Stonepark Zn-Pb Deposit, SW Ireland

Philip Rieger[1], Murray W. Hitzman[1], Victoria Figueira Susin[1], Mark Holdstock[2], Eoin Dunlevy[3], Aline Melo[1]
1. iCRAG - SFI Centre for Research in Applied Geosciences, University College Dublin, Dublin, Ireland, 2. Group Eleven Mining & Exploration Ltd, Dublin, Ireland, 3. Terranta GmbH, Aachen, Germany

Base metal demand is projected to increase significantly to sustain the energy transition. Zinc is an essential element for building the infrastructure for renewable energy production. A large proportion of global Zn resources are hosted in sedimentary basins, and to meet the future Zn demand, current mineral system models need to be refined in order to aid exploration programs. The Lower Carboniferous basin in Ireland ("Irish Orefield") is one of the most prospective sedimentary basins for Zn-Pb mineralization. Irish-type Zn-Pb deposits are typically associated with normal faults and form stratabound bodies within fine-grained limestones. Pre- and syn-ore dissolution/brecciation of the host limestones (i.e., Waulsortian Limestone Fm) is a common feature in many Irish-type deposits. The Stonepark and neighbouring Pallas Green deposits share many characteristics with other Irish-type mineralization, but previous studies have noted two major differences: (1) the apparent absence of faults controlling the location of zones of mineralization; and (2) the close spatial association of zones of mineralization and brecciation with igneous rocks (dykes, sills, diatremes). To better constrain the mineral system at Stonepark, we present a 3D model that integrates drill core logging, exploration drilling data, and regional seismic interpretation. There are considerable differences in Waulsortian limestone facies (mudmound micrites, mudmound-flank wackestones, off-mudmound argillaceous wackestones). Our 3D modelling suggests that this initial facies architecture of the Waulsortian limestones had a strong control on the spatial distribution of zones of alteration/mineralisation due to the lithological contrasts (e.g., permeability, reactivity) between the different facies. The 3D distribution of Waulsortian limestone facies may also be a proxy for identifying initial, now blind or obscured, normal faults at the Stonepark deposit. This study highlights how comprehensive 3D geological modelling utilizing multiple data sets can be used to identify controls on the formation of Zn-Pb deposits in highly complex basin-hosted mineral systems.

Hematite Orebodies Formed by Multiple Geological Events: A Case Study from the Akatani Deposit, Niigata Prefecture, Japan

Kotaro Seno[1], Yasushi Watanabe[1], Takuya Echigo[1], Keita Itano[2], Shogo Aoki[1], Yoshiaki Kon[3], Shota Satori[3]
1. Graduate School of International Resouce Sciences, Akita University, Akita city, Japan, 2. Department of Mathematical Science and Electrical-Electronic-Computer Engineering, Akita University, Akita city, Japan, 3. National Institute of Advanced Industrial Science and Technology (AIST), Geological Survey of Japan, Tsukuba city, Japan

The Akatani hematite deposit in northern Japan is associated with Cretaceous skarn and Miocene hydrothermal mineralization and has a complex history of mineralization, showing the occurrences of diverse hydrothermal minerals such as iron-oxide, skarn, dolostone, and hydrothermal altered minerals like chlorite and kaolinite. We conducted petrographic descriptions and trace element analysis of hematite to identify the formation process of the deposit. The host rock of the Akatani deposit consists of Early Jurassic limestone and mudstone covered or intruded by Late Cretaceous granites (LCG, ilmenite-series) and Middle Miocene rhyolitic rocks (MMR). The hematite orebodies associated with LCG are characterized by the presence of skarn minerals such as andradite and clinopyroxene. The hematite in the distal skarn coexists with andradite and formed by replacing clinopyroxene. Most hematite in the skarn is altered to magnetite. In contrast, hematite orebodies present between MMR and dolostone and in Fe chlorite zones between MMR and skarn are characterized by the presence of siderite and/or kaolinite together with chalcopyrite, pyrite, barite, and/or fluorapatite. The hematite associated with LCG is rich in granitophile elements such as Sn and Mo, poor in REEs (~13 ppm), and shows no correlation between Mn+Zn and Mg, whereas that associated with MMR is poor in granitophile elements, rich in REEs (~40 ppm), and shows a strong positive correlation between Mn+Zn and Mg. Based on these observations, we conclude that the Akatani hematite orebodies formed by multiple mineralization processes form Cretaceous to Miocene.

Geological-Mining Characterization of the Sierra Rancheria Using Remote Sensing in the Northern Region of Chihuahua, Mexico

Carlos A. Silva[1], Luis A. Jiménez[2], Perla Rodríguez[1]
1. Universidad Autónoma de Ciudad Juárez, Juárez, CHH, Mexico, 2. Instituto Politécnico Nacional, Mexico City, CMX, Mexico

Remote sensing techniques are one of the most important to carry out more selective mining geological exploration, which leads to a significant reduction in associated times and costs. Combining these techniques with fieldwork is essential to ensure the consistency and validity of the results. Sierra Rancheria, located in the municipality of Ahumada in Chihuahua, Mexico, belongs to the Basin and Range province. According to the Mexican Geological Service, the geology of Sierra Rancheria is characterized by an intrusive body of granitic composition (Eocene) that affects units of the Aurora Group (limestone and shale) from the Lower Cretaceous, giving rise to areas of hornfels and skarn. This region is recognized as part of the metallogenic provinces of Ag, Pb, and Zn in Mexico. The study area focuses on the southeastern portion of Sierra Rancheria, adjacent to a non-metallic mineral mining concession, where indications of a skarn-type deposit have been identified. The region has economic importance in metallic and non-metallic minerals, although there is currently a lack of geological-mining information, which prevents the use of these resources. This study employs remote sensing techniques, such as band algebra and band ratio analysis, using ASTER and PlanetScope imagery. These techniques were complemented by fieldwork and petrographic and geochemical analyses, using a portable X-ray fluorescence (XRF) spectrometer. Using this methodology, lithological contacts and zones of hydrothermal alteration were defined. Targets for exploration were delimited to guide future geological exploration and discover potential mineralization areas more accurately.

Lithogeochemistry and Magnetic Susceptibility for Stratigraphy and Alteration Identification in the Western Foreland Shelf of the DRC

Amanda M. Stoltze[2], Tim Dunnett[1]
1. Ivanhoe Mines Exploration, Sandton, South Africa, 2. Ivanhoe Mines Exploration, Kolwezi, Democratic Republic of the Congo (Kinshasa)

The Western Foreland Shelf (WFS) is located at the western edge of the Central African Copper Belt (CACB) of the Democratic Republic of Congo. Sedimentation onset was triggered by rifting and deposition of conglomerate and mass flow sediments and sandstone (Mwashya) and finalised with the deposition of the Kundelungu Group at 630 Ma. A 1.8-km-thick package of Grand Conglomerate was deposited on the shelf during rifting. The Grand Conglomerate on the WFS is dominated by diamictite, conglomerate, sandstone, and siltstones layers. Variations in diamictite are subtle, making the identification of their stratigraphic position difficult to assign, especially in parts of the shelf with limited prior exploration. Alteration of sediments is also observed associated with mineralisation but is also only subtly manifested.

For this reason, alternative methods of characterising stratigraphy such as magnetic susceptibility, whole-rock geochemistry, and VNIR-SWIR spectroscopy have been investigated. Magnetic susceptibility is carried out on all drill core, with distinct magnetic signatures associated with stratigraphic horizons, which has been invaluable for locating units during regional drilling. Whole-rock geochemistry has also identified changes in sediment provenance with evolution of the shelf, along with identifying alteration.

A distinct high-Fe-P group of samples was identified at a consistent stratigraphic location near the top of the Ng 1.1.3 subgroup. These elevations in Fe-P within layers of the Nguba have been identified at other localities in the CACB. These layers are proposed to have formed by oxidation of iron to precipitate Fe-oxyhydroxides, which absorb phosphate. These distinct layers may be useful for correlation of widely distributed occurrences of diamictite, as changes in oxygenation of the continental sea could be large-scale events.

Alteration has been identified in both whole-rock geochemistry and VNIR-SWIR spectroscopy as being characterised by muscovite proximal to mineralisation with more distal alteration of chlorite.

Gorno and Piani Resinelli: Genesis and Correlations of Two Alpine-type Pb-Zn Districts

Luca Summino[1], Michele Andrea Giorno[2], Luca Barale[3], Carlo Bertok[1], Max Frenzel[2], Marta Gasparrini[4], Luca Martire[1]
1. University of Turin - Department of Earth Sciences, Turin, Italy, 2. Helmholtz Zentrum Dresden-Rossendorf, Helmholtz Institute Freiberg for Resource Technology, Freiberg, Germany, 3. Institute of Geosciences and Earth Resources, CNR-Turin Unit, Turin, Italy, 4. University of Milan - Department of Earth Sciences, Milan, Italy

The strata-bound, carbonate-hosted Pb-Zn-Ag (± fluorite ± baryte) districts of Gorno and Piani Resinelli belong to the Alpine metallogenic province located between Italy, Austria, and Slovenia. The orebodies are hosted in the lower Carnian stratigraphic succession of the Lombardian Basin (Southalpine Domain, N Italy), mainly composed of peritidal limestones. The aim of this study is to constrain the ore-forming conditions using petrography, geochemistry, fluid inclusion microthermometry, and radiometric datings on both ore minerals and associated diagenetic products.

Petrographic studies reveal that the mineralization was preceded by dolomitization, silicification, and brecciation events. Thermometric data from primary fluid inclusions indicate the involvement of moderately hot (T between 70° and 150°C) and high-salinity fluids (up to ~20 eq% NaCl) during ore precipitation. Radiometric dating performed on associated diagenetic minerals that pre- and postdate the sulfide mineralization reveals that the precipitation of the ore minerals occurred during the LateTriassic. Moreover, in situ analyses (EPMA, LA-ICP-MS) of trace elements on sphalerite will be presented here, including their implications for ore-forming conditions.

Several similarities between Gorno and Piani Resinelli in terms of diagenetic evolution of the host rock, temperature and composition of the fluids, and timing of the ore precipitation highlight the possibility that both districts could be related to a unique, widespread Late Triassic hydrothermal system, in which metal-rich fluids flowed upward through faults and associated fractures. This circulation in the Upper Triassic stratigraphic succession caused major modifications in the host rocks such as multiphase dolomitization, silicification, dissolution, and brecciation, as well as the precipitation of ore minerals in a shallow burial setting.

Unveiling Geometallurgical Potential: Insights from the Soledad Cu-Au-Ag Mineralized Tourmaline Breccias, Central Peru

Victor J. Torres[1], David R. Cooke[1], Lejun Zhang[1], Julie Hunt[1], David L. Kelley[2]
1. University of Tasmania, Hobart, TAS, Australia, 2. Chakana Copper Corporation, Vancouver, BC, Canada

In Peru's Cordillera Negra, within the prolific Miocene metallogenic belt, the Soledad Project was acquired by Chakana in 2017 to advance the exploration of multiple mineralized tourmaline breccias in the area. An initial inferred resource was released in 2022, encompassing seven breccia pipes from surface to an average depth of 330 meters. These pipes yielded 6.7 Mt grading 0.88 g/t Au, 0.88% Cu, and 54.2 g/t Ag combined in both open pit and underground resources. Geometallurgical studies on three breccia pipes to optimize mineral processing and waste management strategies are in progress.

We have developed a geometallurgical workflow involving several key steps:

• Domain Definition: Utilizing geochemical analysis, including machine learning algorithms, to correlate with visible geological features such as breccia type, alteration type, and mineralogy

• Sample Selection: Focus to ensure representativity

• Sample Composite Preparation

• Analysis: Bulk mineralogy assessment was characterized through various techniques such as whole-rock geochemistry, petrography, scanning electron microscopy, electron microprobe, X-ray diffraction, and Automated Mineral Identification and Characterization System (AMICS) imaging. Gold characterization included visible gold grain separation and heavy mineral concentrates analysis using multiple size fractions, as well as LA-ICPMS transects on core slabs and individual mineral LA-ICPMS maps

Results from Breccia 05 defined four geometallurgical domains, from which eight samples were selected. The main sulfides are chalcopyrite-pyrite, and the main gangue minerals are tourmaline and quartz with over 30% each, then carbonates and white micas with less than 5% each. Most of the copper, ~92.8%, is carried by the chalcopyrite; ~45.7% of the silver is hosted by tetrahedrite. Gold deportment studies are in progress, but visible gold grains have been observed, ranging from 25 to 125 microns in length. Liberation sizes, particle locking, and mineral associations have been determined for each domain in the three pipes.

Mineral Chemistry as a Potential Exploration Tool Applied to the Lusale Basin and Fishtie Deposit, Central African Copperbelt (Zambia)

Veronica G. Trevisan[1], Jamie J. Wilkinson[1, 2], John Spratt[1]
1. Natural History Museum, London, United Kingdom, 2. Department of Earth Science and Engineering, Imperial College London, London, United Kingdom

The Central African Copperbelt (CACB) is the world's largest repository of sediment-hosted copper and cobalt, which are critical metals for the energy transition. A key knowledge gap regarding this mineral system is a lack of detailed understanding of the larger-scale alteration mineral assemblage zonation and major and trace element chemistry patterns that are likely to exist around mineralized domains. A detailed characterization of samples from the CACB can address this gap and also enable the development of tools to guide exploration for technology metals across different sedimentary basins. Here, we report microprobe results on a key alteration mineral (chlorite) from (meta)sedimentary rocks from background areas in the Lusale Basin and from distal to proximal ore zones in the Fishtie deposit. This deposit contains approximately 55 Mt of 1.04% Cu at a 0.5 % cutoff in oxide, sulfide, and mixed oxide-sulfide ores, hosted in Neoproterozoic diamictites and siltstones of the Grand Conglomerat Formation that overlies basement schists and quartzites. Our initial findings show that most chlorites classify as ripidolite, with a few in the pycnochlorite field and one in the clinochlore field. Basement (meta-sandstones and schist) and Grand Conglomerat (diamictite) samples present variable Fe/Fe+Mg ratios depending on location within Lusale and Fishtie. All samples demonstrate a negative correlation between Si and Al, which is attributed to the variable substitution of Si^{IV} and Mg^{VI}/Fe^{VI} by $Al^{IV}+Al^{VI}$ via the Tschermak substitution. However, a negative correlation is not observed between Fe+Mg and Mn, which could indicate the presence of other cations in the M^{2+} octahedral site. Future work will expand microprobe and LA-ICP-MS analyses on chlorite from distal to ore-proximal zones in the Fishtie deposit and Lusale Basin, as well as extend analyses to other key alteration (muscovite and biotite) and accessory (apatite, rutile, and zircon) phases in this part of the CACB.

Highway to Grade: The Importance of an Early Quartz Vein Stockwork for Channeling Late Ore-Bearing Fluids While Favoring Sulfide Precipitation

Martin G. Vallejos[1], Laura Maydagán[1], David Selby[2], Kevin B. Heather[3], Mariano Poodts[4]
1. INGEOSUR-CONICET, Departamento de Geología, Universidad Nacional del Sur, San Juan 670, Bahía Blanca 8000, Argentina, 2. Department of Earth Sciences, Durham University, Durham DH1 3LE, United Kingdom, 3. Aldebaran Resources Inc., La Serena, Chile, 4. Aldebaran Resources Inc., Santa Lucía, San Juan, Argentina

At the Radio Cu-Au-(Mo) porphyry deposit, located three kilometers west of the Altar Cu-Mo-(Au) porphyry (San Juan, Argentina), mineralization is closely associated with a dense quartz vein stockwork. Copper-Fe sulfides are mainly observed within A- or B-type veins or disseminated along the vein periphery, as well as within quartz UST layers in the core of the system, whereas molybdenite occurs exclusively within B-type veins.

Conventional petro-chalcographic studies and cathodoluminescence (CL) imagery indicate that Cu-Fe-Mo sulfide precipitation within quartz veins (and UST) post-date volumetrically dominant first-generation barren quartz veins. Moreover, Cu-Fe and Mo sulfides are associated with a second fine-grained quartz generation that is commonly accompanied with variable amounts of chlorite, calcite, sericite, anhydrite, and/or hematite, typically filling cracks, bands, centerlines, and/or interstitial spaces between the first-generation quartz, reflecting a multi-episodic vein formation.

Microprobe titanium (Ti) analyses support these observations, considering that Ti concentrations are positively correlated with CL bright intensity and temperature of quartz formation in several porphyry deposits. The first-generation quartz unrelated to ore from UST layers and A- and B-type veins is characterized by bright-CL quartz, with Ti concentrations varying from 100 to 180 ppm, 50 to 140 ppm, and 55 to 150 ppm, respectively. In contrast, the second-generation quartz bearing Cu-Fe sulfides and molybdenite possesses lower Ti concentrations—35 to 65 ppm and 10 to 30 ppm, respectively—and is characterized by dull-CL quartz with wavy growth zonation.

The early development of a quartz vein stockwork may be analogously compared to a network of "highways." As quartz undergoes thermal contraction during cooling and partial dissolution as it reaches its area of retrograde solubility, it enhances the vein permeability, eventually channeling later ore-bearing fluids, while favoring the generation of spaces for sulfide precipitation.

Analysis of Visual Grading of the Hydrothermal Alteration of the Vazante Zn-Pb Deposit, Minas Gerais, Brazil: A New Vector for Exploration?

Poliana Vidal Salgado, Luiz G. S. Correia, Bruno A. A. C. Arruda, Rafael F. de Freitas, Samuel Boucas do Lago
NEXA Resources, Vazante, MG, Brazil

The Vazante Zn-Pb deposit, located in the northwest of Minas Gerais, Brazil, is considered one of the largest hypogene willemite (Zn_2SiO_4) deposits in the world. Its ore is structurally controlled by the Vazante Fault and hosted by dolomitic rocks from the Vazante Belt. The hydrothermal alteration is marked by strong silicification, dolomitization, and hematitization, composing a diverse brecciated zone. The colour variations induced by hydrothermal alteration has been a visual indicator for the superficial grading of ore in the Vazante and North Extension mines. Looking for new vectors of prospection and to validate this visual guide, Vazante's exploration team analysed the brecciated system through six cross sections within the mines. Five main lithotypes were defined in this evaluation by colour and composition for the brecciated zone: iron-carbonated dolomitic breccias, divided in weak, moderate and strong alteration type; hematitic breccia; and the willemitic ore. The Principal Component Analysis (PCA) of whole-rock samples from the drill holes in the cross sections shows element clusters that distinct the unaltered host rocks, the iron-carbonated dolomitic breccias, and display a progression of the alteration in between the five lithotypes towards the hematitic breccia and the willemite ore with the association of mineralization-related elements (Zn, Pb, Fe, Ag, As, Sb, Cd, Mo). PCA graphics also suggest provenance of determined elements and pathfinder elements that contribute to mineral exploration. The cross sections, however, exhibit a complex placement of the defined lithotypes, where the envelope progression expected for the brecciated zone and mineralization is inconsistently followed. This variability is attributed to interactions between host rock and fluids, or syn-post-mineralization displacement of the hydrothermal zone. The analysis bolstered the classification of Vazante's hydrothermal breccia with technical evidence, introducing efficient tools for pinpointing zinc ore deposits and enhancing predictability in mineral exploration in the region.

Influence of Mafic Units on the Mineralising System Around the Redmoor Project, Cornwall

Georgina C. Woodward
Camborne School of Mines - University of Exeter, Penryn, United Kingdom

Within Southwest England, it is understood that four stages of mineralisation have occurred, both pre- and syn-granite emplacement with mining of these mineral systems occurring since the 17th century. These magmatic-hydrothermal systems are host to multiple polymetallic minerals, including Sn, W, Cu, As, Ni, Co, Pb, and Zn. These include the Redmoor sheeted vein system (SVS) located in East Cornwall, which has been identified as a promising Sn, W, Cu deposit associated with the Kit Hill Granite intrusion. More recently however, research has been undertaken into the influence of mafic units identified within the system, a topic that is highly understudied within SW England.

This was undertaken through detailed core logging of mineralisation, alteration, and structures within the mafic units, creating a comparison of mafic mineralisation to the country rock and granite mineralisation, with specific detail to precipitation controls of Cu, Ni, Pb, Zn, and Co. Through integration of core logging, ore microscopy, and geochemical analysis of the mafic units, an understanding of whether these units may have contributed to controls on mineralisation at Redmoor could be interpreted. An in-depth petrographic study was conducted to understand relationships and microstructures within the unit to determine the role mafic mineralisation has on the wider deposit. Subsequently, a new mineral systems concept has been interpreted for the area to develop the geological model and further aid exploration.

This research has allowed for a better understanding of mineralisation that predates the emplacement of granite intrusions and the role mafic units have in potentially increasing economic grade of the deposit. Interpreting the system as polymetallic in nature has further expanded the scope of potential areas of economic interest and increased the opportunity for exploitation of multiple minerals needed for the energy transition.

Texture, Trace Elements, and Sulfur Isotopes of Pyrite from the Zhao'anzhuang Iron Deposit, Southern Margin of North China Craton: Implications for Ore Genesis

Hengfeng Zhang[1], Hao Hu[2], Jianwei Li[1, 2]
1. School of Earth Resources, China University of Geosciences, Wuhan 430074, China, Wuhan, China, 2. State Key Laboratory of Geological Processes and Mineral Resources, China University of geosciences, Wuhan, China

The Zhao'anzhuang iron deposit (160 Mt at ~40% Fe) in the southern margin of the North China Craton, is hosted in an early Paleoproterozoic metavolcanic sequence with intercalations of evaporite-bearing, Mg-rich carbonate, but its genesis remains poorly understood. Here we present trace element and sulfur isotope data for three distinct types of pyrite to provide new insights into ore genesis. These include pre-ore Py1 disseminated within serpentinized olivine in the altered host rock, syn-ore Py2 closely associated with magnetite-apatite ores, and post-ore Py3 veins intersecting iron ores. Py1 has low $\delta^{34}S_{V\text{-}CDT}$ values (3.55~5.37‰), elevated Co (112-2,516 ppm), Ni (27-1,329 ppm), and Se (265-2,458 ppm), and low Cu (29-1,609 ppm), Pb (29-1,647 ppm), and Zn (3.3-17.8 ppm). The trace element compositions and sulfur isotope data of Py1 indicate its precipitation from a high-temperature magmatic-hydrothermal fluid. Py2 has variable Co (11.8-25,951 ppm), Ni (48-27,958 ppm), and Se (<971 ppm) and exhibits distinctly different sulfur isotopes with $\delta^{34}S_{V\text{-}CDT}$ values ranging from 9.84‰ to 16.91‰. These trace element and isotopic signatures are interpreted in terms of the assimilation of evaporites by a magmatic-hydrothermal fluid, with temperature decreasing from the pre- to syn-ore stage. The contents of Co (52.8-1,168 ppm), Ni (198.4-998 ppm), and Se (<86 ppm) are much lower in Py3, which is further enriched in heavy sulfur isotopes with $\delta^{34}S_{V\text{-}CDT}$ of 14.24~23.53‰. This observation indicates continuous assimilation of the evaporites by the ore fluids, whose temperatures progressively decreased. Given the close association of the three pyrite generations with magnetite and other mineral assemblages, we suggest that the Zhao'anzhuang iron deposit is of magmatic-hydrothermal origin and formed through interaction of the magmatically derived fluid with evaporite-rich carbonates. Assimilation of evaporites into the fluid facilitated the oxidation of ferrous iron that led to the precipitation of magnetite, during which process the Py1 and Py2 formed.

Unlocking Gold Potential: Remote Sensing Insights and Applied Analysis of the Qena-Safaga Area in Egypt's Eastern Desert

Mohamed Badawi
Miskolc Egyetem, Miskolc, Hungary

The eastern desert of Egypt has undergone prolonged deformation as a result of the cratonization of the juvenile crust known as the Arabian Nubian Shield (ANS). This geological phenomenon has raised prospects for potential gold discoveries associated with the corresponding tectonic events. Within this context, the Qena-Safaga zone, located on the border of Egypt's Northern and Central Eastern Deserts, stands out for its metavolcanic rocks, which have been subjected to significant twisting due to the movement along the Qena-Safaga Shear Zone (QSSZ) relative to the Barud magmatic body.

In our recent study, we conducted a thorough examination of numerous deformed post-orogenic granitic intrusions and felsite dikes within the Qena-Safaga area. Our investigation aimed to identify promising mineralization sites, with a particular focus on orogenic gold deposits. Applying a combined approach involving detailed field surveys, advanced analysis of remote sensing data—including band combinations from ASTER and Sentinel 2—and interpretation of aeromagnetic enhanced maps, we systematically distinguished various rock types, structural features, and hydrothermal alterations across the study area.

By integrating insights from multi-sorce data, including the development of remote sensing band combinations and aeromagnetic maps, we were able to validate previously hypothesized mineralized zones within the Qena and Safaga areas, while identifying additional areas with similar mineralization potential. Notably, our analysis, which included sophisticated methods such as the utilization of ASTER and Sentinel 2 data alongside hydrothermal alteration and orientation entropy heat maps, revealed a strong correlation between mineral-rich regions and major structures associated with late-stage deformation of the QSSZ.

Our study stands out for its applied analysis, which provided valuable insights into the geological processes at the Egyptian eastern desert. By elucidating the spatial distribution of mineralization and its relationship with structural features through advanced analytical techniques, our findings offer important guidance for future exploration efforts in this promising area.

Reuse of Granite Quarry Waste for REEs Recovery Along the Industrial Production Chain: Results from Preconcentration Tests

Silvia Gioiello[1], Licia Santoro[1], Alberto Cazzaniga[2]
1. Department of Earth Sciences, University of Torino, Torino, Italy, 2. Minerali Industriali Engineering S.r.l., Masserano (BI), Italy

Granite quarry waste from the "Graniti dei Laghi" plutons in Piedmont, Italy, underwent several lab-scale beneficiation tests to enhance the recovery of critical raw materials (CRMs) in the context of circular economy and sustainable mining. Preliminary analyses by ICP-MS, XRPD, SEM-EDS, and automated mineralogy highlighted that the magnetic waste from granite industrial reprocessing for feldspar production is notably enriched in rare earth elements (REEs), paving the way for simulating and refining a lab-scale treatment to optimize the concentration of REE-bearing minerals. Monazite-(Ce) was identified as the primary target mineral, predominantly locked within Fe-rich biotite (siderophyllite). The feed material was milled to various grain sizes and subjected to multiple magnetic separation stages to identify the optimal conditions for REE mineral beneficiation. Results indicate that approximately 90% of REEs are associated with paramagnetic fractions of 75-150 μm and 150-250 μm grain sizes. The 75-150 μm paramagnetic concentrate recovered at 1 A of current intensity shows the highest enrichment to ~1800 ppm REEs + Y and Sc compared to ~200 ppm in the raw granite. However, significant REE amounts remaining in the paramagnetic material at 0.4 A suggest possible incomplete liberation of monazite. Although grinding below 600 μm favours the rejection of diamagnetic gangue minerals from the monazite-bearing siderophyllite, SEM-EDS observations reveal liberation issues for monazite particles above 100 μm due to their smaller size compared to siderophyllite. Consequently, additional grinding below 100 μm is required to isolate monazite. However, dry electromagnetic separation on fine material demonstrated poor monazite recovery, indicating the need for wet magnetic separation and flotation for effective recovery from the fine-grained preconcentrate. This work underscores the potential of improving beneficiation techniques to maximise the recovery of CRMs as by-products along the industrial production chain, advocating the valorisation and sustainable management of mineral processing waste through its reuse.

Chemical Composition of Garnets and the Relation to Mineralizing Fluids in the Tucano Gold Deposit, Amapá State, Brazil

Deniro F. Gonçalves Costa[1], Gabriel A. Rodrigues Soares[1], Rosaline C. Figueiredo e Silva[1], Mahyra F. Tedeschi[1], Steffen G. Hagemann[2]
1. Graduate Program in Geology, Federal University of Minas Gerais (UFMG), Belo Horizonte, MG, Brazil,
2. School of Earth Sciences/Centre for Exploration Targeting, The University of Western Australia, Crawley, WA, Australia

The Tucano gold deposit, located in the northeastern segment of the Amazon Craton, within the Guiana Shield (Amapá State, Brazil) is a structurally controlled deposit hosted in amphibolite facies metasedimentary rocks of the Paleoproterozoic Serra do Navio greenstone belt, mainly in marble and banded iron formation (BIF). Mineralization is controlled by the N-S trending brittle-ductile Urucum shear zone and associated subsidiary structures. Hydrothermal alteration is characterized by skarn-type calc-silicate assemblages replacing metamorphic minerals and overprinting precursors' fabric. Since garnets are important minerals to consider further relation with mineralizing fluids, this work aims to investigate their chemical composition indicating similarities with gold mineralization. Petrographic investigation identified gold (< 100 μm) as interstitial or fracture-filling in silicates (amphibole, clinopyroxene, garnet). A petrological study was conducted on select diamond drill core samples analyzed by electron microprobe (EPMA), aiming to investigate the compositional variation between minerals from the metamorphic, igneous, and hydrothermal rocks on the area, with focus on garnet. X-ray semi-quantitative maps were reduced with XMapTools. Hydrothermal garnet shows a large compositional range, but in general it contains higher grossular (4 – 34%) and some minor andradite (up to 8%), when compared to metamorphic and magmatic ones. A Garnet-biotite geothermometry yielded the estimate for peak metamorphic temperature: 592 ± 25 °C (1 σ, n = 5). The same garnet crystals resulted in pressure estimates of 4.1 ± 0.6 kbars (1 σ, n = 5) using the geobarometer. Orogenic hypozonal gold mineralization at Tucano is characterized by high temperature hydrothermal minerals such as clinopyroxene, garnet, amphiboles, phlogopite and loellingite. Wide distal alteration zones suggest a high fluid/rock ratio, which is further manifested in the alteration mineralogy – clinopyroxene-garnet-amphiboles-phlogopite-magnetite-pyrrhotite – indicating an intense fluid-buffered composition.

Mapping Cryptic Telescoping Events in Porphyry-Epithermal Systems Through Integrated Microtextural-Chemical Pyrite Analysis at the Bonanza-Type Brucejack Au-Ag Deposit

Duncan F. McLeish[1, 2], Anthony E. Williams-Jones[1], James R. Clark[1]
1. McGill University, Montreal, QC, Canada, 2. Newmont Corporation, Vancouver, BC, Canada

Telescoping, the mechanism by which shallow, low-temperature hydrothermal alteration and mineralisation are superimposed onto deep, high-temperature hydrothermal assemblages, is thought to have played a key role in forming some of the world's largest porphyry-epithermal gold deposits. Although the causes of telescoping, namely rapid, erosion-driven syn-hydrothermal degradation of paleosurfaces and sector collapse of volcanic edifices, are now well recognised, relatively little is known about (1) the prevalence of telescoping during the different stages of porphyry-epithermal deposit formation, and (2) the evolution of telescoped architectural characteristics, especially in multi-pulsed hydrothermal systems. These knowledge gaps exist because mineral assemblages related to early alteration events are commonly obliterated during later overprinting. Recent research has, however, shown that the intra-grain microtextural and trace element record from hydrothermal pyrite can provide a window onto the early, pre-telescoped alteration history of a deposit, in addition to capturing information on the nature of the telescoping process. We have investigated hydrothermal pyrite from Brucejack, a strongly telescoped, bonanza-type epithermal Au-Ag deposit in northwest British Columbia, Canada, to gain insight into the early history of alteration. Backscatter SEM imaging and LA-ICP-MS analyses of pyrite from two vertical sample transects reveal an evolution from magmatic- (elevated Co-Ni; cubic growth zonation) to epithermal- (elevated As-Au; colloform, oscillatory growth zonation) type fluids. We show that this evolution, which is otherwise cryptic due to overprinting by subsequent alteration and emplacement of bonanza-style epithermal veins, was caused by telescoping. Three-dimensional modelling of the microtextural and trace element signatures in pyrite demonstrates that this telescoping led to >1-km vertical and lateral compression of the Brucejack hydrothermal system. Furthermore, we document a previously unrecognised telescoping-reversal event in the deposit, as evidenced by localised mantling of Co-Ni on As-Au growth zones.

Satellite-Based Mapping for Lithium Pegmatite Mineral Exploration in Namibia: A Criteria-Based Approach

Olivia Rhind, Christopher Lambert, Alfie Baines, Mark Broadley
CGG, Crawley, United Kingdom

The growing global demand for lithium (Li) requires innovative mineral exploration methods. Lithium pegmatites, a significant source due to their grade and conventional mining methods, represent a key target. In many cases, Li exploration methods involve observation of surface exposure or investigation through geophysical surveys. In this study we introduce targeting of Li pegmatites in Namibia using satellite-based mapping within a mineral system criteria framework. It integrates expert knowledge with remote sensing techniques to identify and prioritise exploration targets based on geological characteristics, such as source magma types, structural settings and host-rock metamorphic grade.

The study utilises a Bare Earth+ mosaic derived from the entire archive of ASTER satellite imagery. Every image is co-registered at the pixel level with atmospheric and vegetation suppression applied and the "barest" pixel chosen. The result is an enhanced representation of Earth's surface with optimal signal to noise ratio and minimum cover, allowing the most representative spectral signatures to be derived and targeted. The Bare Earth+ mosaic is the foundation for the regional multispectral lithological and alteration mapping, with processing tailored to the Li-bearing granitic pegmatite system and resulting mineral maps supporting target generation, spectral lithological mapping and prospectivity analysis.

Hyperspectral EMIT data are also integrated in the study, enabling enhanced mineral mapping and lithological discrimination. The additional spectral bands are used to further identify and map possible Li-rich pegmatites. Several key proxy clay minerals are used as vectors to highlight candidate pegmatite target areas. Variability in host-rock composition can be spectrally discriminated using both multi- and hyper-spectral data through careful use of Principal Component Analysis, and from the rocks and alteration styles that they highlight. This workflow facilitates rapid regional assessments of Li-pegmatite potential, enhancing target identification accuracy and reducing exploration risk.

The Mineralogy of PGE in the Deep Critical Zone at Sandsloot, Northern Bushveld Complex, South Africa: PPGE Reef Underlain by an IPGE + Rh-Rich Base Metal Zone

Kate Canham[1], David Holwell[1], Iain McDonald[2], Hannah Hughes[3], Katie McFall[4], Andy Lloyd[5], Erin Thompson[1], Kofi Acheampong[5], Lara Du Preez[5], Matthew Loader[5]
1. Centre for Sustainable Resource Extraction, University of Leicester, Leicester, United Kingdom, 2. School of Earth and Environmental Sciences, Cardiff University, Cardiff, United Kingdom, 3. Camborne School of Mines, University of Exeter, Penryn, United Kingdom, 4. Earth Sciences Department, University College London, London, United Kingdom, 5. Anglo American South Africa, Johannesburg, South Africa

The Bushveld Complex, South Africa, is home to some of the world's largest magmatic sulfide deposits. Recent discoveries, including the down-dip extension of the Critical Zone (CZ) hosted Platreef PGE-Ni-Cu deposit, have positioned the Northern Limb of the Bushveld Complex as a frontier for PGE+Ni+Cu+Co exploration and mining in the future decades. In the Sandsloot area, the near-surface resources of the CZ are well characterised, but the recently discovered down-dip extension is not; where multiple high-grade PGE-rich zones on the scale of 10s of meters are underlain by thick base metal sulfide-rich zones. In this study we utilise PGE geochemistry, sulfide and PGM studies and S isotopes to establish the relative controls on the distinctive metal budgets of these two zones.

The PGE reef has 1-2 modal.% disseminated sulfides and is Pt+Pd+Rh(PPGE)-rich. The PGE is dominantly contained within Pt-Fe-alloys, Pt-Pd-(Pb)-bismuthotellurides and Pt-Pd-Pb-alloys, along with high Pd (<3500 ppm), Rh (<270 ppm)-bearing pentlandite. In contrast, the lower base metal (BM) zone contains higher volume, blebby to semi-massive sulfides and lower Pt+Pd grades, but significantly, high Os+Ir,+Ru(IPGE)+Rh. The PGE is predominately contained within sulfide hosted Ir-(Os)-arseno-sulfides and Ru-sulfide (62%), as well as Ru, Rh-rich (<55, 300 ppm) pentlandite and Ru-rich (<45 ppm) pyrrhotite. Within the PGE reef $\delta^{34}S$ values (+0.7-+2.4 ‰), in addition to S/Se ratios (3100-1200), reveal a magmatic control, with subsequent S-loss through fluid mobilisation. In contrast, the underlying BM zone displays an isotopically heavier, contaminated, signature (+4.2-+7.5 ‰).

The difference in PPGE to IPGE+Rh between the PGE reef and BM zone represents a disconnect between the zones. Understanding the PGE mass balance of the deep CZ at Sandsloot, along with the association of PGM and sulfide to surrounding minerals aids not only with understanding of the ore's genesis, but additionally, early understanding of geo-metallurgy has implications for processing.

Partitioning and Mobility of Deleterious Elements from Sulphide Oxidation in the Historical Gold Mine Tailings

Steve J. Chingwaru, Bjorn Von der Heyden, Margreth Tadie
Stellenbosch University, Stellenbosch, South Africa

The Witwatersrand Goldfields, responsible for 27% of global gold production, face environmental challenges due to over a century of extraction, resulting in 6 billion tons of historical tailings. These tailings directly contribute to acid mine drainage and release deleterious elements from unrecovered sulphides. Iron oxyhydroxide alteration products, formed during sulphide oxidation, intricately regulate the mobility of deleterious elements through absorption and incorporation. Despite extensive research on the immobilization properties of iron oxyhydroxides, little attention has been given to understanding the mobility and distribution of deleterious elements during sulphide oxidation, especially regarding the sulphide core and secondary mineral alteration layers, and their impact on environmental contamination. This study extensively characterizes the 70-year-old Klerksdorp tailings dump within the Witwatersrand, focusing on deleterious element partitioning and mobility between sulphide grains and iron oxyhydroxide alteration rims. Using a multi-method approach, including laser ablation inductively coupled plasma mass spectrometry, automated mineralogy, and wet chemistry data integration, metal release and retention across phases were quantified. Results indicate that under net acid-producing conditions, iron oxyhydroxide alteration rims retain a significant proportion (3–15%) of As, Ni, Cu, and Zn (>250 ppm), whereas Co, Au, and Pb exhibit limited compatibility with iron oxyhydroxide alteration (<3% deportment). The majority of remaining deleterious elements reside in mobile, unstable phases, exceeding 88%. Projections highlight a significant partitioning of elements (As, Ni, Cu, Zn, and Pb) into the environment, exceeding 9000 tons, underscoring the urgency of addressing these challenges. Remedial measures for these historical Witwatersrand tailings dumps are imperative, potentially involving metallurgical reprocessing flowsheets targeting sulphides and effective mechanisms for capturing deleterious elements. Such comprehensive approaches are vital for mitigating risks to human health, aquatic systems, and local ecosystems. Additionally, capturing metals such as Cu, Co, and Ni could serve as secondary sources to support the green energy transition.

United Nations Resource Management Systems as a Guide for Sustainable Resource Management: Case Example from Copperbelt Province, Zambia

Michael M. Musialike[1], Eva Marquis[3], Gabriel Ziwa[1], Mubanga Mutale[2], Karen Hudson-Edwards[3], Frances Wall[3]
1. The Copperbelt University, Kitwe, Zambia, 2. Geological Survey Department of Zambia, Lusaka, Zambia, 3. Camborne School of Mines, University of Exeter, Penryn, United Kingdom

In this contribution, we present the results of an initial scoping study of the application of the United Nations Resource Management System (UNRMS) to the critical raw materials (CRM) sector in the Copperbelt and Northwestern provinces of Zambia. As the global economy transitions to use of clean energy technologies, the demand for CRM needed to develop these technologies is rising. The surge in demand for CRM presents real economic benefits to producer countries, especially those that are resource rich and developing. To sustainably benefit from these resources, producers should build legislative and policy frameworks that ensure that operations across the value chain in the sector leave minimal climate footprints and have high environmental standards and quality social circumstances. Zambia is a major producer of CRMs such as copper and cobalt, with major nickel projects underway. Like the rest of the world, Zambia has realised the need to adopt a more sustainable way of managing its mineral resources in order to deliver the socio-economic transformation necessary for improved well-being for its people and the planet. As the country is in the process of reforming its mining policy to align it with the sustainable development agenda, it is timely to investigate how sustainable management of the CRM sector can support this objective. Published in 2019, the UNRMS aims to provide a comprehensive framework for sustainable resource management by government, industry, investors and the civil society. It links to the African Mining Vision, African Minerals and Energy Resource Classification (AMREC) and the Sustainable Development Goals.

A Supergene-Hydrothermal Origin of the BIF-Hosted High-Grade Iron Ores in the Mbarga Prospect, Mbalam Iron Ore District, Southern Cameroon, Congo Craton

George L. Ngiamte[1, 2], Eleanor Green[2], Olugbenga A. Okunlola[1], Roland Maas[2], Alan Greig[2], Cheo E. Suh[3,]
1. Mineral Exploration Geoscience Program, Pan African University Life and Earth Sciences Institute (including Health and Agriculture), University of Ibadan, P.M.B 20, Ibadan, Nigeria, Ibadan, Nigeria, 2. School of Earth Sciences, The University of Melbourne, Parkville, Victoria 3010, Australia, Melbourne, VIC, Australia, 3. Department of Geology, Mining and Environmental Science, The University of Bamenda, P.O. Box 39, Bambili, North West Region, Cameroon, Bambili-Bamenda, Cameroon, 4. Economic Geology Unit, Department of Geology, University of Buea, P.O. Box 63 Buea, South West Region, Cameroon, Buea, Cameroon

The Mbarga prospect lies within the Mbalam iron ore district of the Ntem Complex on the Congo Craton (CC). It is characterized by Banded Iron Formations (BIFs), high-grade supergene, and specularite ores. This study presents mineralogical, geochemical, and isotopic datasets on these ores to determine their origin. Ore microscopic studies indicate that the BIFs are of the oxide facies type, with magnetite showing alteration to hematite-martite. The supergene ores consist of hematite + martite + goethite ± gibbsite ± magnetite ± quartz, while the specularite ores are composed of hematite + martite ± quartz. Magnetite microchemistry suggests formation under low-T hydrothermal conditions (~200–300 °C) with high fO_2. Geochemical analyses suggest that the supergene and specularite ores have higher Fe_2O_3 (88.27 to ~100 wt%) and lower SiO_2 (<0.01 to 0.18 wt%) contents than the BIFs (31.95 wt% Fe_2O_3, 67.16 wt% SiO_2). The enrichment of Fe in the supergene ores is attributed to the depletion of major oxides and trace elements due to weathering and supergene enrichment, while the high Fe content in the specularite ores stems from the precipitation of iron-rich, but trace- and REE-deficient hydrothermal fluids. The slightly higher Al_2O_3 content and positive Ce anomalies in the supergene ores suggest retention of Al-bearing minerals (gibbsite) and reveal highly oxidative conditions during martitization. Stable isotope analyses reveal that the supergene and specularite ores have $\delta^{18}O$ values of -2.5 to -0.3‰ and -2.0 to -3.4‰, and δ^2H values of -75 to -123‰ and -70 to -119‰, respectively, suggesting the involvement of isotopically light-evolved meteoric water in their formation. In contrast, the BIFs exhibit heavier $\delta^{18}O$ (8.5 to 10.2‰) and δ^2H (-85 to -91‰) values, suggesting formation from mixed magmatic-metamorphic fluid sources. A "polygenic-supergene-hydrothermal" model is suggested for the formation of the Mbarga BIF-hosted iron ores.

Resilience and Governance in Artisanal Gold Mining: Addressing Challenges and Opportunities in Western Kenya

Christine A. Omuombo[1], Joy A. Owidhi[2]
1. The Technical University of Kenya, Nairobi, Kenya, 2. Independent Researcher, Nairobi, Kenya

The artisanal and small-scale gold mining (ASGM) sector plays a critical role in global supply chains. Kakamega experienced growth during the 'colonial gold rush' in the 1930s but faced decline after the closure of the Rosterman mine in the 1940s. Since 2019, there has been renewed interest in the Artisanal Gold Mining Sector in Western Kenya due to significant deposit discoveries by mining companies. This surge has resulted in increased engagement in ASM activities by local communities and small-scale investors, leading to notable challenges in environmental and social governance. This study focuses on the role of the local Kakamega ASM committee's resilience team in promoting responsible mining practices and addressing the sector's opportunities and challenges. Kenya is currently in the process of formalizing the ASM sector. Key challenges include a high dependence on artisanal mining and quarrying for livelihoods, a low uptake of modern sustainable resource utilization technologies, and a general lack of awareness about natural resource technologies. Key areas of support identified include access to engineering solutions, environmental data, and geoscience information, which provide a foundation for future mining activities. Additionally, the organization of miners into cooperatives presents an opportunity for the local community to benefit from available financial resources for user-needs-based activities. Miners find these benefits appealing as they seek to mitigate physical and regulatory risks associated with their operations. Moving towards a formalized and resilient ASGM sector requires capacity building and technical assistance from both governmental and non-governmental sector players. This study highlights the pivotal role of the ASM committee's resilience team in facilitating sustainable practices and governance improvements within the ASGM sector in Western Kenya.

The Control of Biotite on the Lithium, Fluorine and REE Budget of Crustal Silicate Magmas

Charles D. Beard[1, 2, 4], Vincent J. van Hinsberg[2], John Stix[2], Madeleine C. Humphreys[3], Owen M. Weller[4]
1. Utrecht University, Utrecht, Netherlands, 2. McGill University, Montreal, QC, Canada, 3. Durham University, Durham, United Kingdom, 4. University of Cambridge, Cambridge, United Kingdom

Of the common rock-forming minerals, biotite contains the highest concentration of lithium and fluorine in both alkaline and calc-alkaline magmatic suites globally. Its control on the distribution of minor and trace-elements, including Li and F, and its impact on metal enrichment, however, is not well understood. Here, we use experiments to quantify that control in sodic (per)alkaline H_2O-saturated magmas with variable F-content at 650–800 °C, 200 MPa total pressure, and log fO_2 ~FMQ +1. Biotite-glass pairs from (tephri)phonolite fall deposits from Tenerife, Canary Islands complement our experimental data set. Our results, alongside a literature compilation covering a wide compositional range constrain empirical models that: (1) describe the exchange of F and OH between the silicate melt and the biotite W-site; (2) predict the partitioning of 1+ cations Li–Cs between silicate melt and the biotite A- and M-sites. The models use the major-element composition of biotite and melt, pressure and temperature as input. Lithium partition coefficients vary from 0.06 to 33 and are lowest in Al-rich, Si-poor biotite formed at high temperature. Residual enrichment of Li should therefore be most efficient during crystallisation of high Al/Si magmas, a finding consistent with the association between peraluminous granites and lithium-caesium-tin pegmatites. Further, biotite-melt $Kd_{F/OH}$ exchange coefficients are 1.2–51, with minima for Al-rich, Mg-poor biotite. Where $Kd_{F/OH}$ values are high, crystallisation of even small amounts of biotite will strip fluorine from the melt. In silica undersaturated peralkaline magmatic systems, biotite $Kd_{F/OH}$ values are low and fluorine can be residually enriched during crystallisation, complexing with REE in the melt and facilitating their enrichment too. Our partitioning models can be used to interrogate biotite from natural systems to determine the composition of their source melt, or to forward model the metal budgets of mafic to evolved igneous systems at upper-mantle to crustal pressures.

Future Lithium Supply: Towards a Robust Quantitative Assessment

Laura Buarque Andrade[1], Max Frenzel[1], Carolin Kresse[2], Britta Bookhagen[2], Michael Schmidt[2], Dirk Sandmann[3]
1. Helmholtz Institute Freiberg for Resource Technology, Freiberg, Germany, 2. Deutsche Rohstoffagentur (DERA) in der Bundesanstalt für Geowissenschaften und Rohstoffe (BGR), Berlin, Germany, 3. ERZLABOR Advanced Solutions GmbH, Freiberg, Germany

There has recently been considerable controversy over whether primary lithium (Li) sources will be able to supply the rapidly growing needs of the electric mobility transition. Unfortunately, existing assessments of future supply are hampered by a lack of consideration for the decision-making processes occurring at the level of individual mining projects and how these respond dynamically to changing market conditions. In the present contribution, we demonstrate how these processes, as well as associated uncertainties, can be incorporated into an assessment of the likely future evolution of global primary Li supply, including probable price trends.

Our method uses Monte-Carlo simulations to achieve this goal. A global database of existing Li mining projects (all development stages, including case histories) is used to build models that can estimate the likelihood of each project proceeding to the next development stage in any given year, depending on specific project characteristics such as location, deposit type, and ore grade, as well as market conditions. A series of stochastic simulations (1,000) are then run, in which projects are moved through the development pipeline according to these estimated likelihoods, year-by-year, up to 2050. New discoveries are also included to achieve realistic results over the relatively long timespan covered. A few simple demand scenarios (low, intermediate, high) are used as fixed inputs to these simulations to allow for easier interpretation.

Results show increasing divergence between demand and supply for demand scenarios incorporating more rapid growth. Some of this divergence may be offset by increasing recycling contributions to overall supply or through demand destruction as Li prices increase due to the divergence.

Hydrothermal Rare Earth Element (REE)-Enrichment in an Iron-Oxide-Apatite (IOA) Type Deposit at Kwyjibo, Québec, Canada: A Fluid Inclusion and Stable Isotope Study

Gary Fung, Anthony E. Williams-Jones
McGill University, Montreal, QC, Canada

The Kwyjibo deposit, Québec (Canada) is an unusual rare earth element (REE)-enriched iron-oxide-apatite (IOA)-type deposit. Hydrothermal processes led to the mobilization of the REE from fluorapatite in a magnetite-apatite host to britholite and andradite, followed by the replacement of andradite by allanite. The physio-chemical conditions of the REE mobilization, however, are poorly understood. Microthermometric measurements identified four distinctly different types of aqueous fluid inclusions in minerals coeval with the mineralization, representing a NaCl-KCl-H_2O-dominated (NaCl) fluid, an intermediate (NaCl-$CaCl_2$)-H_2O fluid, a CaCl2-NaCl-H_2O-dominated ($CaCl_2$) fluid, and a CO_2-NaCl-H_2O (CO_2-bearing) fluid. The NaCl-$CaCl_2$-H_2O fluid inclusions, hosted by fluorapatite, have a higher Na/Ca ratio than the $CaCl_2$-NaCl-H_2O fluid inclusions (quantified by laser ablation inductively coupled plasma mass spectrometry), and some of them are anhydrite-bearing (identified by Raman spectroscopy). The inclusions of medium-salinity (18-24% NaCl eqv.) sulphate-bearing NaCl-$CaCl_2$ composition homogenize to liquid at 185-290 °C. The NaCl-KCl-H_2O fluid inclusions, hosted dominantly by andradite, and the CO2-bearing fluid inclusions, hosted by quartz and fluorite, have low salinity (9-16% NaCl equiv). However, whereas the aqueous NaCl-KCl-H_2O inclusions homogenize to liquid at 290 to 365°C, the CO_2-bearing inclusions decrepitated upon heating. The exclusively secondary $CaCl_2$-NaCl-H_2O inclusions have high salinity (17-27% NaCl equiv) and homogenize at low temperatures (110-170°C). The $\delta^{13}C$ and $\delta^{18}O$ values of calcite, which was crystallized in the REE deposition stage, range from -4.61‰ to -7.6‰ and from +8.86‰ to +17.32‰, respectively. We interpret a magmatic-origin fluid unmixing event occurred, from NaCl-$CaCl_2$ fluid to NaCl and CO_2-bearing fluids, and $CaCl_2$ fluid. The transport of REE were attributed to the predominant chloride and sulfate complexes, given the elevated sulphate activity in the NaCl-$CaCl_2$-H_2O fluid. The deposition was facilitated by the precipitation of fluoride, phosphate, and carbonate ions, reflected by the spatial proximity of fluorite, tertiary fluorapatite, and calcite with respect to the REE-bearing minerals.

Mineral Chemistry, Fluid Inclusion, and Isotopic Characteristics of the Enargite Orebodies at Lepanto, Mankayan, Philippines

Pearlyn Manalo[1], Ryohei Takahashi[1], Manuel Nopeia[1], Kalin Kouzmanov[2], Nigel Blamey[3], Andrea Agangi[1], Akira Imai[4], Leo Subang[5], Marlina Elburg[6], Henriette Ueckermann[6], Linda Iaccheri[7]
1. Akita University, Akita, Japan, 2. University of Geneva, Geneva, Switzerland, 3. Western University, London, ON, Canada, 4. Kyushu University, Fukuoka, Japan, 5. Lepanto Consolidated Mining Co., Quezon City, Philippines, 6. University of Johannesburg, Johannesburg, South Africa, 7. University of the Witwatersrand, Johannesburg, South Africa

Enargite and luzonite are major ore minerals in high-sulfidation epithermal Cu-Au deposits. These minerals have also recently gained interest due to their enrichment in some critical metals, such as Sb, Te, and Ge. In this study, we report new results on mineral chemistry, fluid inclusion microthermometry, sulfur and lead isotope signatures of enargite from the Lepanto high-sulfidation epithermal Cu-Au deposit in the Philippines, and their implications for deciphering the ore-forming processes in the deposit.

Elemental mapping shows that enargite hosts native tellurium, goldfieldite and Au-Ag tellurides inclusions. Enargite and luzonite contain up to 5 wt % Sb and up to 5,000 ppm Te. Variations in Sb and Te contents generally follow the crystal growth planes. Near-infrared microscopic observations revealed that enargite hosts two-phase liquid-vapor fluid inclusions that homogenize to liquid between 205 to 240°C. Fluid salinity ranges from 2.5 to 5.7 wt % NaCl equiv. Bulk gas analysis indicates that the fluid inclusions dominantly contain H_2O with 0.1 to 3 mol % CO_2 and 0.0006 to 0.02 mol % H_2S. N_2/Ar ratio widely varies between 2 and 120. Sulfur isotopic ratios ($\delta^{34}S_{CDT}$) of enargite range from -6.0 to -3.0 ‰, typical of sulfide minerals that were formed from H_2S produced from magmatic SO_2 disproportionation at magmatic-hydrothermal temperatures. Lead isotope ratios of enargite range from 38.48 to 38.53 for $^{208}Pb/^{204}Pb$, 15.57 to 15.59 for $^{207}Pb/^{204}Pb$, and 18.41 to 18.43 for $^{206}Pb/^{204}Pb$, which are similar to the Pb isotopic composition of the Pleistocene dacite in northern Luzon.

AUTHOR INDEX